PEST CONTROL:
An Assessment of
Present and Alternative
Technologies

VOLUME I

Contemporary Pest Control Practices and Prospects:

THE REPORT OF THE EXECUTIVE COMMITTEE

Study on Problems of Pest Control
Environmental Studies Board
National Research Council

NATIONAL ACADEMY OF SCIENCES
Washington, D.C. 1975

This is Volume I of *Pest Control: An Assessment of Present and Alternative Technologies*. The other volumes are:

Volume II *Corn/Soybeans Pest Control*
Volume III *Cotton Pest Control*
Volume IV *Forest Pest Control*
Volume V *Pest Control and Public Health*

NOTICE: The project that is the subject of this report was approved by the Governing Board of the National Research Council, whose members are drawn from the Councils of the National Academy of Sciences, the National Academy of Engineering, and the Institute of Medicine. The members of the Committee responsible for the report were chosen for their special competences and with regard for appropriate balance.

This report has been reviewed by a group other than the authors according to procedures approved by a Report Review Committee consisting of members of the National Academy of Sciences, the National Academy of Engineering, and the Institute of Medicine.

Library of Congress Catalog Card No. 75-43468

International Standard Book No. 0-309-02410-2 (this volume)
0-309-02409-9 (entire set)

Available from

Printing and Publishing Office
National Academy of Sciences
2101 Constitution Avenue, N.W.
Washington, D.C. 20418

Printed in the United States of America

PREFACE

This study originated in discussions in the Executive Offices of the National Academy of Sciences in March 1971. The Environmental Studies Board (ESB)* subsequently continued the discussions and appointed a Planning Committee to develop a detailed proposal. The Planning Committee, sponsored by funds from the Scaife Family Charitable Trusts, submitted its report to the ESB in April 1972. The ESB approved the report and appointed Dr. Donald Kennedy of Stanford University as Chairman of the Executive Committee of the Study on Problems of Pest Control: A Technology Assessment. Financial support for the study was secured from the U.S. Department of Agriculture, the U.S. Environmental Protection Agency, and the Ford Foundation. We thank these public and private agencies for their assistance.

The Committee's charge was to assess the diverse effects on society of efforts to control pests, and to evaluate the long-range alternatives available for accomplishing pest control. Our mission explicitly included an examination of the social and institutional factors--as opposed

---

*The National Academy of Sciences complex (National Academy of Sciences, National Academy of Engineering, and Institute of Medicine) functions as an official but independent advisor to the federal government in matters relating to science and technology through its operating agency, the National Research Council. The National Research Council comprises four assemblies and four commissions, each of which deals with specific disciplines or societal problems. The Commission on Natural Resources, formed in 1973, oversees the work of five boards, among them the Environmental Studies Board. Thus the pest control study, having originated as described above, completed its activities under the aegis of the Commission on Natural Resources.

to the strictly scientific ones--that are important to the conduct of pest control practices and that must be considered in any recommendations for changes in pest control policy. Our charge further directed us to include as "pests" all organisms that man considers detrimental in some way, except those that are directly pathogenic to humans. Thus, pests include insects and other arthropods, weeds, bacteria, fungi, viruses, nematodes, and some vertebrates.

Because a number of evaluations have been made of the external costs of contemporary pest control practices, we did not delve in great detail into an analysis of ecological impacts and health risks associated with the use of pesticides and biological pest control materials. Instead, we accepted the recommendations of other careful investigations of this problem, in particular those produced by the Secretary's Commission.* Where appropriate, we used these findings to make comparisons of the costs of current and prospective technologies.

The Planning Committee concluded and the Executive Committee agreed that the best way to examine pest control problems would be through a series of case studies. In our judgment, units so defined would be more amenable to precise analysis than any others that might have been selected. In addition, because pesticide use is heavily concentrated on a few pests and a few crops, even a limited number of such analyses could, if successful, have a major impact on pest control practices.

Four case studies were selected. Three centered on specific agroecosystems: cotton, corn and soybeans, and timber. The fourth centered on a nexus of issues related to the control of pests of public health significance in the United States and abroad. Each case study examined the pest species involved, current and alternative control measures, and the comparative benefits and costs of controlling the pests by the various methods. The types of alternatives examined included, but were not limited to, genetic methods, biological control, "third-generation" chemicals, cultural techniques, and "integrated" pest management systems.

Our method of comparing benefits and costs of the different control technologies was to ask about each of them this series of questions: What are the prospects for

---

*Report of the Secretary's Commission on Pesticides and Their Relationship to Environmental Health (1969). Washington, D.C.: Government Printing Office.

long-term success in performing the intended functions? What will be the positive and negative impacts on the economy, the environment, and society as a whole? For the new alternatives, we asked these further questions: What incentives and disincentives affect research on their development? What social and economic factors govern their adoption into practice?*

The initial judgment of the Planning Committee was that the role of social and economic factors in governing the adoption of new pest control technologies was of paramount importance. The economic structure of the industries involved with the currently dominant chemical technology affects their ability to produce new control materials. A system of governmental regulatory constraints governs the registration of new chemicals and biological pest control materials. Both public and private institutions are involved in research and development of new technologies and practices and in providing the education necessary for the adoption of new methods. The socioeconomic structure of the agricultural enterprise and the socioeconomic parameters of public health problems can vitally affect the probability of adoption of new techniques and the subsequent impact of any adoption. For these reasons, the study has strongly emphasized an interdisciplinary approach in the organization of each of the study teams assigned to case studies and in the Executive Committee.

The Executive Committee was responsible for organizing the overall plan of the study; for appointing, charging, and coordinating the efforts of the four study teams; and, finally, for developing those areas of this report that are not specific to the work of any given team. Each working group included biologists with a grasp of the ecological and environmental factors involved in pest control, agricultural scientists knowledgable about the system within which pest control is practiced, and social scientists (economists, lawyers, sociologists) familiar with the institutional framework. The rosters of membership for the Planning Committee, the Executive Committee, and the study teams follow this preface.

The reports of the Executive Committee and the study teams are all available from the National Academy of

---

*For a variety of reasons, however, we did not attempt benefit/cost analyses technology by technology. The data bank on which such analyses might be constructed is simply not adequate to yield results sufficiently trustworthy and informative to serve as a basis for public policy formation.

Sciences in five volumes under the general title *Pest Control: An Assessment of Present and Alternative Technologies*. Volume I comprises the report of the Executive Committee and all the summaries and recommendations of the four study team reports. The report of the Executive Committee, in addition to drawing on the findings of the individual study teams, covers other areas not specific to those teams. The remaining four volumes are bound separately as follows: Volume II, *Corn/Soybeans Pest Control*; Volume III, *Cotton Pest Control*; Volume IV, *Forest Pest Control*; and Volume V, *Pest Control and Public Health*.

Donald Kennedy
Stanford University
August 1975

ACKNOWLEDGMENTS

We are grateful to many people who helped in the study through consultation at meetings of the Executive Committee or the study teams, or by recommending persons to serve on the study. They include:

Paul Andrilenas, U.S. Department of Agriculture; Arnold Aspelin, Environmental Protection Agency; Leonard R. Axelrod, Environmental Protection Agency; Helen C. Barry, Food and Drug Administration; Gus Bell, farmer; Carl Bauer, farmer; George Beal, University of Iowa; Bernard L. Berger, U.S. Department of Interior; Parke C. Brinkley, National Agricultural Chemical Association; Ruben Brown, Council on Environment of New York City; Jim Burns, Texas Department of Health; Jack Butler, National Institute of Occupational Safety and Health; Theodore C. Byerly, U.S. Department of Agriculture; Stanley A. Cain, University of California, Santa Cruz; Billy Caldwell, Office of Science and Technology; John Copplestone, World Health Organization; David E. Davis, University of North Carolina; Velmar Davis, U.S. Department of Agriculture; James A. Deer, Texas A&M University; Simon Diaz, University of California, Santa Barbara; Peter Dolinger, Stanford University; Donald Donahoo, U.S. Department of Interior; Thomas Dukes, Environmental Protection Agency; Douglas Early, Texas A&M University; L. E. Ensminger, Auburn University; Donald Farner, University of Washington; William E. Felling, Ford Foundation; E. M. Foster, University of Wisconsin; D. Lee Fowler, U.S. Department of Agriculture; Austin Fox, U.S. Department of Agriculture; John P. Frawley, Hercules, Inc.; Glenn V. Fuguitt, University of Wisconsin; Richard Garcia, University of California; Wayne S. Gardner, University of South Dakota; Leonard C. Gibbs, U.S. Department of Agriculture; Lowell S. Hardin, Ford Foundation; George H. Hepting (Retired), U.S. Forest Service; Fred Honing, U.S. Forest Service; Alton Hornsby, Jr., Morehouse College;

James Horsfall, Connecticut Agricultural Experiment Station; Paul L. Howard, U.S. Department of Interior; Anthony Inglis, Federal Working Group on Pest Management; George W. Irving, Federation of American Societies for Experimental Biology; Edwin Johnson, Environmental Protection Agency; Thomas Jukes, University of California--Berkeley; Ephram Kahn, California Department of Health; Charles D. Kelley, Alabama Department of Conservation and Natural Resources; H. C. Kennett, U.S. Department of Agriculture; Waldemar Klassen, U.S. Department of Agriculture; Henry Korp, Environmental Protection Agency; C. Max Larg, Penn State, Hershey Medical Center; Dale R. Lindsay, Duke University Medical Center; Keith R. Long, University of Iowa; Jerry Longcore, U.S. Department of Interior; George L. McNew, Boyce Thompson Institute of Plant Research; Fowden G. Maxwell, University of Mississippi; Jean Mayer, Harvard University; Fred Miller, Texas Department of Health; Lowell Miller, Environmental Protection Agency; William Mitchell, Environmental Protection Agency; Warren Muir, Council on Environmental Quality; G. A. Niles, Texas A&M University; Lloyd Nolan; T. J. Oleson, National Center for Toxicological Research; Ray Parker, Texas A&M University; Dwight Peabody, Washington State University; David Phillipson, National Agricultural Chemicals Association; Moshe J. Printhus, Hebrew University; Robert L. Rabb, University of North Carolina; John Ritch, Environmental Protection Agency; Jack S. Robbins, U.S. Department of Agriculture; Clark Row, U.S. Department of Agriculture; Peter Rudd, United Farm Workers; R. I. Sailer, University of Florida; Richard A. Schoettger, U.S. Department of the Interior; Nevin Scrimshaw, Massachusetts Institute of Technology; John Siddall, Zoecon Corporation; James Silver, University of Florida; Edgar Smith, Agency for International Development; Max Sobelman, Montrose Chemical Corporation of California; Russell Stevens, National Research Council; Lucile F. Stickel, U.S. Department of the Interior; William Stickel, U.S. Department of the Interior; James H. Street, Rutgers University; Gregory Thomas, Atomic Energy Commission; John G. Thomas, University of Texas; George B. Tindall, University of North Carolina; William Upholt, Environmental Protection Agency; Rosemarie von Rumker, RVR, Inc.; Ralph Wands, National Research Council; Pat Watters, Southern Regional Council, Inc.; R. H. Wellman, Union Carbide; Bruce Wiersma, Environmental Protection Agency; John T. Wilkin, Doane Agricultural Service; and Jerry Young.

Richard A. Carpenter was Executive Secretary of the Environmental Studies Board in the initial stages of the project. He was succeeded by Theodore M. Schad, who administered the Board during the subsequent phases of the study. From the inception of the work of the Planning Committee, John Perkins served as principal staff officer at the National Research Council for the Study on Problems of Pest Control. He participated fully and effectively in the work of the Executive Committee and the study teams and in the writing of the final report. The Chairman and members of the Executive Committee deeply appreciate his efforts. Judith Hough and Howard Engstrom assisted Dr. Perkins and served as staff officers for the study teams on Forests, Public Health, Cotton, and Corn/Soybeans. During the final phases of the study, Bette Lou Fields served as principal staff officer and Robert C. Rooney as editor. Linda Jones, Lucette Comer, Mary Dong, Susan Lesser, Christina Neumann, and Estelle Miller provided able and tireless secretarial support. It is a pleasure to acknowledge the vital contributions made by these and other National Research Council personnel to the completion of the study.

PLANNING COMMITTEE

Donald Kennedy, *Chairman*
Benjamin Scott Crocker
Professor of Human Biology
Department of Biological Sciences
Stanford University
Stanford, California

Wilbur Downs
Yale School of Medicine
New Haven, Connecticut

E. F. Knipling
Plant Industry Station
U.S. Department of Agriculture
Beltsville, Maryland

James E. Krier
Professor of Law
405 Hilgard Avenue
University of California
Los Angeles, California

Jerald Meinwald
Professor of Chemistry
University of California at San Diego
La Jolla, California

William W. Murdoch
Professor of Biology
University of California
Santa Barbara, California

Kusum Nair
Visiting Researcher
East-West Food Institute
1777 East-West Road
Honolulu, Hawaii

Vernon W. Ruttan
President
Agricultural Development Council, Incorporated
630 Fifth Avenue
New York, New York

EXECUTIVE COMMITTEE

Kusum Nair
Visiting Researcher
East-West Food Institute
1777 East-West Road
Honolulu, Hawaii

Charles E. Palm
(ex officio member)
Liberty Hyde Bailey
Professor of Agricultural
Sciences
Cornell University
Ithaca, New York

Vernon W. Ruttan
President
Agricultural Development
Council, Incorporated
630 Fifth Avenue
New York, New York

Roy A. Young
Vice-President for Research
and Graduate Studies
Oregon State University
Corvallis, Oregon

CONSULTATIVE PANEL ON HEALTH HAZARDS OF CHEMICAL PESTICIDES

Matthew S. Meselson, *Chairman*
Professor of Biochemistry
Harvard University
Cambridge, Massachusetts

Bruce N. Ames
Professor of Biochemistry
Department of Biochemistry
University of California
Berkeley, California

Peter M. Dolinger
Research Fellow
Department of Chemistry
Stanford University
Stanford, California

Virgil H. Freed
Professor of Chemistry
Head, Department of Agricultural Chemistry
Director, Environmental Health Sciences Center
Oregon State University
Corvallis, Oregon

John Kolojeski
Special Assistant for Public Health Regulatory Affairs
Office of the Administrator for Water and Hazardous Materials
U.S. Environmental Protection Agency
Washington, D.C.

Robert L. Metcalf
Professor of Entomology, Agricultural Entomology, and Environmental Studies
320 Morrill Hall
University of Illinois
Urbana-Champaign
Urbana, Illinois

Marvin A. Schneiderman
Associate Director for Field Studies and Statistics
National Cancer Institute
National Institute of Health
Bethesda, Maryland

Elizabeth K. Weisburger
Chief, Carcinogen Metabolism and Toxicology Branch
National Cancer Institute
National Institute of Health
Bethesda, Maryland

Charles F. Wurster
Associate Professor of Environmental Sciences
Marine Sciences Research Center
State University of New York
Stony Brook, New York

Roy Hansberry
Retired
217 Stewart Road
Modesto, California

CONSULTATIVE PANEL ON NEW TECHNOLOGIES

Boysie E. Day, *Chairman*
Professor of Plant Physiology
University of California
Berkeley, California

James L. Brann, Jr.
Professor of Entomology
Cornell University
Ithaca, New York

E. Dale Threadgill
Associate Professor
Mississippi State University
P.O. Box 5465
Mississippi State, Mississippi

J. Wade Van Valkenburg
Patent Liaison
3M Company
Central Research Laboratories
P.O. Box 33221
St. Paul, Minnesota

Wesley E. Yates
Professor of Agricultural Engineering
University of California
Davis, California

CORN/SOYBEANS STUDY TEAM

James P. Houck, *Chairman*
Professor
Department of Agriculture and Applied Economics
University of Minnesota
St. Paul, Minnesota

Samuel R. Aldrich
Assistant Director
Agricultural Experiment Station
University of Illinois
Urbana, Illinois

Richard Behrens
Professor
Department of Agronomy and Plant Genetics
University of Minnesota
St. Paul, Minnesota

Walter R. Fehr
Professor
Iowa State University
Ames, Iowa

Arthur L. Hooker
Professor of Plant Pathology and Plant Genetics
University of Illinois
Urbana, Illinois

Richard Levins
John Rock Professor of Population Science
Harvard School of Public Health
Boston, Massachusetts

Eldon E. Ortman
Professor and Head
Department of Entomology
Purdue University
West Lafayette, Indiana

Jerry D. Stockdale
Associate Professor and Head
Department of Sociology, Anthropology, and Social Work
University of Northern Iowa
Cedar Falls, Iowa

Earl R. Swanson
Professor
Department of Agricultural Economics
University of Illinois
Urbana-Champaign
Urbana, Illinois

COTTON STUDY TEAM

Stanley D. Beck, *Chairman*
W. A. Henry Professor
Department of Entomology
University of Wisconsin
Madison, Wisconsin

Perry Adkisson
Professor and Head
Department of Entomology
Texas A&M University
College Station, Texas

Donald C. Erwin
Professor of Plant Pathology
University of California
Riverside, California

Lewis W. Jones
Professor of Sociology
Director, Research Manpower Training and Research Project
Tuskegee Institute
Alabama

William W. Murdoch
Professor of Biology
University of California
Santa Barbara, California

Richard B. Norgaard
Assistant Professor of Agricultural Economics
University of California
Berkeley, California

James S. Plaxico
Professor and Head
Department of Agricultural Economics
Oklahoma State University
Stillwater, Oklahoma

Paul W. Santelmann
Regant's Professor
Agronomy Department
Oklahoma State University
Stillwater, Oklahoma

Christine A. Choemaker
Assistant Professor of Environmental Engineering
Cornell University
Ithaca, New York

A. Dan Tarlock
Professor of Law
Indiana University School of Law
Bloomington, Indiana

Bradford Waddle
Distinguished Professor of Agronomy
University of Arkansas
Fayetteville, Arkansas

FOREST STUDY TEAM

Richard S. Miller, *Chairman*
Professor
School of Forestry and Environmental Studies
Yale University
New Haven, Connecticut

Donald L. Dahlsten
Professor of Entomology
University of California
Berkeley, California

Fred B. Knight
Director
School of Forest Resources
Nutting Hall, U.M.O.
University of Maine
Orono, Maine

Michael Newton
Associate Professor
School of Forestry
Oregon State University
Corvallis, Oregon

William A. Niering
Professor of Botany
Department of Botany
Connecticut College
New London, Connecticut

David E. Pesonen
Garry, Dreyfus, McTernan, Botsky, Herndon & Pesonen, Incorporated
Attorneys at Law
San Francisco, California

William H. Smith
Assistant Dean and Associate Professor of Forest Pathology
School of Forestry and Environmental Studies
New Haven, Connecticut

Earl L. Stone
Charles Lothrop Pack Professor of Forest Soils
Cornell University
Ithaca, New York

Henry J. Vaux
Department of Forestry and Conservation
University of California
Berkeley, California

PUBLIC HEALTH STUDY TEAM

Ray F. Smith, *Chairman*
Professor of Entomology
University of California, Berkeley
Berkeley, California

A. W. A. Brown
Director
Department of Entomology
Michigan State University
East Lansing, Michigan

H. Rouse Caffey
Associate Director
Louisiana Agricultural Experiment Station
Baton Rouge, Louisiana

John E. Davies
Professor and Chairman
Department of Epidemiology and Public Health
University of Miami School of Medicine
Miami, Florida

Walter P. Falcon
Director
Food Research Institute
Stanford University
Stanford, California

Maurice W. Provost
Director
Florida Medical Entomology Laboratory
Florida Division of Health
Vero Beach, Florida

Lloyd E. Rozeboom
Professor of Medical Entomology
The Johns Hopkins University
Baltimore, Maryland

John E. Scanlon
Professor of Medical Zoology
The University of Texas Health Science Center at Houston School of Public Health
Houston, Texas

Thomas H. Weller
Richard Pearson Strong Professor of Tropical Public Health and Chairman of the Department
Harvard School of Public Health
Boston, Massachusetts

CONTENTS

Contemporary Pest Control Practices and Prospects:

THE REPORT OF THE EXECUTIVE COMMITTEE

SUMMARY AND RECOMMENDATIONS

CONTEMPORARY PEST CONTROL

Contemporary pest control in agriculture, forestry, and public health has been a successful technology by a variety of criteria: It has dramatically reduced arthropod-borne diseases on a worldwide basis and has been one of the main factors in achieving high productivity in U.S. agriculture. Toxic chemicals provide the main foundation for contemporary pest control practices; pest control today employs large amounts of toxic chemicals--more than a billion pounds of such materials, worth some $3.4 billion, were used in the world in 1971. Calculations indicate that marginal rates of return per dollar invested in agricultural pest control are in the neighborhood of $4.* Although loss estimates vary widely and are difficult to evaluate, pests clearly cause large annual decreases in yield; in some cases, they could destroy entire crops if control efforts lapsed. There is general agreement that insect control alone annually prevents more than $2 billion in losses in the United States.

Problems Associated with Contemporary Pest Control

The very success of pesticidal chemicals has led to a heavy reliance on them; yet their efficacy--indeed, their continued use--is challenged by several developments. First, public concern about impacts of toxic chemicals on health and the environment has led to their more stringent regulation and, in several cases, to outright prohibitions of

*Headley, J. C. (February 1968) Estimating the productivity of agricultural pesticides, Am. J. Agric. Econ. 50:13.

specific compounds. Some regulations of this type are necessary to protect society against external costs; but where they require the substitution of more-expensive or less-effective alternatives or cause more reliance on a few compounds, they produce problems for pest control decision makers. Second, we have come to recognize that the induction of genetic resistance in pest species and the disruption of natural control mechanisms constitute rising threats to the continued success of our current control technologies. We believe these problems will become more acute, because we predict that incentives for increasing agricultural production will continue over the next decade (see Part II). In this report we draw attention to effects of pesticides on health and the environment-the "external costs" to society--that have been identified by others. We do not, however, attempt a detailed analysis of those costs in relation to benefits conferred by pesticide use. We contend that the problems of genetic resistance and disruption of natural control, lying entirely within the enterprises of agriculture, forestry, and public health, warrant substantial expansion of present efforts to promote alternative technologies, including integrated control strategies. Efforts to combat these problems are handicapped by a lack of knowledge of the basic biology of pest species. Less serious, apparently, is the problem of intrinsic vulnerability of high-yield crop systems, especially to insect pests. Available evidence, though fragmentary, does not clearly support a conclusion that there have been significant decreases in the resistance of crops to insects or plant diseases as a result of selection for high yield, although expanded acreages of genetically uniform crops clearly has increased vulnerability to such losses. Much attention has been given to the deliberate increasing of disease resistance through breeding programs, and similar efforts with insects would be desirable.

We recommend that problems internal to chemical pest control technology warrant a high-priority effort on the part of concerned agencies to evaluate and then to further the development of alternative technologies. At the same time, attention should be given to institutional and technological means by which present methods can be made more effective until the alternatives are ready to provide substantial relief.

In recommendations given below, we consider specific areas, both institutional and scientific, in which such efforts should be mounted.

Problems internal to the enterprises in which pest control is employed are sufficient to support our general recommendation. We recognize, however, that much of the social concern about chemical pest control rests upon perceived costs that, to some degree, lie outside the enterprise. These include, prominently, hazards to human health and to the environment. The Secretary's Commission on Pesticides and Their Relationship to Environmental Health made a detailed survey of these problems and recommended termination of the use of DDT and restrictions on such other chlorinated hydrocarbon insecticides as aldrin, dieldrin, endrin, heptachlor, chlordane, and lindane, as well as compounds containing arsenic, lead, or mercury:

> Recommendation 3
>
> Eliminate within two years all uses of DDT and DDD in the United States excepting those uses essential to the preservation of human health or welfare and approved unanimously by the Secretaries of the Departments of Health, Education, and Welfare, Agriculture, and Interior.
>
> Recommendation 4
>
> Restrict the usage of certain persistent pesticides in the United States to specific essential uses which create no known hazard to human health or to the quality of the environment and which are unanimously approved by the Secretaries of the Departments of Health, Education, and Welfare, Agriculture, and Interior.*

Since that report, the domestic uses of DDT (1970), aldrin and dieldrin (1974), and chlordane and heptachlor (1974-1975) for certain purposes have been terminated by the Environmental Protection Agency (EPA).

Environmental Effects

Hazards to the environment may be posed by pesticides of widely differing chemistry and intended use. Both acute (short-term) and chronic (long-term) effects have been noted across nearly the full spectrum of compounds, though large numbers of chemicals are relatively free of such hazards. The Secretary's Commission reviewed this subject

---

*Mrak, Emil M. (Chairman) (1969) Report of the Secretary's Commission on Pesticides and Their Relationship to Environmental Health, Washington, D.C.: U.S. Department of Health Education, and Welfare.

carefully in 1969. Subsequent information has not led to major changes in our view of the kind or extent of hazards posed, but vigilance is obviously required in maintaining our knowledge of (1) which specific compounds present dangers, and (2) the properties or processes, such as persistence and biological magnification, that might amplify these dangers. In particular, we note that problems of synergism between different kinds of pesticides have surfaced; this area is poorly understood, and new approaches to it need to be developed. Recent lessons have underscored the importance of knowing whether particular cases of environmental contamination result from normal use of pesticides, industrial accidents, or failure to eliminate pesticide contamination from industrial or other wastes.

Research is continuing on the properties of molecules (capacity for biomagnification, transportability, and persistence) that make them likely to produce certain kinds of threats to the environment. We did not identify specific new areas in which policy changes are indicated, but we urge continued attention to the findings and recommendations of the Secretary's Commission.

Health Hazards

A panel on health hazards, The Consultative Panel on Health Hazards of Chemical Pesticides, appointed by the Executive Committee, has examined two aspects of the problem of hazard detection: (1) the use of epidemiological data; and (2) studies on laboratory animals. Both were considered with special reference to the detection of carcinogenic hazard, because this is the problem that has created most concern in connection with pesticides.

The Panel found that no major or obvious alterations in nationwide mortality data can now be associated with the period of onset of major use of modern pesticides. However, more detailed evaluation should be encouraged, including the analysis of the most recent trends of cancer mortality by age, occupation, site, and geographic region.

We recommend that careful and detailed epidemiological studies on the relationship between cancer and other potential risks and exposure to pesticides on the part of the human population are essential to improving our judgment of their potential risk; the gathering and analysis of such data should be supported at higher than present levels. (See Chapter 1.)

A problem of considerable concern is that chronic exposure to a carcinogenic substance may not manifest itself as a noticeable increase in cancer mortality until many years after the substance is introduced into use. In this situation, tests on laboratory mammals afford the only available method for evaluating the risk of chronic human exposure to a potential carcinogen. Laboratory evaluations of carcinogenic risk depend upon assumptions regarding the shape of the dose-response curve and the degree of similarity between experimental animal and man. For several known agents, current data indicate that animals may be no more sensitive than man. It remains possible that some compounds carcinogenic to animals will not be so to man. In National Cancer Institute tests conducted in 1969, 11 of 120 pesticides and other chemicals were positive, and 20 were designated for further evaluation.

We conclude that the carcinogenic risks, and other health hazards, of pesticides require continuing evaluation by testing with laboratory mammals; and that despite the problems involved in translating the results from such experiments to human risk, the present techniques are sufficiently reliable to justify registration actions based upon such data alone, on an interim basis, until evidence convincingly demonstrates that there is no human risk. (See Chapter 1.)

A special sort of health hazard is posed for farm workers by those pesticides having acute human toxicity. Evaluation is hampered by poor reporting systems. Even in California, which has the only effective system, occupational injuries may be seriously underestimated.

Truly occupational deaths from pesticides numbered about 12 in 1969; but as national data are not available for the years after 1970, there is no way to tell whether the partial ban on DDT and other factors have affected this statistic. Preliminary evaluation of state data suggests that the influence, if any, has been slight. The cost due to morbidity, however, may be quite high.

The responsibility at the federal level for regulations to protect farm workers has been subject to jurisdictional confusion between the Environmental Protection Agency (EPA) and the Department of Labor (DOL). Of the states that are in the most logical position to accomplish sound regulation, only California has effective laws. The scientific base for formulating worker protection regulations *is* inadequate.

We recommend that worker protection data be required as part of the registration application. Both the jurisdictional basis for setting re-entry and other standards, and the standards for national reporting of occupational injury and disease among farm workers, require improvement. Concerned federal agencies such as the Environmental Protection Agency, the Department of Agriculture, the Department of Labor, and the National Institute of Occupational Safety and Health of the Department of Health, Education, and Welfare should undertake a cooperative effort to design a data-gathering system (involving mandatory reporting) for occupational pesticide injury in order to insure safety and to avoid unnecessary constraints. (See Chapter 1.)

We note, finally, that defects in the current system with respect to occupational health matters are at least partly attributable to the fact that farm workers in many regions of the United States have not, by reason of linguistic, economic, and other barriers, been able to play a significant role in the determination of public policy.

We recommend that concerned federal and state agencies encourage their participation in the development of occupational health policies regarding pesticides.

The Enterprise of Pest Control: General Theory

The decision to employ contemporary pest control practices is a complex one. It depends as much on the availability of a technology to control a pest as it does on the expected rate of return from employing that technology. The decision maker may be motivated by any of several considerations, such as risk minimization, labor substitution, cost reduction, or yield increase. The specific decision to apply control at a certain place or time has a threshold of action that depends upon a number of considerations, according to the use made of the resource and the information available to the decision maker. These elements define a framework for pest control that is as important as the biology of pest, crop, or vector in determining the timing, style, and magnitude of pest control practice.

The biological elements also vary widely. Important pest categories include fungi and nematodes, other plant pathogens, insects and related arthropods, and weeds. The dollar volume of herbicides sold in the United States has now become much larger than that of insecticides.

Fungicides, nematocides, and vertebrate biocides involve smaller amounts of chemicals than either of the preceding, but categories of each have been involved in regulatory problems. The special biology of each group of organisms has consequences for pest control. Partly because weeds are so different physiologically from man, herbicides have presented relatively few problems in the occupational health area compared with insecticides; and partly because insects often have several generations per year and possess a high potential for genetic change, resistance to pesticides is a problem with them. With weeds, however, resistance has appeared in only a few cases.

At the level of the agroecosystem, which is the functional complex including major crop plant and associated pests, such as parasites and predators, the relationships are still more complex. Most crops are grown outside their area of origin and have pest assemblages derived partly from the new habitat; the community is simplified, mixed, and frequently disrupted. Moreover, its members are apt to have evolved separately instead of jointly, as in natural communities. Modern theory in population dynamics and genetics is beginning to acquire the power to predict some impacts of these factors on the design of pest control techniques. A substantial part of this development is taking place in nonagricultural institutions or departments. We find that this developing body of theory has not in general been well integrated with the research enterprise in applied control.

We strongly recommend that the U.S. Department of Agriculture (USDA) and the State Agricultural Experiment Stations (SAES) support and encourage additional application of contemporary theory in population biology and ecosystem analysis to pest control problems; and, reciprocally, that the USDA, SAES, and other mission-oriented agencies with responsibilities in pest control support and encourage further interactions between theory development and applied research in pest control. (See Chapter 2.)

Research and Development Capabilities

Our capacity to improve the performance of chemical pest control, or to develop acceptable and efficient alternatives, depends upon the research and development enterprises in the private and public sectors. The motivations of private and public research managers are different: Private managers concentrate their efforts on technologies and markets that can be expected to produce profits for corporate owners, whereas public managers are motivated by a complex variety of forces expressed through legislation, agency policies, tradition, and the reward structure of the several professions.

Private R&D resources in pest control are heavily concentrated in the current chemical technology: About 90 firms are engaged in production, and of these 30 produced 80 percent of the poundage and spent $70 million, or 10 percent of sales, on pesticide research in 1970. This percentage and the number of employees engaged in R&D both increased in 1971. There is evidence, however, that an increasing fraction of R&D expenditures went into maintenance (as compared with new product) research, although the EPA figures on rate of introduction of new chemical entities do not reveal a significant reduction. Both the time and expense necessary for the introduction of a new product have been substantially increased. Some industry representatives feel that a decline in innovative research has occurred, which they attribute to the effect of federal regulation; but restrictions on access to information held by private manufacturers make it difficult to document the view that private R&D is losing its capacity to contribute new vitality to the current pest control technology. We believe that serious problems may exist in this area. We make recommendations regarding the regulatory aspect of this problem below. At present, the EPA lacks adequate mechanisms for judging the impact of its own regulations.

We recommend that the EPA improve its capacity for monitoring rates of introduction of new chemicals or technologies by private firms and from the public sector, especially of those representing new approaches to pest control problems. (See Chapter 3.)

Public research on pest control is largely based in the USDA/land-grant-university complex. The two major divisions of this complex, i.e., the SAES and the USDA, spent a total of more than $600 million on agricultural research in FY 1972. About one-fifth of this amount went to research on pest control. Some mission-oriented agencies (the Department of Defense, the Department of the Interior, and the National Institutes of Health) have specialized responsibilities for pest control research, and related basic-science support is provided by the National Science Foundation (NSF) and the National Institutes of Health (NIH). The public institutions support most of the effort designed to improve such alternative methods as genetic, biological, cultural, and integrated programs. In most cases, public research efforts on specific pest problems and on groups of pest organisms tend to be allocated in proportion to the losses experienced. But in certain areas, e.g., the importation of organisms for biological control, the total allocation seems small in light of previous experience, and internal allocations require complex analysis.

Recent criticisms of the efficacy of the agricultural research enterprise were raised simultaneously with a major reorganization of the Agricultural Research Service (ARS), the main research unit within the USDA. It is too early to measure the effectiveness of this reorganization.

We recommend that by 1978 a critical appraisal of the effect of ARS reorganization on research allocations in pest control be made by a group of evaluators outside the USDA. Such an evaluation should also consider the effectiveness of the various inter- and intra-agency coordinating groups in allocating research priorities and coordinating efforts. (See Chapter 3.)

In research on new pest control techniques, the predominance of a single federal agency and related state organizations raises questions about the appropriate role of other agencies. We cited earlier the undesirable separation between basic theory development in population biology and the agricultural sciences. In part, this results from a lack of interest on the part of some basic scientists in applied problems; but the heavily concentrated structure of the agricultural research enterprise is also at fault. These two scientific cultures should be brought together.

We recommend that the NSF and other agencies responsible for the support of basic sciences encourage the increased use in basic research of organisms and ecosystems that may yield benefits to the science of pest control; the USDA and other agencies with a pest control mission should also be funded so that they can make competitive extramural grants for basic research to qualified investigators wherever they are located, in order to enlarge the scope and perspective of research dealing with pest control problems. (See Chapter 3.)

We considered at some length the significance of the growing status of the EPA as a research-granting agency in the areas of its regulatory responsibility, an authority given under the Federal Environmental Pesticides Control Act (FEPCA). We favor diversity in the research enterprise and recognize the EPA's need to obtain scientific and technical information so that it is not dependent solely on information supplied by those it regulates; but we are concerned about the possibilities this raises for conflict with the regulatory function.

We recommend that the EPA evaluate its research-support function, and adopt operating guidelines to insure that its in-house and extramural research activities do not conflict with its regulatory mission. (See Chapter 3.)

The Production, Transportation, and Use of Pesticides

The national system in which contemporary pest control materials are manufactured and used is diverse and complex. The 80-100 U.S. producers of the 1,500 basic registered compounds made 1.2 billion pounds of material worth $1.34 billion in 1973; a quarter of this total was exported. Many of the domestically used chemicals were further processed and packaged by as many as 1,800 formulating concern who marketed over $800 million worth of material to agriculture alone. The primary production of pesticidal chemicals is a level at which we can acquire basic information about use of these compounds, and hence about the extent to which specific ones are likely to be introduced into the environment. Use data, though more directly relevant to this question, are more difficult and expensive to gather; furthermore, because the United States is a major pesticide exporter, production data contain important inter national implications. Although new legislation provides a partial remedy in that production data must be reported t the EPA, public disclosure of such data is still not required.

We recommend that prompt publication of annual productio figures for each pesticide be required as a matter of law. (See Chapter 4.)

A special problem is represented by the use of pesticide for limited purposes. Because of the substantial development and registration costs of new materials for such small markets, some important enterprises are deprived of pest control tools. Sometimes, one of these "minor uses" will unexpectedly fall victim to the shrinkage of a major market: The use of Paris green as a mosquito control agent in the southeastern United States and of zectran as a forest pest control chemical in the Northwest are two examples that appear in study team reports. The Interregional Project IV program is an effort to survey for such problems.

We recommend that the USDA and EPA undertake analysis whenever it appears that the market may be functioning in a way that will result in loss of valuable pest control tools, and recommend public intervention to correct the situation. (See Chapter 4 and Volumes II-V.)

Producing firms exist in a variety of arrangements: as specialist firms, either in the agricultural chemicals or (in some cases) the pharmaceutical field, or as parts of highly diversified corporations. The public sector is involved in a small way in the distribution of pest control materials as well: Sterile male insects in genetic control programs, organisms for biological control, and pheromone traps for sampling are all supplied by the USDA at public cost. In addition, advisory services organized in the public sector supply information and advice that may overlap similar activities organized privately from rather different motivations.

Pesticides are transported as part of interstate commerce, and large amounts are applied by aircraft. Both activities are regulated by the Department of Transportation. Aircraft accidents constitute part of the cost of the present technology; application by air involves an average of more than 30 fatalities each year and an equivalent number of serious injuries. Transportation accidents over the past 2 years numbered about 200, with 35 injuries and less than $100 thousand in damage; but the reporting system needs evaluation, and cost estimates have not been made for these and other accidents. We note, however, that the EPA has recently developed a Pesticide Accident Surveillance System.

We recommend that the Secretary of Transportation and the Administrator of the EPA give priority attention to an evaluation of the reporting system for accidents involving pesticides, including those in the course of transport and aerial application, and to their recommendations for reducing such accidents and the health and/or environmental hazards resulting from them. (See Chapter 5.)

Estimates of total use of pesticidal materials in the United States vary considerably; 800 million to 1 billion pounds seems a reasonable estimate. The fraction employed in agriculture and forestry (about 60 percent of all poundage used in 1971) is increasingly dominated by herbicides, which rose from 33 percent of the total poundage in 1966 to 46 percent in 1971. A few compounds (atrazine, 2,4-D, and propachlor) and a few crops (corn, soybeans, and cotton) account for most herbicide use. Inorganic copper and sulfur compounds, dithiocarbamates containing various heavy metals, quinones, and captan and dodine have been used widely as general-purpose protectant fungicides; other compounds are important as soil fumigants More recently systemic fungicides such as benomyl have been developed for therapy of disease in plants. Of the 170 million pounds of insecticides used by farmers in 1971, about 100 million pounds were accounted for by five synthetic organic compounds: toxaphene, methyl parathion, DDT, ethyl parathion, and carbaryl. The ban on DDT and the shift away from other chlorinated hydrocarbon compounds has produced a sharp increase in the use of organophosphate compounds (e.g., methyl parathion) and a consequent increase in acute and occupational health hazards. As with herbicides, the pattern is a specialized one, particular crops requiring large amounts of certain insecticides. Cotton, corn, soybeans, fruits, and tobacco receive the largest total insecticide treatment.

Several of the study team reports and sections of this report deal with problems that are special to urban or suburban environments. Rodent control in the cities, mosquito control in residential areas, pest control practices carried out in the "suburban forest," and the structural and garden uses of pest control materials by householders are all examples. Although it is true that the vast bulk of all pesticidal chemicals are applied on croplands, the density of application may be higher in lawn and garden uses; and a surprising 19 percent of the value of *sales* of pesticides represent retail sales to individual householders, most of whom are urban or suburban. Pesticides so applied may be especially hazardous, because they are dispersed in a locally concentrated way by untrained persons. A substantial additional volume of material is applied by structural pest control operators, other contract firms, city or parks managers, and industries. It is, therefore, not appropriate to view pest control, as it is sometimes seen, as an exclusively agricultural and rural problem.

It is clearly important to know more about each of the major use categories in order to make good estimates for pest control demand, given various futures. For example, there are indications that the intensity of pesticide use in agriculture may be related to such variables as size of farm and value of crop. We may be able to make projections of these other variables; but unless data are available that relate them quantitatively to level of pesticide use, these projections will yield no information on pest control problems. The Economic Research Service within the USDA, some private firms, and at least one state (California) attempt use surveys. The ERS surveys formerly were done every 2 years; but funding difficulties and delays resulted in such lags that, until early in 1974, the most recent data available were for 1966. Use figures for 1971 have now been published*; but, where opportunities for comparison exist, these do not agree closely with data for California (gathered annually since 1971) or with the results of a private survey commissioned for the Council on Environmental Quality. We believe the state of federal knowledge in this critical area is highly inadequate: The pest control enterprise places a billion pounds of toxic materials into the environment each year, but it is "normal" for us to have only the vaguest idea of how much of each compound was used and where, and even then only after half a decade's lag.

We recommend that the Secretary of Agriculture and the Administrator of the EPA cooperate in planning and implementing efforts to improve both the methodologies and the frequency of surveys on pesticide use. Specifically, we believe that these surveys should be annual, should be published not more than 6 months after the end of the subject year, and should be extended to include pesticides bought by nonfarm firms. We also recommend the use of some mandatory reporting system, such as is incorporated into the California surveys. (See Chapter 7.)

Pest Control and the Law

A complex regulatory structure, engendered in part by concerns about the environmental and health effects of pesticidal chemicals, now affects the registration, sale, and use of these compounds and alternate methods. Although regulation entails costs, it has not been subjected to any formal cost/benefit analysis.

Major responsibilities in the regulatory area fall to the EPA under provisions of FEPCA; these include the formulation of standards, the setting of residues tolerances and

---

*Farmers' use of pesticides in 1971 (July 1974) ERS Report for USDA #252.

registration of materials, the storage of data, the monitoring of some residues, and the investigation of accidents These activities, carried out through the Office of Pesticide Programs, cost $21 million in FY 1974.

The EPA is also responsible for registering pesticides if they meet labeling and performance requirements without producing "unreasonable adverse effects" on the environment; a court decision has held that the administrator must make a calculation of risks and benefits in reaching this conclusion. The law explicitly exempts essentiality as a criterion for registration, which the EPA has taken to mean that a pesticide will be registered entirely on the basis of the information that the manufacturer submits. Hence the availability of alternative control strategies does not affect the registration of a new chemical control agent. The legislation does, however, require that the costs and benefits of alternatives be taken into account in evaluating registrations where risk has already been demonstrated.

The most important aspect of the EPA's role in the regulatory process as it affects innovation is the setting of experimental protocols for demonstrating safety and the procedures for issuing experimental use permits and temporary tolerances. Complaints from registrants have included the following: Signals from the EPA are often conflicting or confusing about what the agency actually requires; the agency is not adequately staffed to deal with peak applications loads, especially where research requirements have a critical seasonal character; and experimental use permit requirements are too stringent. Because these regulatory arrangements strongly influence the incentives for developing new materials and new technologies in pest control, the *administration* of the regulations--as distinct from the provisions themselves--should impose minimal hardships on registrants.

We recommend that the EPA maintain a continuous evaluation of its own effectiveness in processing applications for tolerances, experimental use permits, and registrations, and that it actively seek the assistance of experts outside the Agency, including registrants, in attempting to make these processes more efficient and to minimize their burden to registrants; and further, that detailed studies on the costs and benefits of the regulations themselves be undertaken. (See Chapter 8.)

The impact of regulation is allegedly more severe on the development of new technologies than it is on the development of new chemical entities within the present technology. Specific issues are discussed, and recommendations made, in Part II of this report.

Where possible, regulatory constraints should be applied in such a way as to preserve incentives for innovation. They are, however, not the instruments to use in the primary achievement of this purpose, and we believe it inappropriate to regulate access to scientific data in order to protect innovators by creating barriers to entry. It is our belief that the patent system is the proper device for the protection of proprietary interest.

We recommend that the mandatory licensing-of-data provisions of FEPCA be retained. (See Chapter 8.)

A variety of other areas of federal and state regulatory authority exist. We have already made recommendations in the area of occupational health and safety regarding jurisdictional confusions and the inadequacy of data. The Food and Drug Administration (FDA) is now proposing regulations regarding the chemical content of new genetic strains of crop plants; if these are put in force, substantial delays in the introduction of new varieties resistant to pest attack may be imposed.

We recommend that the USDA and EPA confer with FDA officials regarding the proposed regulations, and attempt to estimate their possible negative impact upon genetically based pest control strategies. (See Chapter 8.)

Other areas of important federal regulation include seed-purity laws, water-quality legislation, and regulations that affect the ability of pest controllers to develop mechanisms for ensuring collective action. Tolerances established by the FDA for pest parts in food directly affect pest control practice. Although originally designed to protect consumers, these regulations now act as strong motivations for chemical pest control in some instances.

We recommend that the FDA revise its regulations regarding pest parts and products in food so as to require that tolerances be established on the basis of health considerations.

Monitoring

Monitoring of pesticide residues in food is the responsibility of the FDA; the USDA carries out monitoring in meat, poultry, and broken egg products. Monitoring activities in the physical environment, in man, and in nonhuman species are carried out by a number of agencies working cooperatively through the National Pesticide Monitoring Program. The operation of the program is in a state of flux; but we can identify defects in a number of specific areas, e.g., the lack of monitoring of air, groundwater, and estuarine and ocean water, the inadequacy of the range of species selected for monitoring, the sample sizes involved in human monitoring, and the uneven relationship between monitoring and research in the various agencies involved in the program.

Other inadequacies are apparent in the choice of organisms to be monitored, the method of capture used, the specific chemicals measured, and the regions sampled, as well as in the organization of monitoring efforts. These suggest inadequate coordination between the several agencies involved, but we recognize that the program is still being shaped.

We recommend that the operating policies of the National Pesticide Monitoring Program continue to be the primary responsibility of a single agency, the EPA; that there be mandatory publication of monitoring results; and that financial support be made adequate to extend the range of species, environments, and human samples. As the program evolves, we believe special attention should be given by the EPA to the connection between monitoring and research and to the integration of monitoring efforts with accurate production and use data. (See Chapter 9.)

PROSPECTS FOR PEST CONTROL

We have attempted to analyze the prospects for changes over the next decade in (1) the U.S. agricultural economy, (2) the patterns of agricultural production that affect pest control practices, (3) the demands for forest products and consequent changes in forest pest control, and (4) the requirements for public health applications. Although predictions are obviously risky, it seems highly likely that there will be continued increases in export demand for agricultural commodities and timber products, and a resulting continuation of domestic policy moves to intensify production. Fluctuating international price structures, continuing fuel shortages, and expected technological innovations in the consumption industries are likely to lead to year-to-year variations in cropping patterns. Changes in such institutional features as average size of holding will also affect pest control practices, as will the price of the petrochemical feedstocks used in pesticide manufacture.

The general effect of these changes will probably be in the direction of more regionally homogeneous, more intensively managed, and more valuable crops. At the same time, because of regulatory constraints and other factors, the number of kinds of chemicals applied to these crops for insect control may decrease. Many of these changes will lead to accelerated loss of effectiveness of chemical control through the more rapid buildup of genetic resistance in insects.

Similar projections can be made for forests, where increases in the demand for recreational uses and possible conversion into cropland is likely to reduce the acreage base slightly, while management is sharply intensified to meet expanded demands for timber products. The impact on pest control will depend on the precise form taken by the response to this demand. Accelerated harvesting of old-growth forests on public lands would reduce some pest problems; intensificaion of even-aged management would increase others. In general, the expanding value of standing timber will provide added incentive for all kinds of pest control measures, but it might also result in improvements in silvicultural practice that would make more chemical control unnecessary.

We foresee the enterprise of vector control in public health as continuing to be essential, but staying about the same size. Increasing problems associated with local losses of control should be approximately balanced by the stabilization of formerly large programs. It will still be necessary to guarantee supplies of critical control tools under circumstances in which all of these are primarily developed and marketed in the context of a variable and expanding agricultural sector.

In summary, the task of pest control over the next 10 years will almost certainly become larger rather than smaller. Contemporary chemical control of some insects (though not of weeds and diseases) is decreasing rather than increasing in effectiveness; and we believe we can identify conditions that will accelerate that loss in the future. Much importance is thus attached to the alternatives to this chemical technology and to the scientific and institutional factors that influence their prospects for further development.

Costs and Benefits of Present and Alternative Technologies

The benefits of using contemporary technology on a wide scale are difficult to measure precisely. Estimates of loss vary widely, and losses frequently are counted more than once. Furthermore, some important nonlinearities are difficult to measure: A change in practice may have little impact on control unless it falls below a certain level or covers an area of critical size. For this reason, we have experienced difficulty in evaluating such institutional recommendations as limitations (whether by regulation or taxation) in the amount of contemporary pest control chemicals used.

We believe that current experimental data on actual losses in relation to intensity of pest control practices are highly inadequate. This is but one aspect of a defect in our knowledge that is reflected in the following analysis of alternative strategies. The agricultural research enterprise now lacks adequate capacity to conduct experiments on the scale necessary to evaluate the impact of changes in practice on a particular crop over a wide area, except for forests. The efficacy of new methods is often rightly suspected by practitioners unless it has been shown on a scale that is seen to belong to the real world. As we show below, the only projects of such scale to have drawn significant support to date, unfortunately, are eradication programs.

We recommend that high-priority consideration be given by Congress and the USDA to the design and funding of large-scale (i.e., practical) experiments (1) to measure the economic consequences of altering the intensiveness of present pest control practices on major crops and (2) to evaluate the impact of substituting alternative strategies.

Such tests are required for more accurate cost-benefit assessments of present and substitute technologies and also for mounting effective and convincing programs that may lead to their adoption.

"Third-Generation" Chemicals

Substantial publicly funded basic research, as well as some private developmental activity, has been invested in the synthesis and testing of so-called "third-generation" compounds that intercept hormonally controlled developmental or reproductive processes in insects. Unlike contemporary pesticides, which were selected for their general toxicity, these compounds affect biochemical processes that are unique to insects; hence, the chemicals are more specific and pose fewer hazards for human workers and consumers. Temporary tolerances have been established for one mosquito larvicide that shows promise for public health applications. Unfortunately, there is reason to be pessimistic about the prospects for controlling major (especially lepidopteran) crop pests with these compounds. Resistance is expected to develop in response to them, just as it has to contemporary pesticides. Further advances in this area depend, among other factors, upon (a) the willingness of private firms to invest in relatively narrow markets in the face of substantial regulatory constraints designed for a more dangerous group of chemicals and (b) a continued flow of research findings on basic insect biochemistry and physiology from publicly and privately funded university and government laboratories. Although we expect the third-generation compounds to be highly useful in certain applications, it is doubtful whether they can significantly replace contemporary methods on major agricultural crops.

Control by Pathogens

Both bacteria *(Bacillus thuringiensis)* and insect viruses have been used against crop pests on a commercial scale in the United States and other countries. *B. thuringiensis* is marketed for use on cotton and other crops by at least two private firms in this country, and temporary tolerances have been granted for one nuclear polyhedrosis virus (NPV). The NPV's and granulosus viruses (bacculoviruses), because of their specificity and variety, offer great promise as control agents. Further development is inhibited by two factors. First, the language that defines a pesticide in the Federal Insecticide, Fungicide, and Rodenticide Act (FIFRA) (as amended by FEPCA) includes, probably unintentionally, microbial organisms and perhaps other "biologicals" with chemicals. Caution is well-advised when dealing with a potentially virulent organism that can replicate itself in nature; but all present evidence indicates that the NPV's and the related granulosus viruses--the bacculoviruses--cannot infect vertebrates and most invertebrates. Second, the large-scale growth of viruses is dependent upon improved methods for culturing insect host tissues.

In view of the promise of the bacculoviruses as agents for insect control, we recommend that the EPA and USDA extend the efforts already begun to undertake a comprehensive review of the potential threats of these organisms to man and the environment. If, as present data suggest, they present relatively slight risk of virulence to man and to nontarget species, we believe that consideration should be given to revising registration protocols so as to substitute tests appropriate for the properties of living pathogens. (See Chapter 11.)

As with the case of the third-generation chemicals, the development of new agents depends critically upon basic research advances--in this case, from the fields of insect pathology and tissue culture. These needs exemplify the importance of funding of basic research by agencies having applied missions in pest control practice--a recommendation made in other parts of this report.

Genetic Techniques

Genetic methods for pest control include two disparate technologies: manipulating, by breeding, the resistance of crop plants to pests and introducing genetically modified (sterile or autolethal) pests into natural populations for control purposes. The first method includes the oldest and still among the most effective techniques for the control of plant pathogens. Resistance to insects can also be manipulated genetically in some instances; an important aspect is the possibility of changes in vulnerability to insect attack as a consequence of genetic improvement for other purposes. Though private-sector activity in this area is increasing, continued support of research programs in the public sector will be essential for the future success of these methods.

Insect control by the so-called "sterile male" technique has been effective in some situations, but it is limited to circumstances in which the population has been reduced to low density by other techniques. The large-scale rearing facilities required and the narrowness of potential markets make it likely that developments in this area will remain in the public sector. Where these and other methods involve production activities by federal agencies, we make the following recommendation:

We recommend that the costs of production and distribution, and a share of the research costs, be charged to the beneficiaries of the technique wherever they can be identified. (See Chapter 11.)

Cultural Practices

This term embraces a variety of techniques used by growers to achieve crop production goals, of which pest reduction may be an important part. Weeds, insects, and pathogens can all be affected by such practices, which may involve adjustment of planting times, irrigation methods, destruction of plant remains, use of trap crops, and other methods. The practice of "organic" farming represents an application of these techniques to the exclusion of practices that depend upon manufactured agricultural chemicals; a market for such produce that may have some influence on pest control practices now exists, and some "organic" gardening strategies may be worthy of study, although at present their impact appears likely to be greater for the home gardener than the large-scale grower.

Research on cultural practices is largely located within the USDA/land-grant-university complex, and adoption depends heavily upon the Cooperative Extension Service system. Many of these methods require uniformity of practice if they are to be successful, and they thus need enforcement mechanisms. Pest control districts provide a means of employing the legal power of the state or federal governments to enforce practices. For the success of certain kinds of cultural practices, we regard it as important that access to such devices be ready.

Formulation and Application

There is substantial room for innovation in the methodologies used to formulate and apply contemporary pest control chemicals; they could result in lower dosages per unit of control, lower residues, and improved worker safety. In general, for example, little attention has been paid to integration of the overall design of aircraft with that of the spray equipment; the situation is better with ground spray rigs. Atomization technology has still been unable to produce uniform and controllable drop size; further advances in ultra-low volume application will depend on improvements in this area. In the area of formulation, much remains to be done with microencapsulation methods, systemic formulations, and surfactant carriers for herbicides.

We recommend that the area of formulation and application technology be given a higher research priority in the USDA, because it is a promising way to reduce the cost of pest control as well as the unwanted side effects of pesticide use. (See Chapter 11.)

Integrated Control

More modern schemes for pest management take advantage of ecosystem features to maintain pest populations at economically acceptable levels. These strategies may involve altering conditions so as to augment populations of parasites or predators (natural control), rearing and releasing parasites and predators in cases where the pest is exotic (classical biological control), or combining these forms of density-dependent regulation of pest populations with a variety of other management tools, including genetic cultural, and chemical methods (integrated control). The common theme is an emphasis on agroecosystem ecology and on containment rather than catastrophic reduction of pest populations.

The rearing of predators and parasites is a complex and expensive task now located almost entirely within the public sector and likely to remain there. Substantial investments in facilities and research are involved. It is possible that the FEPCA definition of *pesticide* (so troublesome in the area of microbial control) may also induce inappropriate regulation in this area. We therefore recommend, in this context as well, that the definition be clarified to avoid this result.

The ecologically based methods of pest control are coming into increasingly frequent use, but they are not yet as widely employed as they might be. Among the several barriers to their further expansion are: (1) they depend heavily on research specific to each managed agroecosystem; and (2) they often require that the grower have access to rather specialized advice. We reiterate:

We recommend that the agencies concerned with pest control problems recognize their dual responsibility to the support of basic research on ecosystem dynamics and the biology of pest organisms, and to the training of the pest management specialists who will be needed to guide growers in the application of the new knowledge. (See Chapter 11.)

Eradication as a Control Strategy

In the reports of each of the study teams, reference is made to attempts to eradicate particular agricultural insect pests or arthropod disease vectors. Such efforts often attract support for expensive ventures, because, if successful, they yield a permanent benefit without continuing control costs. Experience shows that, although this goal has been achieved in the case of some isolated pest populations, it is difficult to attain. Even such successes as the screwworm in the southwestern United States have involved continuing control costs. The strategy is still being strongly recommended, as in the recent proposal to mount a program costing nearly $1 billion to eliminate the boll weevil as a cotton pest in the southern United States, a project evaluated in some detail elsewhere in this report. We believe that such programs, in the event of failure, may endanger public confidence in the alternative strategies that are often employed in eradication attempts.

We recommend that agencies concerned with agricultural pest and vector control, support eradication programs only where the geographic isolation of the area or other features give especially strong reasons for believing that they will be successful; and that when they are undertaken, care be taken to advise the public fairly of the chances and the consequences of failure. (See Chapter 11.)

Assisting Technology Transfer

It is difficult to determine whether or not fear that alternative pest control strategies will fail is a major barrier to their adoption. On the assumption that it is, others have argued for various schemes designed to facilitate the transfer of new technologies, including insurance devices. Although there are substantial problems in the identification of risks and the estimation of losses, we believe that such institutional mechanisms show promise.

We recommend that the USDA and EPA sponsor additional research on insurance and other institutional devices for assisting the adoption of new pest control technologies. (See Chapter 11.)

Finally, we recognize that the development of new and multidisciplinary approaches to pest management may outstrip the capacity of traditional systems to deliver the technology. Improvements in extension and other public services may require new kinds of trained personnel and supporting research enterprises.

We recommend that the USDA and Cooperative Extension Services (CES), as part of the effort described in the preceding recommendation, develop estimates of the need for future pest management services and programs for training the personnel necessary to implement them.

PART I

CONTEMPORARY PEST CONTROL PRACTICES

1

ACHIEVEMENTS AND PROBLEMS IN CONTEMPORARY PEST CONTROL

PESTS

Pests are organisms that diminish the value of resources in which man is interested. They may interfere with the production and utilization of crops and livestock he uses for food and fibers, adversely affect his health by acting as vectors for disease, or reduce the perceived quality of an environment he chooses to live in. Biologically, pests span the range of living complexity, from the viral pathogens that affect crops and livestock to the rodents that rob the granary; ecologically, their relationship can be that of competitor (weeds), consumer (insect herbivores), pathogen (fungi), or vector (e.g., malaria-bearing mosquitoes). For the purposes of this study, we have included as "pests" all organisms that man considers detrimental in some way, except those that are directly pathogenic to humans.

The biologic role of pest organisms is as variable as man's definition of them, and so are the motivations that cause human societies to adopt control procedures. Some pests, including various fungal root diseases, competing weeds, and others, reduce crop yield in a linear fashion over a wide range of population densities. In such cases, control will increase continuously as long as the crop price guarantees that the marginal gain in yield will pay for the incremental cost of control. More commonly, an agricultural pest causes little or no damage when scarce, but can potentially erupt to population densities that cause devastation of entire crops. In this case, uncertainty about the likelihood and extent of pest eruption can cause managers to employ control as *insurance*, but the rational manager will limit his control efforts to actions that yield a positive economic return over a period of years. Arthropod disease vectors are likely to be considered an unacceptable threat at whatever population density they occur. Serious efforts at eradication, in this case, are likely to occur on a continuous basis.

These illustrations do not begin to cover the range of human motivation in pest control. They do, however, suggest that the activities of pests--and our attitudes toward them--are sufficiently diverse to have produced marked categorical differences in control practices. In designing or evaluating alternative strategies, these motivational differences must be dealt with just as realistically as the differences in the biology of pest organisms.

ACHIEVEMENTS OF PEST CONTROL

The technologies available for controlling unwanted organisms are far more diverse and efficient today than they were 50 or even 25 years ago. Significant biologic advances have occurred in the development of genetically resistant plants and animals, in the introduction of exotic organisms, in autocidal control techniques, and also in increasing the efficacy of contemporary methods. The formation of legal pest control districts and the increasing importance of cultural practices are notable examples of socially spawned contributions to the solution of pest control problems. Nevertheless, the general toxicity and application efficiency of chemical control agents has led to their widespread adoption; and the capacity of toxic materials to deal at low cost with pest problems in agricultural ecosystems has constituted a disincentive for the development of alternative technologies of control. Man's effort to control pests with chemicals is a significant activity: In 1971, $3.4 billion worth (retail) of such materials was applied on a nationwide basis for agricultural (including forestry), industrial, and household use--about half in the United States (Where Are the Markets? 1973).

The returns for pest control expenditures in agriculture, forestry, and public health come in a variety of forms, ranging from increased yields to peace of mind. Secondary returns have included the increased ability of users to adopt other new technologies that could not have been implemented without the simultaneous use of pesticides.

The steady improvement in worldwide agricultural yields is directly attributable to the use of pesticides and such other chemical inputs as fertilizers. Chemicals, however, represent only one class of increased input that contributes to higher yields. Other factors include irrigation, improved germ plasm, mechanical devices, and cultural practices that increase and stabilize overall land fertility. Humans, however, are currently dependent on pesticidal chemicals to maintain present levels of agricultural production; their removal from use would lead to an immediate drop in food supplies.

It is difficult to estimate the levels of loss that would occur without pesticides. The USDA (1965) estimated that in the early 1960s insects alone did $3.8 billion worth of damage per year. Such estimates depend heavily upon educated guesses and assumptions about commodity prices under conditions of changing supplies. But there is no doubt that in certain agricultural sectors, economically disabling losses would result if the chemicals were not available. Furthermore, the calculated marginal return for the dollar invested in pesticidal chemicals is strongly positive for agriculture as a whole: Estimates by Headly (1968) indicated that the value was approximately $4 in the mid-1960s.

Pest control for public health purposes is no less dependent upon chemical technology. The eradication of malaria and other devastating disease problems in many parts of the world is properly credited to DDT and other chemicals that made it possible to control arthropod vectors efficiently and cheaply.

The unqualified successes of pesticides have created a dilemma: Agriculture, forestry, and public health are now quite dependent upon chemical control technology. Yet, contemporary pesticide technology shows signs of vulnerability that distress those concerned with the future of crop protection and vector control. Apart from rising concern about the environmental or health hazards of pesticidal chemicals, a subject touched upon only briefly in this report, *the efficacy of some of the existing compounds is decaying* as a result of genetic resistance in pest organisms and other factors. Man's commitment to the enterprise of pest control makes it of critical importance that limitations on our current technology be carefully explored before turning to prospects for alternatives.

PROBLEMS WITH CONTEMPORARY PRACTICES

The Impacts of Public Debate and Public Policy

The intense public attention focused on pest control practice during the past decade has precipitated instances of sharp and emotional debate. The publication of Rachel Carson's *Silent Spring*, the use of herbicides in Vietnam, and the campaign of the United Farm Workers against occupational hazards posed by certain pesticides are familiar examples of controversial foci involving pesticides and pest control. Each of these events spawned polarized groups that engaged in adversary activity; in addition, the Congress has passed new laws governing the conduct of pest control practices. The National Environmental Policy Act (P.L. 91-190) (NEPA), the Occupational Safety and Health Act (P.L. 91-596) (OSHA), and the Federal Environmental Pest Control Act (P.L. 92-516) (FEPCA), are the most prominent additions. Administrative proceedings and improved toxicological methods have led to the virtual banning of a variety of formerly widely used pesticides, including DDT, aldrin, dieldrin, Compound 1080, and certain mercurial fungicides. Still other commonly used products, including chlorinated hydrocarbon insecticides, arsenicals, and certain phenoxy herbicides, could also eventually pass from common use.

We do not intend, at this point, to assess in detail these debates and policy changes or describe their effects on different groups of people. For the present, we merely note that all of them tend to constrain the uses of pesticides and that they therefore can have an immediate and sometimes significant impact on how business must be conducted. The constraints of FEPCA and of the administrative prohibitions on chemicals, for example, may have affected the willingness of the private chemical industry to undertake research and development activities to develop new materials for pest control. Similarly, occupational safety requirements of OSHA may have influenced the way in which farmers conduct pest control operations. Emergency control of mosquitoes and other nuisance or disease-vector organisms has been affected by the requirements of NEPA, FEPCA, and the administrative bannings of chemicals. Indeed, it can be argued that pest control activities of all types have been made suspect by the acrimonious nature of the public debate.

The constraints on pest control emanating from the body politic are not likely to be temporary. Whether one applauds or decries the constraints, they will continue to constitute a "problem" in that the status quo has been disrupted and a variety of economic, environmental, and social impacts will result.

We will now consider certain biological and medical aspects of contemporary pest control problems.

Genetic Resistance of Target Pests

Target Pests

One of the most troublesome impediments to the continued use of chemical pest control has been the tendency of target pest organisms to evolve resistance to the compounds used. Such resistance may depend on a variety of mechanisms. Perhaps most commonly, it involves the biochemical capacity of the pest organism to metabolize the chemical so that it is converted into nontoxic products. But other modifications may accomplish the same purpose. The sensitivity of critical target enzymes may change, the permeability of the organism or its internal organs may be altered, and changes in behavioral mechanisms may affect the contact of the organism with the chemical. Natural populations of pest organisms are apt to be diverse with respect to one or all of these features. Upon repeated exposure of the population to chemical stress, those individuals possessing the resistance mechanisms survive and leave relatively larger numbers of progeny than nonresistant individuals; and the frequency of resistant genes is thereby increased in the population. (The problem is quite analogous to that of drug resistance in human microbial pathogens.)

In theory, genetic resistance could be shown by any group of pests subject to control by toxic chemicals. In practice, however, the most serious problems of this kind have arisen among the arthropod pests. During the quarter century since the first discovery of resistance to DDT in the housefly, some 200 other species of arthropods have been shown to exhibit genetic resistance to chemical insecticides (Georghiou 1972). Included in the roster of resistant organisms are most of the major pests of agricultural and public health importance, at least in some regions and to some chemicals. For example, programs for the control of malaria are in serious jeopardy in some areas because of the resistance of mosquitoes to DDT and to other effective and economical means of chemical control. A typical illustration of resistance in an agricultural pest is offered by the history of chemical control of mites in Michigan apple orchards. Spray recommendations of the Cooperative Extension Service for the period 1952-1973 have included 20 different miticides that were recommended for 2 or more consecutive years. Of these, only five are still recommended for 1973; all of the rest have been discontinued due to the development of resistance.

Although the development of resistance in a pest population treated for control is a severe problem, there are now well-documented examples of the creation of resistance in species that shared the same geographic range with a control target and hence came under similar selective pressures. Resistance development, like the inadvertent killing of beneficial species, can thus result in the creation of "secondary pests." The threat to man posed by such cases of compound resistance is well illustrated by the situation in Central America. Insects attacking cotton have developed resistance to insecticides sufficient to make production of the crop uneconomical over wide regions. Moreover, the insecticides applied to cotton insects have inadvertently produced resistance in arthropod disease vectors (see Volume III, *Cotton Pest Control*, for details). Should the latter resistance result in a breakdown of efforts to control the disease, severe harm could result to local human populations.

Considerable research has been devoted to understanding the mechanisms by which genetic resistance may be conferred upon pest organisms. Studies on resistance to the synthetic organic insecticides, for example, have established that either novel or increased levels of detoxifying enzymes often characterize resistant individuals. Insect populations are quite likely polymorphic for different forms of these enzymes. DDT resistance may depend on the enzyme DDT dehydrochlorinase, whereas resistance to organophosphate and carbamate insecticides may involve several types of enzymes, most prominently mixed-function oxidases, alkyltransferases, and esterases. Differences in penetration rate between resistant and nonresistant individuals have also been noted. In some instances a reduced susceptibility of acetylcholinesterase to inhibition by the toxin is involved. These mechanisms have recently been reviewed by Georghiou (1972).

Unfortunately, our understanding of selection for pesticide resistance does not begin to match our knowledge of the biochemical mechanisms underlying it. We do know, at least, that nearly every species of insect is potentially capable of developing resistance to any kind of insecticide. At one time, it was believed that resistant genes were extremely rare in wild populations of insects that had never been exposed to chemicals as selective agents; but this belief was probably caused by a general tendency on the part of biologists to underestimate the degree of genetic polymorphism present in natural populations. At present, it seems sensible to assume that most natural populations have the genetic diversity required for the development of resistance. The rate at which resistance develops will depend upon the stringency of the selective regime. Obviously, concentrations of chemical treatment so high that all pests die can lead to no selection; neither can concentrations so low that they fail to affect the reproductive potential of individuals of normal genotype. At some level below the former, the rate of response to selection should be highest. In practice, the point of maximal response rate depends on the genetic nature of the system controlling resistance. For example, in the resistance of mosquitoes to dieldrin, a single dominant gene confers the trait. In such a case, very high selection pressures would produce the fastest rates of evolutionary change (Brown 1971). In the case of most chemicals, the genetic basis for resistance is much more complex: Inheritance is apt to be multifactorial, and the genes involved are often recessive. In these instances, more moderate levels of selection actually appear to produce faster responses.

Most of the rather scanty data on relations between selection rate and selection pressure have been obtained from experiments performed on insects under laboratory conditions. In the field, information is even sparser; and the situation is vastly more complex in view of the heterogeneity of the selective environment to which the population is likely to be exposed. Moisture, soil conditions, temperature, developmental stage, and exposure to other stresses can all modify the pests' sensitivity to the toxin by orders of magnitude. Furthermore, the effective concentration in different microenvironments--even within one field of a single crop treated in a single application--will vary widely.

How an insect population will respond to an unevenly distributed selective agent also depends upon the extent to which genes are able to be transferred through the population. Little is known about this feature for most pest species of insects. But in some lepidoptera it has been shown that subpopulations separated by only a few tens of meters can show considerable isolation from one another--despite the fact that individual insects are strong fliers and appear capable of moving freely from one region to another. Yet even so, the degree of *genetic similarity* between the subpopulations may be quite high. Thus, although it is possible that in some pest species behavioral or other barriers provide effective isolating mechanisms for subpopulations having the range of a single field or cultivated area, it is difficult to predict the actual magnitude of gene flow; because gene flow itself depends (among other things) on the intensity of selection. One might expect great regional heterogeneity in resistance, with the consequence that excessive control programs mounted by individual growers would result in relatively low costs to others.

In cases in which gene flow is relatively unconstrained, local variation in dosage and sensitivity is likely to insure that at least some pockets of individuals are always exposed to concentrations of the chemical that result in maximum rates of resistance development. Each of these local pockets will respond in its own way, depending on the particular set of genes it possesses; when it expands to recolonize areas left bare by the death of nonresistant individuals, recombination will accelerate the adaptation process.

Thus, although we can make some rather general theoretical predictions about the development of resistance in pests and its relation to the concentration of the chemicals applied, data from actual experiments are not readily applicable to agroecosystems. Furthermore, not enough is known about the population genetics of major pest species to allow us to apply the existing theory as carefully as we would like to. About all that can be said, in summary, is that some dosage rate will produce the maximum rate of resistance development. Because of the inconsistency with which chemicals are applied, this optimum relationship is likely to be achieved somewhere within the treated area, and resistant "pockets" of pest organisms will develop. How effectively these spread will depend upon the mobility of the pest, the genetic isolation of the "pocket," and other parameters.

Of course, predators and parasites are also exposed to the selective pressure of chemicals and may be expected to develop resistance, too. It would be reassuring if beneficial species in fact showed the same rate of resistance development as pests; but in practice this does not seem to be the case. The explanation is not exactly clear. Some predators have rather long generation times compared to prey species; but other predators, and most parasites, do not. Because predators and parasites in general are highly mobile compared to pests, the local pockets of resistant individuals are much less likely to develop. In agreement with these suppositions, those predators in which resistance has been found have low dispersal powers and a generation time shorter than that of their prey (Georghiou 1972). Among other possible explanations for the lower incidence of resistance in predators are: (1) that herbivore populations are normally larger than those of predators and hence likelier to harbor any given rare genotype and (2) that herbivores (but not predators) may have been "pre-adapted" by exposure to the selective influences of protective chemicals possessed by the plants.

Weeds, in contrast to insects, tend to have long generation times. Furthermore, many kinds of weed seeds are mobile or are capable of existing in a dormant state in the soil for long periods of time. Thus, evolutionary rates tend to be slow, and resistant genes, even where produced, are effectively diluted by a continuous input of nonresistant individuals. Nevertheless, weed control literature is full of cases in which control is said to be "difficult" because of decreased effectiveness of the herbicides. It is not clear what proportion of these cases involve true genetic resistance. For example, common groundsel shows resistance to simazine and atrazine (Ryan 1970). In several weed species, different clones or subpopulations show markedly different susceptibility to herbicides, and it may be inferred that the underlying mechanism is genetic. In the field, the problem has been serious enough to produce suggestions that herbicides should be rotated to make resistance development less likely. In general, we believe that the potential for resistance to herbicides by weeds has been underestimated and recommend that weed scientists maintain a careful watch for resistance development. The fact that plants can express resistance to herbicides, of course, may be potentially important in a positive sense: In principle, crop plants with high herbicide resistance could be selected, permitting more selective action of those herbicides against weeds (Pinthus et al. 1972).

Pathogens

Changes in the pathogenicity of plant disease organisms may result from mutation and genetic recombination. In general, development of resistance to the newer fungicides, which are inhibitors of specific enzymatic processes, has been more rapid than to the older compounds, which primarily are general inhibitors of protein synthesis. New races occur through hybridization in the sexual stage of fungal reproduction. This phenomenon has been studied intensively in several fungi, particularly the organisms that cause flax rust, black stem rust of wheat, certain smuts, and the apple scab disease. The rise and fall of resistant wheat varieties has reflected the constant change in races of *Puccinia graminis tritici*, most of which have arisen as a result of hybridization on the alternate host. Genetic recombination in the black stem rust fungus may also result from other interchanges between mycelia. The continual development of new races of rust and smut fungi make it essential that reservoirs of resistance to different pathogens be maintained in crop-breeding programs so that seed can be developed quickly to avoid severe disease outbreaks when a new race of rust or smut fungus develops.

Summary

Genetic resistance is a mechanism of crucial importance: It is a source of built-in obsolescence for many of the foreseeable strategies of chemical control; and, because of it, the cost of achieving a unit of protection rises continually. Yet, we do not know in any detailed way the relationship between the concentration of a toxic chemical and the rate of development of resistance to it, especially when the spatial heterogeneity characteristic of natural situations is a factor. The form of the relationship undoubtedly depends upon mode of inheritance of the resistance, its biochemical mechanism, and, in particular, the extent of genetic interchange in the pest population. Typically, exposure of the pests to chemicals is intensive, and selection is of the kind that population geneticists term *catastrophic*. The resistant character is often negatively correlated with fitness; but, in the insecticide-free environment, continued selection usually integrates the genetic innovation into its background, so that overall fitness returns to the level of the nonresistant population. Furthermore, the resistance often involves biochemical modifications that have fairly wide applicability, so that cross-resistance to compounds other than the selected agents is often found. These two characteristics, of course, exacerbate the control problem by guaranteeing that the pest organism is more likely to have decreased than increased vulnerability to other methods of chemical attack.

Finally, we know less than we would like to know about "natural" resistance. Plants synthesize compounds that are toxic to herbivorous animals of various kinds and that may function as protective agents in nature. There are indications that plants having a variety of these compound with no single one dominating, enjoy more effective protection than plants employing large concentrations of only one substance. Studies on the biology of such natural defense systems may help us formulate the theory we need to mount effective defenses ourselves.

In short, there are few areas in which sound experiments and good hypotheses would be more valuable to the conduct of the pest control enterprise than in genetic resistance.

Disruption of Natural Control

Whenever chemicals are employed against pests, there is a likelihood that nontarget species will be affected as well. When these nontarget species have little or no effect on the population density of the target or other pests, the costs depend entirely on the value of the inadvertently killed organisms and are reckoned as a part of the environmental hazards of pesticide use. But frequently chemicals uses in pest control affect natural enemies of the target pests or of organisms that, if relieved of predation, become pests. Thus there is a cost of chemical pest control that is largely internal to the agricultural enterprise; this cost results either from the resurgence of a target pest or from the activities of what in effect is a pest of human manufacture--a secondary pest.

Most of our knowledge of the effects of disruption of natural control involves insects and mites. The predator and prey are often reasonably closely related in these cases--at least belonging to the same biological class--and most of the compounds that now dominate the chemical insecticide market have a broad spectrum of toxicity. The result is that in a great many instances, the grower who decides to treat a pest outbreak with these compounds runs a heavy risk. A broad coverage of this issue is clearly beyond the scope of this volume, but an example should serve to illustrate the point.

In the San Joaquin Valley, in California, two cotton pests of importance are *Lygus* and *Heliothis*. The former, a plant-feeding bug that appears in early- and mid-season, is often subjected to treatment with organophosphate insecticides. *Heliothis*, the bollworm, is a lepidopteran pest that appears in late season; and its outbreaks are particularly severe when earlier chemical treatments for *Lygus* have been heavy. The bollworm is normally kept in check by a complex of natural enemies that include several species of predaceous bugs (*Nabis* and others) and the lacewing fly (*Chrysopa*); when these beneficial insects are destroyed by insecticide applications, bollworm populations build up rapidly in the absence of predation (van den Bosch and Hagen 1966). Entomologists at the University of California, Berkeley, have conducted experiments on the biology of these predators and also on economic damage thresholds of *Lygus*. These findings have enabled them to recommend for control measures against *Lygus* more conservative procedures, which allow growers to maintain better late-season natural control over *Heliothis*. As a result of their experiences and these efforts, growers in the San Joaquin Valley are widely aware of the relationship between early-season spraying and late-season bollworm infestations; and, for them, the decision whether to spray for *Lygus* involves a calculation of these costs.

Similar problems have been well documented in other insect groups and for other chemicals. The problem is not restricted to the organophosphates, although these compounds are especially broad in their toxicity; early cases, involving the relationship between the codling moth and outbreaks of several different secondary pests, resulted from the use of DDT control in fruit orchards.

Disruption problems have not as yet been identified for weeds, even though it is known that a number of economically important weed species are controlled by natural populations of phytophagous insects. It seems reasonable to suppose that the use of chemical insecticides might in some cases induce certain species of weeds as secondary pests, but to our knowledge no documented case has yet appeared. A somewhat analogous problem, however, does occur with the use of herbicides against weeds. As documented in Volume II of this series, *Corn/Soybeans Pest Control*, the extensive use of herbicides in corn frequently produces species displacement, whereby the target species is replaced by another that may pose equal or, in some instances, more severe problems. In this case, the replacement weed can be thought of as a secondary pest. It does not, however, result from the removal of a predatory check, but rather from the displacement of a competitor. The responses are similar in that they represent predictable adjustments of an agroecosystem to the removal of agents in the control of numbers. The costs, though hard to reckon, are clearly often high.

Vulnerability of High-Yield Systems

The vulnerability of food production systems has received much recent attention in both scientific and popular publications (NAS 1972a). Increased use of fertilizers and pesticides, both in countries with advanced technologies and in developing countries where the Green Revolution is an important factor in food production, has led many to assume that high-yield systems per se are more fragile and thus must be protected to a greater extent than their lower-yielding counterparts.

The vulnerability issue and its possible implications for pesticide requirements has three components:

1. a narrow germ plasm base within species;
2. lack of regional crop diversity; and
3. possible intrinsic susceptibility of high-yield systems to pests.

Narrow Germ Plasm Base

This subject was thoroughly reviewed in the National Academy of Sciences monograph (NAS 1972a) and is discussed in Volume II, *Corn/Soybeans Pest Control*. The genetic base of crops should be as broad as is compatible with practical needs for uniformity as to size, maturity, and quality attributes for the intended use; there is no need for uniformity in the genes for pest resistance (provided, of course, that there are available diverse sources of resistance) (Hooker 1972). Fortunately, some factors militate against the probability that the potential for disaster from a narrow genetic base will be realized. The potential threat is only from new diseases and insects (or modifications in pests previously encountered), because modern high-yield types undergo rigorous selection for resistance or tolerance to existing pest problems.

Modern corn breeders, for example, not only subject their nurseries to natural levels of insects and diseases, but also frequently inoculate materials under test with leaf (*Helminthosporium turcicum*, *H. maydis*) and stalk rot (*Gibberella* and *Diplodia*) disease organisms. Furthermore, they frequently send breed material to other regions and other countries to determine their susceptibility to diseases that are not indigenous to the geographic region for which the phenotypes are being developed. (It is obvious that diseases and insects should not be imported for testing purposes unless fail-safe procedures are followed to prevent their escape. Even in exchanging genetic materials for testing, it is not possible to eliminate entirely the risk that an unwanted organism may be introduced.)

Although we cannot thoroughly evaluate the adequacy of subjecting genetic material to present and potential diseases and insects, the matter has received a great deal of attention from practical plant breeders.

The maintenance of germ plasm for commercial corn hybrids is unique, in the sense that after the inbred lines have been developed they may be propagated by a hybrid corn company for many years--during which time, conceivably, some deviation from the original line could develop, for example, in resistance to a disease. This could come about by mutation or long-term natural selection in some characteristics for which the line was not homozygous. This seems an unlikely potential problem because the useful life of inbreds without intentional modification is usually no more than 10-20 years, and the lines are continuously being altered by crossing, backcrossing, and selecting among the progeny.

Whereas the vulnerability of a narrow germ plasm base to new pests has received much publicity, too little has been said of its uniformity of resistance to many prevalent diseases and to some insects.

New plant genotypes should, of course, be thoroughly tested before being widely distributed in new geographic areas. This is sometimes ignored; but if, as a result, disease occurs, the fault lies in the testing program. A case in point was the original expansion of short wheats in Turkey and Morocco, where *Septoria tritici* is prevalent. [Personal communication (1973) from R. Glenn Anderson, Associate Director, International Maize and Wheat Improvement Center, Mexico.]

Lack of Crop Diversity

Subsistence agriculture usually involves a wide range of crops. There is a strong tendency in modern agriculture to develop intensified areas of production that often involve only one or two crops. This arrangement concentrates each crop species in the most suitable edaphic areas, and fosters the development of efficient production support services as well as facilities for marketing, processing, and transporting the end-product. The implications for pesticide requirements are mixed. Some insects, diseases, and weeds are favored by diverse cropping systems; others are disadvantaged. We believe that, on balance, the need for pesticides per unit of land area is increased in narrow-based cropping systems. There are, however, tradeoffs in terms of higher yields per unit of land area and lessened environmental impact because of the reduced arable area.

Intrinsic Susceptibility

We next examine some possible mechanisms for differential pest susceptibility or tolerance between high-yielding and low-yielding crops. There is a common impression that high-yielding types are more succulent, hence more susceptible to attacks by insects and pathogens. Succulence implies higher moisture, thinner cell walls, and lower lignin content. These characteristics could trace to a genetic base or could be caused by fertilizer, irrigation, or planting patterns. There is no reason to believe that high-yielding genotypes tend inherently to differ from lower-yielding types in moisture content, cell wall thickness, or lignin except as selection for standability would bias in favor of thick rather than thin cell walls and more rather than less lignin content.*,†

Optimum fertility, especially for nitrogen, promotes vigorous plant growth; but plant moisture content, thickness of cell walls, and lignin content are unlikely to be much affected. Lush growth always favors the incidence of foliar diseases, because it raises the humidity within the crop canopy. In addition, it frequently causes lodging (of small grains and forage crops) that exacerbates the situation. The implications of lush growth for the use of pesticides may be made academic by practical approaches to the control of excessive disease attack through genetic resistance, combined with proper management

---

*Crushing strength on two early corn inbred lines--Ohio 41 and WF9--were 313 and 283 kg compared to 516 and 514 kg for two recently developed lines--B14A and B36 (G. F. Sprague, personal communication, 1972).

†Anderson of CIMMYT (personal communication) in comparing introduced dwarf wheats with local types stated that:

> The moisture content will be essentially equal in both types and, if anything, would be lower in dwarfs because of greater tillering effects: The cells are known to be smaller in size in the dwarf plants as compared with the tall, since cell numbers are about equal--dwarfing of plant is largely the result of shortened cells. One could expect therefore, *more lignin per unit length of straw*. Again one would have to say that variation exists within both groups from genetic viewpoints, and be careful not to compare varieties belonging to different backgrounds.

decisions as to planting density, fertility level, and amount of irrigation rather than application of chemicals. Some foliar-feeding insects are favored by excessive vegetative growth; others are not much affected, but the crop shows less damage because the plants produce new growth rapidly enough to compensate for lost tissue.

High-yielding crops are usually less affected by weeds than lower-yielding ones because their vigorous growth competes more effectively under field conditions, especially during the latter part of the growing season. Doubling of the corn plant population since 1930 has greatly reduced late season weeds in the Corn Belt. In the early part of the season, fertilizer placed near the row gives the crop a competitive advantage over inter-row weeds. High yields, when obtained through improved varieties, more timely planting, and improved fertilization, require less pesticide per unit of crop produced.

Does the germ plasm associated with high yield produce plants that are inherently different in chemical composition (more attractive, better nutritionally, lacking in toxic compounds), hence more suitable to the growth of insects or diseases? There are, of course, examples of wild phenotypes that contain substances toxic or distasteful to humans and perhaps to certain insects. Once these plants are domesticated, these traits are largely lost; and further improvement does not affect the substances.

We find no reason to suspect that the multiplicity of genes directly contributing to high-yield potential are other than randomly associated with the genes for resistance to specific insects and diseases. Where this is true, crop-breeding programs that focus on high-yield performance should not coincidentally influence susceptibility to insects or diseases to which the crop may later be subjected. To the extent that breeding programs are conducted within the producing areas, selection will favor resistance to local pests even though the principal focus is on high yield. On the other hand, if selection and testing are conducted in a protected environment (as in the case of strawberries on chlorpicrin-fumigated soil in

California), a genotype may be produced that depends upon continued protection.*

Some plant breeders feel that attention to the incorporation of resistance to certain diseases was relaxed when fully effective seed treatments became generally available at reasonable cost. Such cases are unfortunate, especially when the treatment has involved chemicals subsequently found to cause environmental problems; but it is logical that since research dollars and manpower are limited, researchers have placed top priority on moisture problems rather than on those that had already been solved by alternative strategies--including pesticides.

We conclude that the capability of high-yield systems (whether genetic or based upon superior management) to respond economically to greater use of technological inputs--including pesticides--has been mainly responsible for the notion that such systems have a high pesticide requirement because they are inherently fragile and vulnerable.

Losses in Agriculture from Pesticide Use

Unintended effects associated with pesticide use (especially that of insecticides and herbicides) have caused losses to farm firms from time to time, and some of these have been substantial. As long as the chemicals continue to be used, there is every reason to expect such losses to recur. Well-publicized examples have included the destruction of commercial beehives by drifting or misapplied insecticides, the contamination of dairy products with unacceptable chemical residues, and the destruction of cotton plants by drifting phenoxy herbicides. Congress has established indemnity programs to compensate beekeepers, dairy producers, and dairy manufacturers for

---

*Other exceptions have been described by Dr. A. L. Hooker (personal communication):

> There are a few diseases of senescing tissue, such as the corn stalk rots, where high grain yield is associated with loss of vegetative vigor of the plant and as a consequence greater susceptibility to disease. But even here, breeders have been able to combine high yield and an adequate amount of disease resistance. The usage of fungicides is not feasible in these instances.

such losses. The beekeeper program alone paid 3,190 claims totaling $6.5 million in the period 1967-1972. The dairy indemnification program paid a total of $1.3 million between 1964 and the end of fiscal year 1973. No other government programs cover pesticide-related losses, although, of course, individuals suffering damage may seek relief in the courts.*

The damages caused by accidental contamination are not large compared to the benefits received from pesticide use, but they also have not been acknowledged for what they are, i.e., episodes likely to recur at unpredictable times and places in amounts difficult to estimate. Hence they pose a hazard to individual farmers against which little can be done in advance. Individual firms could be destroyed economically by an accident, especially in cases with no government indemnification and unclear liability. Furthermore, the accidents pose a threat to consumer health, because accidentally contaminated commodities may pass through the marketplace undetected.

The indemnification programs themselves are not without drawbacks. The public subsidy may be warranted by the benefit associated with the continued activity of producers at risk. But the indemnification program may actually discourage more careful handling of pesticides. Equally, they might discourage the adoption of precautions by those at risk. The latter possibility should be carefully examined before the existing programs are renewed or before any new programs are proposed.

Environmental Effects

Although there is wide agreement that pesticides have some detrimental effects on the environment, the significance of these is in controversy. This Committee made a deliberate decision that it would not examine or review in detail the problems of environmental contamination by pesticides. The Secretary's Commission (1969) made such a study--the Mrak Report--and the amount of new knowledge about the effects of environmental contamination is

*An attempt was made in the United States Senate in 1974 to indemnify poultry producers and workers for losses sustained when a large batch of chickens was found to be contaminated with illegal residues of dieldrin. Total losses were estimated at about $7 million. The bill died for lack of support in the House of Representatives.

too limited to make a lengthy review profitable. Also, the problems of genetic resistance in target species and the disruption of natural control discussed in the preceding pages are sufficiently severe that we saw our main task as one of evaluating alternative control possibilities. Some general principles about environmental contamination by pesticides, however, can be formulated.

Types of Organisms Affected

Organisms at all taxonomic levels can be affected by pesticides: Green plants (including angiosperms, gymnosperms, ferns, mosses, and algae), nongreen plants (such as fungi and bacteria), and animals (including mammals, birds, reptiles, amphibians, fishes, insects, crustacea, various types of worms, coelenterates, and protozoans) have all been shown to be affected by one compound or another. Species vary in their sensitivity to a given chemical; conversely, there is variation in toxicity of different chemicals to a given species. Furthermore, for any species, sensitivity to a given chemical varies with age, sex, nutritional intake, stress, general health status, and local environment. This complexity is important in the evaluation of the precise effect of a given chemical on a given species or group of species; but it should not obscure the basic principle that most organisms are affected at some level of exposure by one or more of the pesticides currently in use. Unfortunately, our systematic knowledge of the different effects is very sparse.

Ecological Level of Effects

Pesticides can affect all levels of ecological organization. At the simplest level, they affect individuals: A bird in a cotton field may be killed during an aerial spraying operation, or dragonfly nymphs in a pond sprayed for mosquito control may be destroyed. Except in the case of endangered species, having critically low population densities, the reproductive capacity of organisms normally insures that such losses are replaced quickly. The pest control decision maker, however, may well have cause for concern over the inadvertent killing of nontarget species because the unintended victim may be an organism of direct use in the control effort. For example, birds around cotton fields may eat sizable quantities of cotton insects, and dragonfly nymphs may destroy considerable numbers of mosquito larvae.

Pesticides can also affect an entire population of a given species in a given area. For example, there is considerable evidence (Blus et al. 1971, Snyder et al. 1973, Cooke 1973) that DDT pollution has been one important agent in the decline of breeding brown pelican (*Pelecanus occidentalis*) populations off the California coast. Similarly, reduction in numbers of several raptorial and other types of birds in the United States has been attributed to residues of DDT and its metabolites. Declines of this sort are of concern to society whenever the demise of an entire species is threatened--as in the case of the bald eagle, which has been kept on the endangered species list by pesticide pollution, among other factors. Extinction represents an irreversible change in the ecosystem. Although human beings clearly can survive without the full complement of species that co-inhabit the world with them, no one knows how many, or which ones, are especially essential for survival. The extinction of any species must, therefore, be considered a cause for concern. Beyond these long-term and somewhat abstract costs, however, we must count two more specific ones. First, each species represents a scientific resource that can never be replaced. Second, the aesthetic joy caused to man by the varied life forms on earth is a resource of considerable value. We cannot quantify either of these values, but extinctions of species surely represent serious impoverishments of man's environment.

At the highest level of ecological organization, pesticides affect communities; their influence in altering species composition is, in fact, their intended function--just as it is the function of agriculture itself. Vast areas of North America and the rest of the earth have been cleared of their native vegetation and planted with agricultural crops, millions of other acres have been permanently altered with asphalt and concrete. These activities have clearly had an immense impact on the species that formerly lived in these areas, and mankind has benefited from the changes in that his food has become more plentiful and his transportation systems more efficient. The unintended changing of species compositions, however, is quite another matter. Deforestation of parts of North America was undertaken intentionally to alter the species composition of the landscape; but involuntary changes, such as the induction of algal blooms and consequent reduction in predaceous insect populations in freshwater ponds by an organophosphate insecticide (Hurlbert et al. 1972), are both unintentional and undesirable. We know much too little of such unintentional alterations of species composition and its reversibility. In the face of such ignorance, prudence must clearly be the guiding principle.

Types of Compounds Involved

In part because recent disputes have attracted such notoriety to the chlorinated hydrocarbon insecticides, there has been a tendency to presume that some other categories of pesticides are not hazardous at all. Current evidence does not, however, warrant the assumption that any compound can be presumed free of negative environmental impact just because it belongs to a particular class of compounds. Pesticides of every class have been implicated in an environmental poisoning episode. For example, among the herbicides, trifluralin, commonly used in the cultivation of cotton and soybeans, is highly toxic to fish; among the carbamate insecticides, carbaryl, a valuable compound applied to fruits and vegetables, has effects on reproduction in the cotton rat; and among the fungicides, ferbam, a dithiocarbamate applied to tobacco and apples, is known to reduce populations of beneficial insects in orchards under some conditions (Pimentel 1971). Only where a single mechanism of action on the unintended target is involved are we likely to be able to exempt compounds by category from toxicological scrutiny. We argue below, however, that for certain classes of compounds related by molecular structure and known mechanism of action, there should be an adjustment of testing protocol requirements.

Types of Effects

Pesticides are biologically active compounds that can produce a myriad of symptoms and effects depending on the specific chemical, the affected organism, the dose, and the conditions under which it is received. Time is of special importance: Some compounds produce rapidly developing symptoms in the affected organism but are themselves relatively quickly destroyed by the organism's own metabolic machinery. Such *acute effects* are characteristically produced by the organophosphate insecticides. Different nontarget species differ in their susceptibility to these compounds. For example, azinphosmethyl is an organophosphate insecticide commonly used on a variety of fruit and vegetable crops; it is highly toxic to a wide variety of vertebrates, including mammals. In contrast, it does not seem to be very toxic to some earthworms (*Eisenia* sp.) (Pimentel 1971).

Some compounds do little or no harm to organisms other than target pests during short-term (3-4 days) experiments. Sustained exposure to such compounds, especially if they are chemically resistant to biological degradation, can result in a gradual increase in their concentration; then deleterious symptoms--chronic effects--may begin to appear following a delay. The symptoms are quite varied and may be so subtle as to be demonstrable only by sophisticated laboratory experiments. Because the exposure is prolonged, there may be sequelae short of death that are crucially important to the population of organisms at risk and to the environment as a whole. Chronic effects have been classified by toxicologists primarily concerned with human health, but the categories used apply equally to nonhuman organisms. They include the capacity to increase mutation rate (mutagenesis), the incidence of tumors (carcinogenesis), and the frequency of malformations occurring during embryonic development (teratogenesis).

Effects of this kind, exerted upon a wide range of nontarget organisms, have been associated with pesticides belonging to all of the major functional and chemical classes. The Secretary's Commission (1969), for example, listed Strobane, a chlorinated hydrocarbon insecticide with low acute mammalian toxicity, as a carcinogen, and held that human exposure to this compound should be limited. Linuron, a herbicide that is commonly used for selective weed control in corn, soybeans, and cotton, was listed as a mutagen in barley; and captan, a fungicide commonly used in the production of cotton, fruits, and vegetables, as a mammalian teratogen. None of these compounds has a very high acute toxicity for mammals.

The principle to be drawn from these examples, and many others that could be cited, is that the toxicological properties of pesticides are so complicated that each compound requires testing in a variety of ways in order to understand its exact properties. Moreover, problems can arise from compounds having various properties and various uses. The hazards to humans posed by the chemicals are treated on pp. 54-92.

The discussion so far has focused on the toxicity problems of chemicals administered or received one at a time. Organisms, however, are often exposed to more than one chemical at a time; and when many compounds reach them simultaneously, the effect may be quite different from that of any one acting alone. In some cases, the effect of a combination may be to negate the effects of each taken separately. For example, atropine (a drug) and methyl parathion (an organophosphate insecticide widely used on a variety of crops) are each poisonous to humans when taken separately; but when they are taken together, atropine acts as an antidote to the insecticide. In other cases, a combination of chemicals may cause greater toxicity than the sum of the effects observed for the chemicals taken separately. Such chemicals are said to exhibit synergism. Synergism is deliberately employed in some pesticidal formulations. For example, pyrethrum, an insecticide extracted from chrysanthemums, is widely used, especially by homeowners. Pyrethrum's toxic effects are enhanced by piperonyl butoxide, a relatively nontoxic substance by itself. This kind of synergism differs from "accidental" synergism, i.e., cooperative interactions between two or more chemicals that are unintended and remain unregulated.

An example of accidental synergism has recently been described by researchers at the University of Wisconsin (Lichtenstein et al. 1973). The herbicides atrazine, simazine, monuron, and 2,4-D (all commonly used in agriculture) enhanced the toxicity of a wide variety of common insecticides (including parathion, diazinon, carbofuran, DDT, and dieldrin) to vinegar flies (*Drosophila melanogaster*), houseflies (*Musca domestica*), and mosquito larvae (*Aedes aegypti*). As might be expected, there was variation in the level and consistency of the enhancement effect between different chemicals acting on different species. The major conclusion of the authors is worth noting:

> This further illustrates the necessity for continued investigation of the interactions of pesticides with other chemicals in biological systems. Although it is imperative to study the fate and behavior of single environmental chemicals, it is also apparent that these problems must be approached within the concept of a whole system in which chemicals can interact with each other.

A case of synergism between organophosphate and organochlorine pesticides in increasing virus susceptibility in laboratory mice was also recently reported (Crocker et al. 1974).

The potential for synergistic action to create environmental hazards for man and other organisms is one that this Committee judges to be of great importance. First, our systematic knowledge of the potential for synergistic action between different chemicals is very meager. Second, and more importantly, the research needed to generate a systematic knowledge of synergism would be immense and would grow exponentially with the addition of new chemicals to the armamentarium of pesticides. If each new chemical were to be tested for toxicity by itself and for synergism with each of the 1,500 other chemicals now available in combinations of two, three, four, and so forth, the number of separate tests needed would be 10^{1050}. Furthermore, compounds added to the environment for other uses may be potential synergists. The problem is, fortunately, less than would be suggested by these figures because in the habitat of any one organism, only a small fraction of the chemicals is likely to be represented.

This Committee does not recommend that tests for synergistic activity should immediately be conducted on a large scale. The numbers involved are prohibitive. Rather, we point out that the existence of synergism between different chemicals creates a potential and unevaluated hazard to man and his environment.

We recommend that scientific criteria be developed on which an economically efficient system for screening for synergism can be constructed--especially for chemicals likely to be present in the same locality.

Sources of Pesticide Pollution

Pesticides may enter the environment from point sources or from a diffuse area. The most common examples of the former involve accidental release, because pesticides are generally intended to be applied in small amounts over extended areas. Spillages of pesticide at the production plant--during transit, storage, or use--can cause high amounts of material to be placed in a small area in a short period of time. Introductions of a chemical in this way create higher concentrations of toxic substance in a small area and are, therefore, more likely to lead to acute toxicity in organisms in the zone of introduction.

By contrast, diffuse sources--such as result from normal agricultural uses--are more likely to create a lower level of toxic substance over a larger area and thus to produce widespread effects. Pesticides applied for control operations in forests, agriculture, home gardens, industry, and elsewhere may move from their target area to places in which they are not wanted. The mechanisms for movement are quite varied and include (a) drift at time of application, (b) adherence to eroding soil particles, (c) runoff with rainwater, and (d) volatilization into the atmosphere.

It is believed by many that pesticides are not injurious to nontarget organisms when used according to the directions on the label. Use according to label is surely important and undoubtedly does help avoid damage to nontarget organisms. We emphasize, however, that pollution problems may result even from uses of a pesticide that are strictly according to label directions. Too many mechanisms exist by which conscientiously applied pesticides can act as a diffuse source of pollution to guarantee that use according to directions will be "safe" in a broad environmental context.

One example of a point-source pesticide pollution problem is worth noting. It now appears quite likely that much of the DDT and its metabolites that contaminated organisms off the coast of southern California during 1969 and early 1970 was the result of spillage from a manufacturing plant. The exact magnitude of the spillage is disputed; but in one measurement made by the Los Angeles County Sanitation Districts, it amounted to 600 pounds per day. On the basis of measurements of DDT content in the Los Angeles County sewer sludges and in oceanic organisms near the outfall through which it passed (Burnett 1971), it constituted for at least a year the main source of chlorinated hydrocarbon residues for the California Current ecosystem. The "leak" was reduced to negligible levels by changes made in the wastewater disposal system at the plant in question.

The point of this example is not that factories sometimes spill chemicals, but that factory spillage of a chemical can be as significant a source of pollution as normal use of the chemical at low rates over large areas. In short, point sources of pollution may be as important as diffuse sources, but they are without redeeming benefits. In any case, the point source may potentially confound a determination of where the actual source of the pollution lies. Such confusion may have important implications for policymaking on pesticides. Had pollution in California coastal waters been incorrectly attributed to the *use* (rather than the *manufacture*) of the chemical, a mistaken policy banning use instead of controlling manufacturing sources might have resulted. Such examples emphasize the importance of identifying the source of pesticide effects before acting or setting policies to correct them.

Changes in Environmental Hazards from Pesticides

Patterns of pesticide use--the kinds of chemicals used, the rates and times of their applications, and the purposes for which they are applied--change rapidly. Changes in any of these parameters may significantly alter the environmental hazards created by the use of the chemicals, and often it is impossible to predict the direction of change without careful examination of the new circumstances. For example, the shift toward the use of organophosphate insecticides from the chlorinated hydrocarbon insecticides resulted in a reduction in the amount of persistent pesticides with low mammalian toxicity introduced into the environment, but increased the amount of less-persistent but considerably more-toxic materials. We know this switch created immediate hazards for the people occupationally exposed to the pesticides, but we cannot accurately evaluate the long-term significance for humans and nontarget organisms. It is presumed to be environmentally sounder to use materials of less persistence instead of those with longer lives. This presumption, however, may be based merely on the fact that we know some of the problems created for wildlife, especially birds and fish, by the persistent chlorinated hydrocarbons. We are more ignorant of the long-term effects of using the organophosphates, and only continued research can tell us whether the change was sound on environmental grounds.

Trade-offs in Evaluating Strategies

Pesticides are usually part of a broader strategy that is being used to accomplish some goal. The use of the pesticides inevitably alters the particular mixture of practices that are being used to accomplish the goal, and it may turn out that the total environmental impact from the activity is less when pesticides are used than when they are not. The environmental impact of pesticides should thus be evaluated in comparison with the impacts of other ways of achieving the same objective. In most cases, there will be a complex trade-off between one type of pollution or environmental impact and another.

For example, in the production of corn on slopes,* the crop can be planted by killing the sod with herbicides

---

*See Volume II, *Corn/Soybeans Pest Control*, for a discussion of no-till corn production.

without tillage of the soil, or by conventional seedbed preparation. Herbicide use results in the introduction of a toxic material into the environment; tillage on a slope results in increased rates of soil erosion and siltation in streams. Both toxic compounds and erosion may have negative environmental effects, but in this case man has a choice of which one he is willing to tolerate. It would be a mistake to evaluate the impact of using the herbicide solely on the basis of environmental toxicology. A better procedure would be to compare the relative impacts of the chemical and the moving soil particles and the difference in long-term soil productivity.*

Another example concerns the mycotoxins produced by fungal infections of plant seeds. Aflatoxins produced by species of *Aspergillus* are known to occur in significant quantities on peanuts, cottonseed, and on several nut crops. The fungi often infect the seed by entering the cotton boll through the tunnel provided by the pink bollworm. It is now known that feed prepared from cottonseed may contain sufficient quantities of aflatoxin to cause hepatoma in animals: As little as one part per billion of aflatoxin B_{17}, for example, can cause liver cancer in rainbow trout (Lee et al. 1971). It has also been reported that a toxin produced by *Phytophthora infestans*, the organism that causes late blight of potatoes, may cause birth abnormalities in children. Storage of grain under conditions that favor development of fungi may also result in accumulation of substances toxic to animals.

It is important that decision makers weigh the costs of continued use of plant disease control chemicals against those posed by human health hazards that may result from an increase in mycotoxins on food products. Decisions concerning insect control, furthermore, may have implications for the control of virus and fungus pathogens.

---

*This companion of the toxicological impact of a herbicide with the damage to streams by siltation assumes that production of corn on the slope is the most desirable use of the land in question. It may be, however, that the most beneficial use of the land would be to keep it under a constant cover crop, such as pasture or forest.

Health Hazards of Chemical Pesticides*

Introduction

Monitoring Human Exposure to Pesticides The U.S. production of pesticides of all kinds is more than 1 billion pounds per year, with synthetic organic compounds making up more than 90 percent of the total. This includes some 900 different chemicals, about 100 of which are produced in quantities of more than a million pounds per year.[†,‡] U.S. pesticide production is roughly half of the worldwide total. Inorganic and natural organic pesticides were used even in the last century. However, the large-scale use of synthetic organic pesticides is a relatively recent development, starting in the 1940s with the commercial production of the insecticide DDT and the herbicide 2,4-D.

Because pesticides must be released into the environment in order to perform their function, it is inevitable that some portion, however small, of these chemicals will reach the general population via the diet and other routes (National Pesticide Monitoring Program 1974). The FDA conducts a program for the analysis of "market basket" samples taken from various parts of the United States in order to monitor amounts of pesticides in the diet. Analyses have been reported for 22 different pesticide chemicals in market basket samples collected during the period FY 1965-1970 (Table 1). The results indicate that the average young adult American male, whose diet the market basket is intended to simulate, consumed about 40 mg of these residues per year, averaged over the 6 years studied. More than half of the total amount consisted of chlorinated hydrocarbon insecticides.

Although the principal route of exposure of the general population to most pesticides appears to be through the diet, some additional exposure results from pesticides in the air. The EPA conducted an air-monitoring program during 1970-1972 and is considering a proposal to resume pesticide monitoring of air on a continuing basis (Stanley et al. 1971).

---

*Members of the Consultative Panel on Health Hazards of Chemical Pesticides are listed at the beginning of this volume. This section constitutes the Report of the Panel.

†See Chapter 4, "The Production of Materials for Pest Control," Estimated U.S Pesticide Production Volume, 1971.

‡A summary of properties of selected pesticides is presented in Appendix A to this volume.

TABLE 1 Pesticide Chemicals in Food[a]

Compound	Daily Intake (mg)						6-year
	FY 1965	FY 1966	FY 1967	FY 1968	FY 1969	FY 1970	Average
Organochlorine insecticides							
DDT	0.031	0.041	0.026	0.019	0.016	0.015	0.025
DDE	0.018	0.028	0.017	0.015	0.011	0.010	0.017
TDE	0.013	0.018	0.013	0.011	0.005	0.004	0.011
SUBTOTAL	0.062	0.087	0.056	0.045	0.032	0.029	0.053
Dieldrin	0.005	0.007	0.004	0.004	0.005	0.005	0.005
Lindane	0.004	0.004	0.005	0.003	0.001	0.001	0.003
Heptachlor epoxide	0.002	0.003	0.001	0.002	0.002	0.001	0.002
BHC	0.002	0.004	0.002	0.003	0.001	0.001	0.002
Aldrin	0.001	0.002	0.001	T	T	T	0.001
Dicofol (kelthane)	0.003	0.002	0.012	0.010	0.007	0.004	0.006
Endrin	T	T	T	0.001	T	T	0.000
Methoxychlor	-	T	0.001	0.001	T	0.001	0.001
Heptachlor	T	-	T	T	T	T	T
Toxaphene	-	0.002	-	0.002	0.004	0.001	0.002
Perthane	T	0.001	-	0.001	0.004	-	0.001
Endosulfan	-	T	T	T	0.001	0.001	0.000
Organophosphate insecticides							
Malathion	-	0.009	0.010	0.003	0.012	0.013	0.008
Diazinon	-	0.001	T	T	T	0.001	0.000
Parathion	-	T	0.001	T	T	T	0.000
Ethion	-	T	0.002	0.001	0.003	0.004	0.002

TABLE 1 (continued)

Compound	Daily Intake (mg)						6-year
	FY 1965	FY 1966	FY 1967	FY 1968	FY 1969	FY 1970	Average
Carbamate insecticide							
Carbaryl	0.15	0.026	0.007	-	0.003	-	0.031
Herbicide							
2,4-D	0.005	0.002	0.001	0.001	T	T	0.002
Fungicide							
PCP	T	0.006	0.001	0.001	0.002	-	0.002

[a]Values are determined from analyses of foods prepared from "market basket" samples collected in five major U.S. cities and designed to simulate the diet of a 16-19-year-old male.
T (trace) indicates an amount less than 0.001 mg. From Duggan, R. E., and P. E. Corneliussen (1972) Pestic. Monit. J. 5:331-341.

Analyses for pesticides in streams and surface waters are conducted by the U.S. Geological Survey and the EPA. The EPA is embarking upon a program to analyze drinking water for pesticides and other chemicals, as is required under the Safe Drinking Water Act (P.L. 93-523). Tap water is monitored to the extent that it is employed in the preparation and cooking of foods for FDA market basket samples.

Another possible route of exposure is adsorption through the skin of pesticides that may be present in soaps and various cosmetic preparations, and in clothing. Also, a definite although not well known degree of exposure of the general population must result from direct contact with insecticides at the site of their use.

The levels of several chlorinated hydrocarbon insecticides in human adipose tissue are monitored by the EPA. Adipose tissue is a principal repository in the body for pesticides that are insoluble in water, soluble in fat, and resistant to metabolic degradation, as are certain of the organochlorine insecticides. Results from the first 5 years of the program, FY 1970-FY 1974, show an average total of about 11 parts per million (ppm) in body fat for the pesticides studied, corresponding to approximately one-tenth of a gram for an average adult (Tables 2 and 3). Considerable variation is observed among individuals, and values tend to be somewhat higher in southern states.

The relationship between the total quantity of a pesticide used and the amount that reaches the human diet depends on several factors, including the pattern of use, environmental mobility, water and lipid solubility, and chemical and metabolic stability. The propensity of a pesticide to reach the diet is relatively great for stable, water-insoluble, lipid-soluble compounds, such as some of the chlorinated insecticides, including DDT and dieldrin. For these compounds, under conditions of rather steady and prolonged application, a few hundredths of a gram enter the U.S. diet for each pound used.

<u>Pesticide Regulations and Public Health</u> Because of the exposure of the general public to pesticide residues and the risk of higher exposures to formulators, applicators, farm workers, and other special groups, intensive efforts are being made by government, industry, and other organizations to test the toxicity of current and proposed pesticides in an attempt to avoid chemicals or practices harmful to human health.

Under the Federal Environmental Pesticide Control Act of 1972 (FEPCA), the registration of a pesticide may be

TABLE 2 Chlorinated Hydrocarbon Pesticides in Human Adipose Tissue, U.S., FY 1970[a]

Pesticide	Concentration in Lipid (ppm)						
	Lim. of Detect-ability	Percent Pos.	Range: Maximum	Range: Minimum	Arith-metic Mean	Geo-metric Mean	Median
Total DDT equivalent[b]	0.02	99.93	270.05	0.00	11.65	7.87	8.10
Alpha-HCH[c]	0.01	2.05	0.14	0.00	0.01	0.02	0.00
Beta-HCH	0.02	99.15	26.93	0.00	0.60	0.37	0.34
Gamma-HCH (Lindane)	0.01	1.77	0.18	0.00	0.01	0.02	0.00
Delta-HCH	0.01	0.07	1.11	0.00	0.01	0.02	0.00
Dieldrin	0.01	96.46	15.20	0.00	0.27	0.18	0.18
Heptachlor epoxide	0.01	94.76	10.62	0.00	0.17	0.09	0.08

[a] Values are determined from 1,412 autopsy and surgical samples systematically collected in 31 U.S. cities. The monitoring program is gradually being expanded to include additional pesticide residues. From Kutz, F. W., A. R. Yobs, W. G. Johnson, and G. B. Wiersma (1974), U.S. Environmental Protection Agency.

[b] 77.15 percent of total DDT equivalent was found to be DDE.

[c] HCH = hexachlorocyclohexane (= BHC).

TABLE 3 Chlorinated Hydrocarbon Pesticides in Human Adipose Tissue, U.S., FY 1970-1974[a]

Pesticide	Concentration in Lipid (Arithmetic Mean) (ppm)				
	FY 1970	FY 1971	FY 1972	FY 1973	FY 1974
Total DDT equivalent	11.65	11.55	9.91	8.91	7.83
Beta-HCH	0.60	0.48	0.40	0.37	0.32
Dieldrin	0.27	0.29	0.24	0.24	0.20
Heptachlor epoxide	0.17	0.12	0.12	0.12	0.10
Oxychlordane[b]	--	--	0.15	0.15	0.15
Sample size	1412	1612	1916	1092	898

[a]From U.S. Environmental Protection Agency (1974).
[b]First full year in which oxychlordane was analyzed was FY 1972.

denied, canceled, or modified by the Administrator of the EPA on the basis of evidence for ". . . any unreasonable risk to man or the environment, taking into account the economical, social, and environmental costs and benefits of the use of any pesticide." Although specific guidelines are still being designed by the EPA, a presumption of risk to human health may be based on evidence of acute, subacute or chronic toxicity, including tests on laboratory animals showing carcinogenicity, teratogenicity, or mutagenicity.

The "Report of the Secretary's Commission on Pesticides and Their Relationship to Environmental Health," issued by the Department of Health, Education, and Welfare (USDHEW) in December 1969, provides a broad overall perspective as well as much information on individual pesticides (USDHEW 1969). The Commission recommended that all uses of DDT be terminated within 2 years except for specific, individual cases unanimously judged essential to the public health or welfare by the Secretaries of the USDHEW, USDA, and USDI (U.S. Department of the Interior). In a separate recommendation, the Commission called for similar restrictions on the use of other pesticides with highly persistent residues, including aldrin, dieldrin, endrin, heptachlor, chlordane, benzene hexachloride, lindane, and compounds containing arsenic, lead, or mercury. The Commission further recommended minimizing human exposure, pending additional studies, to several pesticides for which there existed indications of carcinogenic or teratogenic activity in test animals (see Appendix A).

In 1972 the EPA prohibited nearly all uses of DDT in the United States. The 1974 U.S. production of DDT was largely for export for malaria control and other uses. It was slightly under half of the 1950-1970 average production of 130 million pounds per year. Recently, the EPA suspended registrations for nearly all uses of aldrin and dieldrin (Federal Register 1974). In this case, as well as in that of DDT, the regulatory decision cited human exposure levels and the results of animal studies suggesting adverse health effects to man.

<u>Scope of the Health Hazards Discussion</u> It was not the assignment of the (Health Hazards Panel) to conduct an overall and detailed review of the toxic hazards of pesticides, as did the Secretary's Commission. Instead, our charge was to summarize the current state of epidemiological and laboratory approaches to detecting and averting major environmental health hazards from pesticides, particularly regarding carcinogenesis.

As its first area of inquiry the Panel reviewed national mortality statistics (see pp. 61-65) for the leading causes of death, including cancer, for any recent indication of adverse effects that might be attributed to environmental factors, including the use of pesticides.

The recommendation of the Secretary's Commission that human exposure to certain pesticides be reduced rested in part on tests in which animals receiving relatively high doses of these chemicals developed significantly more tumors than did the controls. Concern regarding possible carcinogenic hazard played a major role in the EPA decisions to withdraw registrations of DDT and aldrin/dieldrin and appears likely to be given prominence in future regulatory actions.

Cancer is a major human illness that is strongly correlated with environmental factors. This is shown by the fact that the incidence of many types of cancer among immigrant groups differs from that found in the country of origin and approaches the incidence prevailing in the adopted country (Haenszel 1961, Kmet 1970, Lilienfeld et al. 1972). From occupational and therapeutic exposures to certain chemicals, it is well known that man is subject to chemical carcinogenesis (IARC 1974). In these cases there is generally a delay of years, or even decades, between the time of first exposure and the occurrence of the disease. Hence, it is possible that major effects resulting from an environmental carcinogen would begin to show up in mortality statistics only many years after its introduction into widespread use. As its second area of inquiry, the Panel therefore considered the present basis of risk estimation from carcinogenicity tests on laboratory animals (see pp. 64-68).

Mortality Statistics

Death rates in the United States at all ages and especially for infants declined rather steadily during the twentieth century until the middle 1950s, when mortality leveled off and then, in the 1960s, showed indications of rising. Although the epidemic of lung cancer in men accounted for some of the change, other causes of death participated as well, including circulatory disease, emphysema, and cirrhosis of the liver. Similar trends were evident in Europe and Japan. This caused some apprehension that changing environmental factors might be adversely affecting man's health on a global scale (USDHEW 1971, 1973, 1974; U.S. Public Health Service 1964).

Statistics for the last few years, however, show that a general upswing of death rates has not materialized (National Center for Health Statistics 1952-1972). In 1970 and 1971 most age groups experienced the lowest death rates on record. Considering the major causes of death, data for the period 1968-1973 show that on an age-adjusted basis there have been small decreases in heart and vascular disease, stroke, diabetes, and in the category recorded as bronchitis, emphysema, and asthma. There has been a continued rise in mortality due to cirrhosis of the liver. The age-adjusted mortality rate for all types of cancer combined has been essentially constant, although there have been definite increases for some types of cancer and decreases for others.

The Panel examined age-adjusted mortality data for the period 1950-1967 for the 10 leading primary sites of cancer (lung, large intestine, breast, lymphoma, prostate, pancreas, leukemia, stomach, bladder, uterus plus cervix), which account for more than two-thirds of the total mortality due to the disease in the United States (Burbank 1971). For cancers for which recorded mortality is clearly increasing, such as lung, pancreas, and leukemia, the upward trends appear to have become established before the widespread introduction of modern pesticides beginning shortly after World War II. No clear trend associated with the rising use of pesticides is apparent for any type of cancer examined in age-adjusted data for the general population.

This of course does not exclude the possibility of less drastic, but still important, trends in mortality statistics. In this regard it would be important to conduct a more detailed epidemiological study than the Panel has carried out, extending the period of observation more nearly up to the present time. Mortality rates among different age groups should be examined for each separately recorded type of cancer. Particularly, a comparison should be made of geographical and other subpopulations for which pesticide exposure may vary. For example, the recent National Cancer Institute compilation of data on cancer mortality by county in the United States should be examined for possible "hot spots," especially among young adults, that might yield clues to important geographically varying carcinogenic factors (Mason and McKay 1974). Finally, studies on occupationally exposed populations would be useful, particularly where large numbers of persons are exposed for long periods of time.

Unfortunately, among other problems with the epidemiological approach, mortality statistics are likely to be very slow to reflect the effects of newly introduced environmental carcinogens. There will generally be a build-up period of several years from the introduction of a new product to the attainment of near-steady levels of production and human exposure. Also, a delay exists between the first diagnosis of a fatal tumor and its contribution to published mortality statistics. Currently this amounts to roughly 6 years (USDHEW 1972, U.S. Public Health Service 1975).* For a few types of cancer, improvements in therapy may obscure important trends in incidence as opposed to mortality. Few states regularly report cancer incidence. The Third National Cancer Survey gives cancer incidence rates for the period 1969-1971 for two states and seven metropolitan areas (Cutler et al. 1974). Although the sample is limited, the results deserve detailed study for trends that may not be apparent in mortality statistics.

An even more serious limitation to the utility of mortality and morbidity statistics for the detection of environmental carcinogens arises from the long delay that can occur between the first exposure and the appearance of tumors. Evidence from animal experiments and from the age distribution of cancer in man suggests that the incidence of cancer induced by continuous exposure to a carcinogen depends very strongly on the duration of exposure.†

---

*This estimate is the sum of the 2-year average survival time following the first diagnosis of a fatal cancer plus the current 4-year lag in publication of mortality statistics.

†Druckrey, H. (1967) Potential Carcinogenic Hazards From Drugs, UICC Monograph Series 7, R. Truhaut, ed., pp. 60-78. New York: Springer-Verlag; Cook, P. J., R. Doll, and S. A. Fellingham (1969) Int. J. Cancer, 4:93-112. The mathematical relationship discussed by the latter authors giving the incidence I of numerous types of cancer in a population continuously exposed to a carcinogen is $I = f(d)(t-k)^n$, where $f(d)$ is a function of the dose rate d alone and t is the exposure time. The quantities k and n are constants characteristic of the carcinogen and k is the minumum delay between first exposure and the expression of cancer. The relation deduced by Druckrey relating time and dose rate at 50-percent accumulated induced incidence is dt^n = constant. This is a special case of the former equation with $k \ll t$ and $f(d)$ equal to d itself. *See also* Peto, R., P. N. Lee, and W. S. Paige (1972) Brit. J. Cancer, 26:258-261.

Indeed, it would appear that, after a delay, the induced incidence of most types of human cancer rises as the fourth power of the time.* Hence, the incidence even as long as approximately 20 years after the introduction into use of a chemical carcinogen may be only a very small fraction of the eventual incidence it induces. Such a time dependence can give the impression of a "latent period" of many years or even several decades before cases begin to appear with noticeable frequency. Thus it is possible that a serious environmental hazard could go undetected in mortality statistics for many years and then appear with seemingly explosive force.

Laboratory Studies on Animals

Because epidemiological surveillance for trends linking pesticides or other environmental factors to major illness in the population might indicate hazard only after serious damage has already occurred, other information must be used to establish standards for the protection of public health. Where widespread damage to health may be long delayed after the introduction into use of a chemical, as with agents that may induce cancer or cause heritable mutations, the establishment of safety standards is particularly dependent upon tests on animals or other laboratory systems.

For carcinogenesis, the principal test systems in current use are laboratory mice and rats. A large proportion of such testing is done or overseen by the National Cancer Institute (NCI) of the U.S. Public Health Service (USPHS), using standard protocols for administration, diagnosis, and evaluation. Approximately 200 compounds are currently under test by the NCI and its contractors, at a cost of approximately $100,000 per test. The testing and evaluation of a compound for carcinogenicity requires approximately 3 years. For pesticides, the compound is administered in the daily diet. In order to simulate the usual conditions of human exposure, the most realistic

---

*It may also be noted that this relation predicts that the age distribution of induced cancer will be independent of dose unless the induced incidence is so great as to deplete the older age ranks. For the present age distribution of the U.S. population and in the case with $n = 4$ and $k = 2$ years, about one-quarter of the incidence would occur before age 55 and about half by the age of 68.

administration of a compound would begin with the mothers of the test animals before conception and continue throughout life. For convenience, however, the current NCI protocol specifies that administration begin at the time of weaning and terminate in late middle age. The animals are divided into several groups, each receiving a different concentration of the compound in the diet. An undosed group of animals serves as a control. There are usually 30-100 animals in each dosage group. Animals are allowed to die spontaneously or are sacrificed shortly after dosing is terminated or earlier if tumors are evident. Postmortem examinations are performed to determine the type and number of cancers. A compound is considered carcinogenic to the test animals if the treated groups show a statistically significant increase in the age-specific frequency of one or more specific types of cancer, as compared to the control group. As a check on the reliability of the experiment and for the quantification of carcinogenicity, a positive test is generally not considered adequate unless the frequency of cancers is found to be dose dependent.

The sensitivity of such tests is inherently limited by the number of animals employed. For example, even if no cancers are found in a group of, say, 100 animals, one can conclude with 95-percent confidence only that the true frequency of cancer induction at the given dose is less than 3 percent. Clearly, in extrapolation to man, a likelihood of inducing cancer in no more than 3 out of 100 individuals would hardly be reassuring. In practice, animal tests may be even less sensitive than the foregoing example suggests. This is because the control animals themselves may develop a substantial frequency of various kinds of cancer, making the problem one of detecting a significant increase above an already appreciable background. In order to increase the sensitivity of animal tests, administered doses are chosen to be considerably higher than the level to which humans usually are exposed. This procedure may fail if the test compound greatly shortens the life span of the animals as a result of toxic effects other than cancer. In that case, the older ages at which carcinogenesis is likely to be most manifest cannot be studied.

Attempts to utilize data from animal experiments in order to quantify carcinogenic risks to man must be based on assumptions for which available evidence is suggestive but far from complete. Two basic kinds of assumptions are generally involved. One concerns the relative susceptibility of man and the test animal to the chemical in question. The other concerns the relationship between the dose and the response. The following is a simplified description of the current basis for estimation of carcinogenic risk for environmental agents such as pesticides.

Carcinogenesis in Man and Laboratory Animals Notwithstanding the difficulty of identifying specific carcinogens affecting man, about a dozen agents are now recognized as human carcinogens. All of these but one are definitely carcinogenic in tests on laboratory animals, as indicated in Table 4. The only exception is arsenic, which is still under test on animals. It may be noted that the organs affected in man are not always the same as those that are found to develop tumors in laboratory animals, nor is the organ specificity the same in different species of rodents. Nevertheless, the evidence, based on a limited number of carcinogens, suggests that most agents that pose a carcinogenic threat to man will be carcinogenic in laboratory tests on animals. However, this leaves open the possibility that such tests may also identify chemicals carcinogenic to rodents that do not pose such a threat to man.

It would be desirable to have some idea of the proportion of useful chemicals that are carcinogenic in standard animal tests. In this regard, it is of interest that only a minority of pesticides that have been subjected to standard rodent tests have been found positive. In a test of 120 different pesticides and industrial chemicals conducted under contract with the NCI, 11 were judged positive, 89 gave no significant indication of carcinogenicity, and 20 were designated for further evaluation (Innes et al. 1969). It should be noted that one of the criteria used in selecting the 120 pesticides to be tested was structural or other prior indication of possible carcinogenicity. This suggests an upper limit of about 25 percent on the proportion of currently used pesticides that might be suspected of human carcinogenicity on the basis of such tests.

Research on molecular mechanisms in carcinogenesis has provided evidence that the actual carcinogen is often not the compound administered but instead is a metabolite of that compound (Miller and Miller 1971). Active carcinogens tend to possess particular features of chemical structure by which they may be recognized. Although apparent exceptions are known, similarities in the metabolic fate of a compound in different species may imply similar carcinogenic properties. Hence, metabolic studies using human and animal tissues can support the case that

TABLE 4 Results of Rodent Tests on Substances Known to be Carcinogenic to Man

Compound		Site	References
Aflatoxin	Mouse	Lung	Wieder et al. *J. Nat. Cancer Inst. 40*, 1195 (1968)
		Liver	Vesselinovitch et al. *Cancer Res. 32*, 2289 (1972)
	Rat	Liver	Newberne and Wogan *Cancer Res. 28*, 770 (1968)
		Stomach	Butler and Barnes *Nature 209*, 90 (1966)
		Kidney	Lee et al. *J. Nat. Cancer Inst. 43*, 1037 (1969)
	Man	Liver	IARC 1, 145 (1972)[a]
4-Aminobiphenyl	Mouse	Liver	Clayson, Lawson, and Pringle *Brit. J. Cancer 21*, 755 (1967)
		Liver, bladder	Clayson et al. *Brit. J. Cancer 19*, 297 (1965)
	Rat	Intestine, injection site	Walpole, Williams, and Roberts *Brit. J. Industr. Med. 9*, 255 (1952)
		Intestine, injection site	Hendry et al. *Nature 175*, 1131 (1955)

TABLE 4 (continued)

Compound		Site	References
	Rabbit	Bladder	Wood *Industr. Med. 39*, 55 (1970)
	Man	Bladder	IARC 1, 74 (1972)
Arsenic	Mouse	Negative in several tests. Recent reports of skin, lung and other tumors await confirmation	IARC 2, 48 (1973)
	Rat	Negative in several tests	IARC 2, 48 (1973)
	Man	Lung, skin	IARC 2, 48 (1973)
Asbestos (Crocidolite)	Mouse	Lung, injection site	Roe et al. *Food Cosmet. Toxicol. 6*, 566 (1968) Wagner et al. *Brit. J. Cancer 29*, 252 (1974)
	Man	Lung	IARC 2, 17 (1973)
Auramine	Mouse	Liver, kidney, intestine	Bonser, Clayson, and Jull *Brit. J. Cancer 10*, 653 (1956)
		Liver, kidney,	Williams and Bonser

		Liver	Green *Brit. Emp. Cancer Campaign Rept. 34*, pt. 2, 434 (1961)
		Liver	Walpole *Acta Un. Int. Cancer 19*, 483 (1963)
	Rat	Liver, injection site	Bonser, Clayson, and Jull *Brit. J. Cancer 10*, 653 (1956)
		Liver, injection site	Williams and Bonser *Brit. J. Cancer 16*, 87 (1962)
	Man	Bladder	IARC 1, 69 (1972)
Benzidine	Mouse	Liver	Bonser *Brit. J. Cancer 10*, 653 (1956)
		Liver	Profofjeva *Vop. Onkol. 17*, 61 (1971)
	Rat	Liver, ear duct, colon, injection site	Spitz et al. *Cancer 3*, 789 (1950)
		Liver, ear duct, cholangioma	Boyland et al. *Brit. J. Cancer 8*, 647 (1954)
		Liver, ear duct, injection site	Pliss *Vop. Onkol, 10*, 50 (1964)
		Mammary carcinomas	Griswold et al. *Cancer Res. 28*, 924 (1968)
	Hamster	Liver, cholangiomas	Saffiotti et al. In *Bladder Cancer, A Symp.* p. 129 (1967)

TABLE 4 (continued)

Compound		Site	References
	Man	Bladder	Zavon, Hoegg and Bingham *Arch. Environ. Health 27*, 1 (1973)
			IARC 1, 80 (1972)
Bis-(chloromethyl)ether	Mouse	Lung	Leong et al. *Arch. Environ. Health 22*, 663 (1971)
		Skin	Van Duuren et al. *Arch. Environ. Health 16*, 472 (1968)
		Skin	Van Duuren et al. *J. Nat. Cancer Inst. 43*, 481 (1969)
	Rat	Injection site	Van Duuren et al. *Arch. Environ. Health 16*, 472 (1968)
		Injection site	Van Duuren et al. *J. Nat. Cancer Inst. 43*, 481 (1969)
		Lung, nasal sinuses	Laskin et al. *Arch. Environ. Health 23*, 135 (1971)
	Man	Lung	IARC 4, 231 (1974)

N,N-bis(2-chloroethyl)-2-naphthylamine			*Brit. J. Cancer 15*, 252 (1961)
		Lung	Shimkin et al. *J. Nat. Cancer Inst. 36*, 915 (1966)
	Rat	Injection site	Koller *Heredity 6*, 181 (1953)
	Man	Bladder	Thiede and Christensen *Acta Med. Scand. 185*, 133 (1969) IARC 4, 119 (1974)
Cigarette smoking	Mouse	Lung	Harris et al. *Intern. Journal of Cancer 14*, 130-6 (1974)
	Hamster	Larynx	Dontenwill et al. *J. Nat. Cancer Inst. 51*, 1781 (1973)
	Man	Lung	Doll and Hill *Brit. Med. J. 1(5396)*, 1460 (1964)
		Lung	Haenszel, Ed. *National Cancer Inst. Monograph No. 19*, 127 (1966)
Diethylstilbestrol (DES)	Mouse	Mammary	Gass et al. *J. Nat. Cancer Inst. 33*, 971 (1964)

TABLE 4 (continued)

Compound		Site	References
		Vagina, cervix and other tumors	Dunn and Green *J. Nat. Cancer Inst. 31*, 425 (1963)
	Rat	Pituitary	Waelbroeck et al. *Compt. Rend. Aca. Sci. (Paris) 262*, 1646 (1966)
		Mammary	Gibson et al. *Toxicol. Appl. Pharmac. 11*, 489 (1967)
	Hamster	Kidney	Polkina *Prob. Oncol. 7(7)*, 969 (1961)
		Kidney	Ward et al. *Cancer Research 24*, 319 (1964)
		Kidney	Horning *Brit. J. Cancer 10*, 678 (1956)
	Man	Vagina and cervix	IARC 6, 55 (1974)
2-Naphthylamine	Mouse	Liver, injection site	Bonser et al. *Brit. J. Cancer 6*, 412 (1952)
		Liver, injection site	Bonser et al. *Brit. J. Cancer 10*, 533 (1956)
	Hamster	Liver, bladder	Saffiotti et al. In *Bladder Cancer*, p. 129

Bis (2-chloroethyl) sulfide (Mustard gas)	Mouse	Lung	Heston *J. Nat. Cancer Inst. 11*, 415 (1950)
		Lung, liver	Heston *J. Nat. Cancer Inst. 14*, 131 (1953)
	Rat	Injection site	Haddow In *Physiopathology of Cancer*, p. 606 (1959) F. Homburger, ed.
	Man	Lung	Wada et al. *Lancet*, 1161 (1968)
Vinyl chloride	Mouse	Lung, mammary, liver	Maltoni and Lefemine *Accad. Nazional dei Lincei Rend Cl Sci. Fis 66*, 1 (1974)
	Rat	Liver, zymbal gland, kidney	Maltoni and Lefemine *Accad. Nazional dei Lincei Rend Cl Sci. Fis 66*, 1 (1974)
		Lung, skin, bones	Viola et al. *Cancer Res. 31*, 516 (1971)
	Man	Liver	*Report of a Working Group on Vinyl chloride, IARC Internat. Tech. Report, No. 74/005 Lyon* (1974)

[a](IARC) references refer to monographs on the Evaluation of Carcinogenic Risk of Chemicals to Man, vols. 1-6, International Agency for Research on Cancer, Lyon, France.

a given compound that is carcinogenic in animals may or may not be carcinogenic in man. At present, however, such information is available only for very few compounds.

Recent evidence suggests that within several broad chemical classes, compounds that are carcinogenic in standard rodent tests are also mutagenic to bacteria in the presence of mammalian activating enzymes and vice versa (Ames et al. 1973, McCann et al. 1975). The test is sensitive, rapid, and inexpensive, and offers high promise as a preliminary screen for the many chemicals in man's environment that have not been evaluated for carcinogenicity. A further important implication of the finding that many carcinogens are also mutagens is that measures taken to protect a population against chemical carcinogenesis may also offer a significant measure of protection against chemical mutagenesis.

The finding that most compounds known to cause human cancer are demonstrably carcinogenic to laboratory animals is a qualitative one. For risk estimation it is necessary to have some quantitative estimate regarding the sensitivity of man relative to that of laboratory animals. In order to simulate conditions of long-term human exposure to environmental carcinogens, it would be of interest to compare the induced cancer incidence as a function of age for man and for laboratory animals exposed to the same carcinogen via comparable routes of administration throughout life. However, reliable data for such comparisons do not exist. For the few carcinogens for which comparisons may be undertaken, the total induced incidence in man and the intensity of exposure are usually very poorly known, and the duration and conditions of exposure are often not comparable in the available studies on animals. Nevertheless, in order to bring together some of the data and to encourage more adequate comparative studies, the Panel reviewed the available data for the small number of carcinogens for which human exposure and induced incidence may be at least roughly estimated. These carcinogens are benzidine, chlornaphazine, DES, aflatoxin B_1, vinyl chloride, and cigarette smoke. The results are outlined in Table 5. Dose is expressed as the total amount of carcinogen ingested, injected, or inhaled per kilogram of body weight. Where several animal tests involving prolonged exposure are published, the ones indicating the highest sensitivity for each species have been chosen. In the case of DES, a single dose experiment on newborn female mice is also included since it more closely simulates the conditions under which prenatal exposure is known to cause cancer in women.

TABLE 5 Degree of Exposure and Reported Cancer Frequencies for Agents Carcinogenic to Man and Laboratory Animals

	Conditions of Exposure	Accumulated Dose (mg/kg)	Cancer Induction	Reference
Benzidine				
Man	13.6 yr avg occupational	200 avg	50% bladder 13/25	Zavon, Hoegg, and Bingham *Arch. Environ. Health 27*, 1 (1973)
	>1 yr occupational	50 avg	22% bladder 47/210	Vigliani and Barsotti *Med. Lavoro 52*, 241 (1961)
Mouse	1/wk for 8-13 mo subcutaneous	about 10,000	67% liver 31/46	Prokofjeva *Vop. Onkol. 17*, 61 (1971)
Rat (female)	1/3 da for 1 mo gastric intubation	100 50 control	80% mammary 7/9 50% mammary 5/10 2% mammary 3/132	Griswold et al. *Cancer Res. 28*, 924 (1968)
Chlornaphazine				
Man	3 yr avg therapeutic	60-6000 2000 avg	16% bladder 10/61	Thiede and Christensen *Acta Med. Scand. 185*, 133 (1969)

TABLE 5 (continued)

	Conditions of Exposure	Accumulated Dose (mg/kg)	Cancer Induction	Reference
Mouse	3/wk for 1 mo intraperitoneal	4800	100% lung 25/25	Shimkin et al. *J. Nat. Cancer Inst.* *36*, 915 (1966)
		1200	86% lung 25/29	
		300	62% lung 18/29	
		75	40% lung 12/30	
		control	38% lung 124/330	
DES				
Man (woman)	During pregnancy therapeutic, oral	0.5-300 <100 avg	0.2% vaginal and cervical adenocarcinoma in daughters	Herbst et al. *New Eng. J. Med.* *287*, 1259 (1972)
Mouse (male)	20 mo in the diet	13	27% mammary 30/112	Gass et al. *J. Nat. Cancer Inst.* *33*, 971 (1964)
		3.5	7% mammary 9/134	
		2	4% mammary 6/139	
		control	0/251 mammary	
Mouse (newborn female)	Single injection subcutaneous	about 400	33% cervix and vagina 9/27	Dunn and Green *J. Nat. Canc. Inst.* *31*, 425-438 (1963)
Aflatoxin B_1				
Man	In the diet	0.1	0.5% liver	Peers and Linsell *Brit. J. Cancer* *27*, 473 (1973)

Mouse	1/3 da, 1-5 times, intra-peritoneal	1.25-6.0 control	25-100% liver 3%	Vesselinovitch et al. *Cancer Res.* *32*, 2289 (1972)
Rat	About 25 mo in the diet	1.5 0.9 0.3 control	100% liver 28/28 80% liver 20/25 19% liver 4/21 0/18	Wogan et al. *Fd. Cosmet. Toxicol.* *12*, (in press)
Vinyl chloride				
Man	18 yr avg occupational	70,000 (based on est. of 150 ppm avg in factory air)	0.2% liver	*Report of a Working Group on Vinyl Chloride*, IARC Internal Tech. Report No. 74/005 Lyon, 1974
Mouse	250 ppm in air 4 h/da, 5 da/wk for 30 wk inhalation	30,000 control	25% lung 4/16 13% mammary 2/16 0/28	Maltoni and Lefemine *Rend. Sci. Fis. Mat. Nat. (Lincei)* *66*, 1 (1974), as quoted in above
Rat	250 ppm in air 4h/da, 5 da/wk for 52 wk inhalation	40,000 control	9% kidney 6/67 6% liver 4/67 0/68	Ibid.

TABLE 5 (continued)

	Conditions of Exposure	Accumulated Dose (mg/kg)	Cancer Induction	Reference
Cigarette smoking				
Man	40 yrs, est	Smoke from 1,000 cigarettes per kg	2.5%	*Smoking and Health*, Report of the Advisory Group to the Surgeon General, 1964
Mouse	12 min/da 2 da/wk for life inhalation	Smoke from 400 cigarettes per kg	4.9% lung 16/326	Harris et al. *Int. J. Cancer* 14, 130 (1974)
		control	1.3% lung 4/319	
Hamster	About 9-26 min/da 5 da/wk for life inhalation	Smoke from 6,000 cigarettes per kg	6% larynx 29/480	Dontenwill et al. *J. Nat. Cancer Inst. 51*, 1781 (1973)
		control	0% 0/160	

NOTES FOR TABLE 5

Benzidine

1. *Human dose*. Estimate 0.05 mg/day average in urine × 70 recovery factor × 250 working days/year × 11 years average employment ÷ 70 kg man = *ca* 150 mg/kg average total intake (Zavon, Hoegg and Bingham 1973).

   Estimate 0.02 mg/day average in urine × 70 recovery factor × 250 working days/year × 10 years assumed average employment ÷ 70 kg man = *ca* 50 mg/kg average total dose (Vigliani and Barsotti 1962).

2. Approximately 1/70 of benzidine ingested by Rhesus monkeys is excreted in the urine as benzidine and monoacetyl benzidine (Rhinde and Troll, *J. Nat. Cancer Inst.*, in press).

3. The estimated doses may be underestimates of the actual averages due to improvements in factory conditions before urine samples were collected.

4. In both the United States and the Italian studies, part of the study population was still at risk at the time of the reports.

5. The periods of observation from the start of exposure in the experiments on mice and rats were 27 months and 10 months, respectively.

Chlornaphazine

1. There were 24 surviving patients still at risk at the time of the 1969 report.

2. Nine of the patients with bladder carcinoma received ^{32}P and three received busulfane in addition to chlornaphazine.

3. Shimkin et al. continued observations on treated mice for 43 weeks after the start of exposure.

DES

1. *Human dose*. Range among mothers of cancer patients 0.03-18 g total. Average is less than 6 g ÷ 60 kg woman (Herbst et al. 1974).

2. *Human incidence*. The annual incidence has increased rather steadily up to 1972, the most recent year

reported. In that year, there were 21 cases with medical records of administration of DES or related drugs to the mother during pregnancy. Several additional cases occurred for which adequate medical records are not available. Assume that the annual rate will not continue to rise as the population at risk continues to age and that it represents a steady state, reflecting the average usage pattern of DES in the United States since the middle 1940s. Assume further the number of women treated with DES during pregnancy giving birth to daughters has been 13,000 per year, the average of the range estimated for the period 1960-1970. Then the incidence is 21/16,000 or 0.16 percent (Herbst et al. 1974; Heinonen 1973 *Cancer*, *31*, 573).

3. The apparent carcinogenic activity of DES in humans is higher than that estimated above in the special case of endometrial carcinoma after DES therapy for gonadal dysgenesis (Cutler et al. 1972 *New England J. Med.*, *287*, 628).

4. The observations of Gass et al. continued for about 20 months and those of Dunn and Green for up to 26 months.

Aflatoxin B_1

1. *Human dose*. Average dietary intake for three areas in Kenya is estimated as 7.7 nanograms/kg/day. Assume average exposure time is 40 yr. Then estimated total dose is *ca* 0.1 mg/kg (Peers and Linsell 1973).

2. *Human incidence*. Average incidence for above population is 7 per 100,000 adults per year. Assume an annual death rate of 1.5 percent. Then the total incidence is *ca* 5×10^{-3} (Peers and Linsell 1973).

3. The observations of Vesselinovitch et al. continued for up to 82 weeks.

Vinyl chloride

1. *Human dose*. Estimate an average of 0.4 mg/liter (150 ppm) in factory air breathed by polymerization workers. Average employment of known cases is 18 yr. Assume 5 years average employment for the exposed population. Assume 40 h/wk, 50 wk/yr work schedule and 10 l/min breathing rate. Then average inspiration for 70 kg

man is *ca* 70,000 mg/kg (Baretta, Stewart and Mutchler 1969 *Amer. Industr. Hyg. Assn. J.*, *30*, 537; IARC 1974).

2. *Human incidence.* Liver angiocarcinoma cases among polymerization workers reported through June 1974:

U.S.	13/20,000	= 0.07%
F.R.G.	3/3,600	= 0.08%*
Sweden	2/790	= 0.25%†
U.K.	1/1,366	= 0.07%
Total	19/25,756	= 0.07%

*One was not a polymerization worker.
†If 790 represents essentially all polymerization workers.

The first reported case occurred 13 years ago. More than half of all reported cases occurred in the past 4 yr. Assume that for the population currently at risk the annual incidence is at or near its peak and will soon decline, giving a total incidence, with complete reporting, about three times that reported thus far. This gives an estimate of *ca* 0.2 percent (IARC 1974).

3. Assume a breathing rate of 0.03 l/min for the mouse and 0.14 l/min for the rat.

Cigarette smoking

1. *Human incidence.* Men smoking approximately 10 cigarettes per day have an age standardized death rate from lung cancer of approximately 0.05 percent. Assume that the annual death rate from all causes among smokers is about 2 percent. Then the total incidence associated with smoking 10 cigarettes per day is about 2.5 percent (Doll 1967).

2. *Human dose.* Assume that the average duration of smoking is 40 yr and that half of the smoke is inhaled. Then ten cigarettes per day represents the smoke from $1/2 \times 10 \times 365 \times 40 \div 70 =$ *ca* 1,000 cigarettes/kg.

3. It is to be noted that questions have been raised as to whether the observed mouse lung tumors were actually caused by smoke and as to whether the nose breathing of mice results in filtering out carcinogens to which the human smokers' lungs are exposed (*Lancet*, 506, 1974).

4. The breathing rate of the hamster is taken as 0.08 l/min.

The carcinogens for which comparisons can be made are those already known to affect humans. In generalizing to other compounds, this selection may impose a bias, exaggerating the sensitivity of man relative to laboratory test systems. Other factors may introduce an opposite bias. For two carcinogens, vinyl chloride and DES, observations on man are for considerably less than a full lifetime so that the reported incidence may be a serious underestimate of the eventual total. Also, these two compounds and aflatoxin are known as human carcinogens because they are associated with types of cancer that are otherwise rare. If these carcinogens also induce more common types of cancer, even at much higher frequency, this could go undetected, again giving rise to an underestimate of their overall carcinogenicity to man.

The limited conclusion that emerges from the comparisons given in Table 5 is that if the data from the most sensitive published test on animals are used to predict lifetime human incidence on a dose per body weight basis, the result seems approximately correct for benzidine, chlornaphazine, and cigarette smoking. For aflatoxin, the predicted human incidence is about ten times greater than estimated from existing epidemiologic studies, while for vinyl chloride it is about 500 times higher. For DES, the human incidence predicted from the result of a single dose administered to newborn female mice is about 50 times higher than that estimated from studies of adenocarcinomas in daughters of women given DES during pregnancy.* Thus, as a working hypothesis, in the absence of countervailing evidence for the specific agent in question, it appears reasonable to assume that the lifetime cancer incidence induced by chronic exposure in man can be approximated by the lifetime incidence induced by similar exposure in laboratory animals at the same total dose per body weight. (For a discussion of interspecies scaling factors for dose, see Rall 1974.)

Dose-Response Relationships In the case of suspected environmental carcinogens, we are usually concerned with

---

*In comparing human and animal sensitivity it is assumed that the response is simply proportional to dose. See the discussion of dose response relationships in the next section.

exposures much lower than those producing a statistically significant increase in tumors in the small number of animals used in laboratory tests. Therefore, we must extrapolate the experimentally measured rate to lower doses. This is the second area in which risk estimation involves a basic assumption, as mentioned earlier.

For estimating the response at doses lower than those tested in laboratory animals, several quite different assumptions have been proposed.

Threshold hypothesis. This hypothesis assumes that there is a dose below which cancer induction cannot occur. An examination of published dose-response data for chemical carcinogenesis in laboratory animals provides no clear indication of a threshold for any carcinogen (Craig and Miller 1974). In a review of 151 dose-response curves, none was found to be clearly inconsistent in a manner suggesting a threshold with both the single-event and the probit hypotheses discussed below. Neither is there any adequate theory of chemical carcinogenesis that would require the general existence of thresholds. Thus, even if a threshold is postulated, there is presently no empirical or theoretical basis for determining the dose at which it may occur. Unless and until this can be done, the threshold concept does not provide a practical basis for risk estimation.

Single-event hypothesis. This assumes that the induced incidence of cancer is directly proportional to the dose all the way from the lower incidence levels that can be measured in animal experiments of practical size down to zero dose and zero response. In other words, below an induced incidence of about 10 percent, the dose-response curve is, for practical purposes, a straight line. This would result, for example, if cancer is induced by a single cellular event, the likelihood of which is directly proportional to the dose of carcinogen. An essentially linear dose-response relationship can also result under much more general assumptions, so long as the carcinogen in question simply adds its effects to those of other carcinogens already present (Crump et al., in press).

The single-event hypothesis is in agreement with the limited data available for man. The induction of leukemia by ionizing radiation from nuclear explosions is compatible with a linear dose response down to an induced incidence of about 0.1 percent, the lowest incidence for which the available data can meaningfully be be analyzed.* Other data on the induction of various types of cancer following therapeutic and occupational exposure to ionizing radiation, although less extensive, are likewise compatible with linearity (NAS 1972b). The dose response relating the incidence of lung cancer in men to the average number of cigarettes smoked per day is also compatible with the single-event hypothesis. In this case, the data can be analyzed down to an induced incidence of approximately 2 percent (Doll and Hill 1964, Doll 1967, Kahn 1966).

Animal experiments are not usually conducted on a scale large enough to measure induced incidence below a few percent. For some carcinogens in some investigations, the dose-response relation is compatible with the single-event hypothesis, while in other cases it is not. However, it is quite possible that a dose-response departing from the single-event hypothesis at high induced incidence may nevertheless converge to a linear relation at lower incidence values (Crump et al. 1975).

The probit and other hypotheses implying a dose-response curve that is concave upwards. This class of hypotheses assumes that there is no threshold for a population but that the incidence at doses below the lowest tested is less than that implied by the single-event hypothesis. Below a response of a few percent, such a relationship between dose and response is described by a smooth curve that is concave upwards. For example, the incidence of skin tumors produced by surface application of benzo[α]pyrene in the mouse has been found to vary as the square of the amount of chemical applied over the dose range examined (Lee and O'Neill 1971).

Another dose-response relation that is concave upwards at low dose levels is the probit curve. It assumes that the sensitivity of individuals in a population to chemical carcinogenesis is a Gaussian function of the logarithm of the dose. The probit dose-response curve is S-shaped, with a slope that at first increases and then decreases as the dose is lowered. Its use requires the choice of an adjustable parameter, the probit slope, that describes the narrowness of the presumed Gaussian distribution of sensitivity to carcinogenesis in the population

---

*It has recently been suggested, however, that while the response for neutrons is linear, γ radiation may vary more nearly as the square of the dose (Jablon 1975).

at risk. The probit slope may be estimated from dose-response data at high incidence, as determined in an animal experiment, so long as the experimental data is compatible with a probit curve. For sufficiently low dose levels, the probit extrapolation always predicts a lower incidence than does the single-event hypothesis. However, using a value* for the slope that is well in the range of values found for various carcinogens in animal experiments at high incidence, the probit extrapolation for lower doses does not differ by more than a factor of 10 from the incidence predicted by the single-event hypothesis down to an induced incidence of about 1 per 100,000 exposed individuals.

Estimates of Carcinogenic Hazard Until more is known about the mechanisms of chemical carcinogenesis, any method of extrapolation to predict cancer rates at doses much lower than tested will remain partly a matter of conjecture. However, the single-event hypothesis probably provides an upper limit for induced incidence estimates at low dose levels and is compatible with the very limited human data on carcinogenesis at intermediate response levels. It would therefore seem prudent to employ the single-event hypothesis in making risk estimates, at least for those carcinogens for which the dose-response curve from animal experiments approaches linearity at the lower response levels that can practically be studied. In that case, the estimated total lifetime incidence in man resulting from continuous exposure to an environmental carcinogen would be the lifetime incidence for continuous exposure to the same total dose per body weight found by extrapolation of the animal data under the single-event assumption. For risk assessment, the resulting estimate would then be subject to adjustment to allow for statistical uncertainty in the input data.

Findings of the Health Hazards Panel

1. For many pesticides in widespread use, it is inevitable that some will appear in the diet of the general population, which may also be exposed, although apparently to a lesser degree, via other routes.

---

*One standard deviation per log dose (see Mantel, N., and W. R. Bryan 1961).

2. There is no clear indication in U.S. mortality statistics through 1967 reviewed by the Panel of any major deterioration in health attributable to the widespread use of synthetic organic pesticides. This does not preclude the possibility of less-extensive, but still important, effects requiring more detailed evaluation of such statistics for their detection. Such evaluation, including the analysis of the most recent site-specific and age-specific trends and of cancer mortality by county, should be encouraged as a means of detecting previously unsuspected environmental effects on health.

3. Chronic exposure to a chemical carcinogen may not manifest itself as an increase in cancer until many years after the agent is introduced into use. However, the induced cancer incidence may rise sharply thereafter. This seriously limits the assurance that may be drawn from short-term observations of mortality or morbidity statistics for the general population and for smaller groups, exposed to unusually high doses, such as applicators and manufacturers.

4. Although there are major uncertainties in extrapolating the results of animal tests to man, this is usually the only available method for quantitative risk estimation. Despite the uncertainties, enough is known to indicate what dependences on dose and time may operate and to provide rough predictions of induced cancer rates in human populations.

Occupational Health Hazards

A special kind of health hazard posed by pesticidal chemicals involves the risk to agricultural workers and others exposed to these compounds in the course of their occupation. Although most of society's concern has been focused on the involuntary consumer of pesticide residues, a substantial population of persons is at much more direct risk. There has been a tendency on the part of many policymakers to ignore the occupational health aspect of the problem, acting perhaps on the common assumption that the risks are "paid for" because they are effectively internalized within the agricultural enterprise. To a degree that is true: The spray-plane pilot may indeed receive higher wages for his services in proportion to the risks he takes. But, as in other sectors of our society, the effectiveness of the mechanisms for internalizing cost depends in large part on the power of the group at risk. It should be remembered that where occupational disease of this kind is highest--in the semiarid croplands of the Far West, especially California--the population at risk is largely isolated by ethnic, linguistic, and traditional factors from the main culture and has up until now lacked an effective and established organization to bargain for its interests. Furthermore, even if the costs associated with health risks were effectively contained within agriculture, the magnitude of the problem and the potential human tragedy involved would make it a legitimate focus for our concern.

A variety of problems make it difficult to obtain an index of the seriousness of pesticide-related occupational health problems on a national basis. The most serious of these is the absence of systematic, statewide surveys of occupation-related injury. The only agricultural state in which such data are obtained as a regular procedure is California. The arrangement there is a potential model for other states. The law requires that a physician treating any occupational injury or disease must file a form entitled "Doctor's First Report of Work Injury," copies of which go to the employee's insurance company, the state Bureau of Labor Statistics and (if an illness) to the state Department of Health. Since agricultural employees have been covered under California workmen's compensation for many years, an accurate tally of the pesticide poisoning cases treated as workmen's compensation is available. Yet even this system misses cases. Some are brought to physicians under Medic-Aid, or privately; others are misdiagnosed or never appear in a doctor's office at all. Indeed, some officials of the California Department of Public Health believe that the actual number of cases is underestimated manyfold. In discussions with the Committee, Dr. Ephraim Kahn, of that department, cited two lines of evidence for this view. First, in follow-up investigations of dramatic mass poisoning episodes, it has been shown that the number of cases was underestimated by about one-third in the doctors' first reports. Second, in a 1969 study done by the department in Tulare County, Mexican-American interviewers--purportedly doing a "Health Survey" in which pesticides were not explicitly mentioned--questioned farm workers about symptoms for which they had sought medical treatment during the previous 12 months. Doctors' first reports from the same region for the same period yielded an occupational illness incidence only *one-hundredth* that reported by the workers themselves. Although the interview design used in this study has been criticized, the disparity is so large that it is difficult to explain on methodological grounds alone.

On a national level, the EPA maintains a Pesticide Accident Surveillance System, but it relies on voluntary reporting and employs a form that permits no analysis of either illness or the occupational context that produced it.

Despite its defects, then, the California system provides the only useful data on the epidemiology of pesticide related health problems in units having the dimensions of states. It is interesting that despite probable underreporting, the California surveys show agriculture to be the most dangerous occupation in the state--except that of supplying pest control and other agricultural services! The occupational disease rates per 1,000 workers for farming and industry as a whole in 1970 were 10.8 and 4.7, respectively. Pesticides accounted for only 5 percent of all occupational disease reports in that year, but were responsible for nearly half the occupational poisonings. In all, there were 1,447 cases of occupational illness attributed to agricultural chemicals; 249 of these were systemic poisonings, mostly by organophosphates and carbamates.

Not all workers, obviously, are at equally serious risk. The majority of field-workers receive an exposure that will depend, among other things, upon their level of contact with the crop during harvesting or other operations, upon the care and frequency with which pesticides are used, and on the amount of rainfall. More seriously exposed workers include "scouts" who assess insect population levels in sprayed fields, or flagmen who direct aerial spraying operations from the ground. Those at greatest risk are mixers and loaders in applicator plants, or pilots of spray aircraft and drivers of ground rigs.

The hazards of these latter occupations are well illustrated by the situation in the Lower Rio Grande Valley of Texas, where cotton requires heavy protection with organophosphate insecticides. In the middle and late 1960s, nearly 70 cases per year of organophosphate pesticide poisoning were reported. Officials of the Community Pesticide Study (CPS) in Brownsville report that the current totals are somewhat lower, but on occasion the cases have been severe--in two instances, severe enough to involve likely loss of life. Among teenage boys employed by applicator companies during the summer as loaders and mixers, the incidence of poisonings has ranged between 10 and 15 percent of the population at risk. CPS personnel have attempted to enforce clothing safety provisions, but the heat and other factors make it difficult to obtain compliance.

In a few instances like this, the scrutiny of a state or federal agency falls upon a particular occupational health problem and results in corrective action. After the announcement that DDT would be banned for general use at the end of 1972, officials of the EPA and USDA drafted a plan for training farmers (largely small cotton, soybean, and peanut growers in the Southeast) in the handling of the more toxic organophosphate compounds that would inevitably be used instead of DDT. Agricultural Extension Service centers in 14 states were asked to submit proposals for training programs in their own localities; these were to be supported partially through reallocation of nearly $3 million in USDA regional funds. At this point the program seems to have been successful in avoiding major occupational health disasters, but an accurate evaluation of its effectiveness would ultimately have to await retrospective epidemiological studies--the data for which are unavailable. It should be noted, however, that although hearings on the DDT ban had been in progress for months preceding Administrator Ruckelshaus' announcement to ban DDT on June 14, 1972, the first planning meeting was not held until September 6 and the first tentative program not issued until October 10. The delay made some tactical approaches to the problem, e.g., the production of training films, impracticable.

In the majority of situations, however, there is little surveillance of pesticide-related occupational health problems. It is somewhat alarming that the most carefully studied state turns out to have by far the largest incidence of such problems; one is tempted to conclude that the situation elsewhere looks better than it really is primarily because of ineffective reporting. Furthermore, even in California there is strongly suggestive evidence that the number of cases is far larger than that reported. Because the data base is so poor, it is impossible to estimate such health parameters as morbidity. Mortality, it would appear, is relatively low: In California, in the period 1951-1970, there were 23 occupational and about 100 nonoccupational deaths due to pesticides. Fatalities in aircraft spray operations may or may not be partially due to occupational exposure to the compounds being sprayed: Most observers believe that some such accidents are contributed to by the physiological symptoms that result from exposure.

Occupational mortality statistics for pesticides on a national basis are not now available for the years since

1970. Analysis of the national data for 1969 (Hayes, in press) and of state data for more recent years (Personal communication from J. Freeman, Department of Epidemiology, North Carolina State Board of Health) shows that these figures are heavily inflated by suicides, homicides, and accidental deaths that are nonoccupational. Truly occupational deaths numbered about 12 nationally for the year 1969 (Hayes, in press), and the number is not likely to have changed significantly since. The much larger cost due to morbidity, however, cannot be calculated.

The transition to organophosphate insecticides from DDT following the ban on the latter compound was expected by many to produce an increase in occupational mortality, especially in cotton- and tobacco-growing areas of the South, where DDT was depended on more heavily than in other sections of the country. While national data are not available to indicate whether such an increase has in fact occurred, spot checks of state data have not revealed it, nor does the evaluation of Project Safeguard indicate a serious new problem.

Both federal and state governments promulgate regulations regarding the safety of agricultural workers exposed to pesticides. At the federal level, there has been jurisdictional confusion about this responsibility. Under FIFRA, as amended by FEPCA, the EPA has the authority to set and enforce reentry standards for all pesticide uses. The DOL, under OSHA (1970), has authority to issue and enforce occupational safety standards in areas where other agencies have not prescribed them. In early 1973, the DOL proposed reentry times for 21 organophosphate pesticides on five crops, which ultimately were rendered inoperative by judicial decision; in the meantime the EPA was proceeding to promulgate its own standards. In November of 1973, a memorandum of agreement between the DOL and EPA stated that the agencies would work jointly, employ combined records, and make sure that their standards were not in conflict. The EPA's preliminary proposals were published in the *Federal Register* in July of 1973; hearings were subsequently held, and the proposed final regulations were published in March 1974. The hearing process produced significant reductions in the reentry times for most pesticides. Moreover, the current regulations exempt from the reentry provision workers wearing "appropriate protective clothing"--coverall, shoes, and a hat. Persons medically experienced with pesticide poisoning doubt the efficacy of such protection. The EPA proposals do, however, contain two important positive features. First, they establish the label instructions--in particular, the preharvest interval set forth on the label--as enforceable standards. Second, they will require registration data regarding the protection of workers from foliar residues.

At the state level, there is of course substantial diversity in regulations. California has the most carefully developed and stringent requirements: For nearly every pesticide and crop, California's reentry times exceed the most recent federal ones. Moreover, California's regulations provide for specific medical monitoring in cases of sustained exposure and define protective clothing more explicitly.

The most obvious feature of these occupational health regulations, taken together, is the complexity of jurisdictions involved. Both state and federal governments are involved; and within the federal apparatus, more than one agency believes it has responsibilities. Furthermore, problems are inherent in setting such standards nationally, especially where hearings and other adversary devices form part of the promulgation process. There is a tendency to settle on the least common denominator of protection--that is, to make federal regulations appropriate to the least, rather than the most, dangerous situation. If all states set standards appropriate to their own regional character, this characteristic of the federal regulations might not be cause for concern; but few states so regulate, and the existence of a federal regulatory body in the same area may be misread as a sign that the situation has been taken care of.

Finally, we consider several difficulties in the setting of worker-protection regulations by the EPA. There is no substantial scientific data base to permit conclusions about how much exposure to a given pesticide on a particular crop in a particular locale is really dangerous. Field tests are not done; indeed, the location-specificity of the problem is so great that it discourages such data gathering.

The level of scientific input going into the protection of farm workers is oddly at variance with that invested in the protection of consumers. We believe that the heterogeneity of the U.S. agricultural enterprise is great enough to suggest that ideally specific regulations should be set at the state level. But for the foreseeable future, it is likely to remain a federal responsibility; and as long as that is so, it should be carried out with the most reliable data that can be obtained. It is difficult to see how a useful data base can be constructed unless the EPA (1) develops stronger policies to require worker-protection data as a part of registration requirements and (2) develops the in-house competence (including that provided by occupational health specialists) to generate the information necessary to set reentry standards more scientifically. Finally, we point out that the jurisdictional confusion between the EPA and DOL is damaging to efforts in this area. While we make no specific recommendations for its resolution, we believe consideration must be given to the allocation of responsibility between these agencies and the National Institute of Occupational Safety and Health, which also has important skills to bring to the problem.

HISTORICAL ROOTS OF THE PROBLEM

Public Policies and the Adoption of New Technologies

Throughout most of this past century, U.S. agriculture has experienced rising farm productivity and declining relative prices. Growth in agricultural productivity increases the supply of farm products that can be produced from a given aggregate amount of resources. At the same time, the overall demand for agricultural products is relatively inelastic. A 1-percent change in supply, relative to the change in demand, is associated on the average with a 5-percent change in price in the opposite direction. Thus, relatively small increases in supply cause much larger decreases in prices.

This combination of factors has resulted in a continuing problem for the individual farmer, who perceives that the only way he can improve his own position is to produce more at the same cost or produce the same amount at a lower cost. The American farmer has had three ways, not mutually exclusive, of changing his production costs per unit of output: (1) He could cultivate more land, (2) he could adopt labor-saving technologies, mostly in the form of new machinery,* and (3) he could adopt land-saving technologies that allowed increased production per acre--examples would be hybrid corn, use of fertilizers, and, significantly, the use of pesticides. Some pesticides especially herbicides, are also labor-saving technologies.

---

*Adoption of labor-saving technologies was historically a very important step, because North America was land rich and labor poor; any device or technique that allowed increased production per man-hour was a potential boon to the farmer, and the adoption of labor-saving technology was a hallmark of the changes in nineteenth-century American agriculture (Benedict 1953).

The government and private sector have long taken an active interest in the exploitation and development of agricultural resources through the development and introduction of new technologies. Pest control research has been a prominent part of both public and private R&D efforts, and, in many cases, there has been close cooperation between them. The target of this research has usually been to lower farmers' production costs. The effect of the resultant technologies has also been to expand output. Technological change thus permitted agricultural producers to meet the nation's food and fiber requirements at lower relative prices and created a downward pressure on total net farm income.*

Severe agricultural price depressions in the 1920s--caused primarily by worldwide disruptions of economies following World War I--stimulated major debate on government intervention in the agricultural economy. The government price-stabilization programs instituted in the 1930s can be viewed as attempts to design institutional structures that would capture a larger share of the social gains from new technology for the agricultural sector of the economy.

A prominent feature of these programs has been limitation of the acreage on which crops may be grown. These restrictions on land use enhanced the attractiveness of land-saving technologies. Substantial increases in the use of improved varieties, fertilizers, and pesticides occurred after the introduction of the acreage restriction programs.

The price-support and acreage-limitation programs were designed to meet supposedly temporary emergencies. Instead, these programs have operated within the framework of a continual flow of new technologies. In fact, to the extent that stabilization programs have been successful in raising or stabilizing prices, they have probably encouraged the introduction of new technology and have helped to assure that the problems they were set up to deal with would not be solved.†

---

*This statement does not contradict the observation that farmers react to higher or lower prices for their commodities by increasing or decreasing their production, respectively. Rather, it points out that the new technologies allow a given farmer to produce a unit of output with lessened inputs of land and labor.

†Price-stabilization programs have been justified primarily on two grounds: stabilization and equity. The equity

The continual introduction of new technologies has caused production to rise steadily both per man-hour and per acre. This increased production has, except during periods of rapid growth of external demand, generated a downward pressure on prices. Over the long run, this has led to an exodus of many farmers and to consolidation of the remaining farms into larger units.* The withdrawal of labor from agriculture has further required growers to adopt those technologies that save labor. The adoption tends to increase production further, to lower prices, and thus to reinitiate the cycle. Viewed in this light, the adoption of the new pesticides and other technologies has not been entirely a matter of free choice for the farmer: He has had to adopt them or run the risk of being slowly forced out of business.†

---

argument has usually been used to support higher than market clearing prices in order to transfer income from consumer to farmer. The equity rhetoric in the United States has usually emphasized the need for price supports to protect the small, family farm. The stabilization rationale, by contrast, is more soundly based in economic fact. A well-run price stabilization policy, backed up by an appropriate stock acquisition and disposal policy, is capable of reducing the burden of excessive price fluctuations--or of surpluses and shortages--on both producers and consumers.

The price support programs have, in practice, concentrated the relative burden at the lower end of the consumer income scale and yielded the greatest benefits for the larger agricultural producers. Hence, the equity argument is largely rhetoric. Stock acquisition and disposal have been more frequently the accidental by-product of pricing decisions made primarily with the "equity" objective in mind. Nevertheless, the stabilization feature of the program is likely to remain important in the future, regardless of whether the "equity" features tend to favor farmers or consumers.

*Higher relative wages in industry have also helped attract labor out of agriculture. Thus farmers and farm laborers have been subjected to both a "push" and a "pull" to encourage them to leave agriculture.

†For a more detailed presentation of the "agricultural treadmill," see Cochrane (1957).

Creation of a Growth Industry

The discovery of the pesticidal properties of certain synthetic chemicals--particularly of DDT, 2,4-D, and dithiocarbamates at the outset of World War II--sparked a large effort to discover other synthetic organic chemicals with pesticidal properties. The pesticide industry launched a vigorous R&D effort in response to the demand for new technologies generated by the economic pressures described above. Within 20 years, a large number of successful compounds had been found, and production had grown tremendously. As a consequence, the industry and the pest control sciences became oriented to chemical technology.

The Growth of Regulatory Policies

Although "overly stringent regulation" is sometimes cited as a significant part of our current set of problems with pest control techniques, it should be recognized that our current regulatory policies developed over an extended period of time and for a variety of reasons. The following summary highlights the more important developments:

- Residues of pesticides in or on food are subjected to regulation.
  - 1906: Food and Drugs Act establishes federal jurisdiction over food treated with pesticides and traded in interstate commerce.
  - 1938: The Federal Food, Drug, and Cosmetic Act (FFDCA) allows tolerances to be established.
  - 1950: The Food and Drug Administration holds public hearings on tolerance levels of chemicals on fresh fruits and vegetables but is unable to set tolerances due to complexity of FFDCA.
  - 1950-1952: The House Select Committee to Investigate the Use of Chemicals in Food and Cosmetics holds hearings. Recommended that FFDCA be amended to require that chemicals employed in or on food be subjected to safety testing such as required then for new drugs and meat products.
  - 1954: The Miller Amendment to FFDCA gives the Food and Drug Administration a new mechanism to set tolerances of pesticidal residues on food.
  - 1958: The Delaney Amendment to FFDCA declares food additives found to induce cancer in man or animal to be unsafe.

- The pesticide consumer receives protection against misbranded and adulterated pesticides.
  - 1910: Insecticide Act of 1910.
  - 1947: The Federal Insecticide, Fungicide, and Rodenticide Act (FIFRA).
- Concern about the environmental impacts of pesticides becomes a significant policy issue.
  - 1962: Rachel Carson's *Silent Spring* sparks a still-lingering controversy.
  - 1963: The President's Science Advisory Committee issues *The Use of Pesticides*. Recommends reducing reliance on pesticides, especially the persistent ones.
  - 1969: The Secretary's Commission on Pesticides and their Relationship to Environmental Health (the "Mrak Commission") recommends limiting uses of DDT and DDD because of adverse environmental effects and restricting other persistent pesticides to "essential" uses.
  - 1970: The National Environmental Policy Act requires a statement of environmental impact for every major federal activity "significantly affecting the quality of the human environment."
  - 1972: The Federal Environmental Pesticide Control Act (FEPCA) amends FIFRA. FEPCA requires registration of a chemical unless the EPA determines the chemical will have an "unreasonable adverse effect" on the environment. Use of pesticides is regulated.
  - 1972: Most uses of DDT are eliminated by the EPA.
- Long-standing occupational hazards associated with the use of pesticides are regulated at the federal level.
  - 1970: The Occupational Safety and Health Act (OSHA) establishes a public policy on the provision of safe and healthful working conditions.
  - 1972: FEPCA provides that pesticides are to be classified for "general use" or for "restricted use." Restricted chemicals must be applied by or under supervision of a certified applicator.

The Irreversibility of Technological Change

Once a technological change has occurred, it is difficult or impossible to return to the status quo without severe disruption of human behavior patterns. New technologies alter the flow of resources into economic activities and change human perceptions of what can and should be done with resources. The former effect leads individuals, organizations, and firms to assume economic and social obligations that they would not have assumed in the absence of the technology. The latter leads to the formation of value judgments that can be reversed only with difficulty. Both create barriers that impede a return to the old way of doing things. Two examples can be given to illustrate these principles.

Agricultural pest control that depends upon herbicides is in part a substitution of capital for labor. The substitution was accompanied by an exodus of labor from agriculture into industrial and service occupations. In addition, there was a physical movement of people from rural to urban areas. Our population is now partitioned into two major segments: a large nonagricultural group and a small agricultural force. The partitioning depends upon the substitution of herbicides and other capital inputs dependent upon herbicides for manual labor. If the pesticides failed or if their risks were seen as too high to bear, the partitioning scheme would be strained and large social disruptions would be likely. Food and fiber could not be produced without extra labor, but the urban-based work force could not move easily into rural areas to do the job. Neither is it likely that U.S. consumers would readily reverse a trend that has given them, over the past few decades, food at remarkably low cost. Thus, alternative pest management schemes that are labor intensive are not likely to be adopted by contemporary U.S. society under present conditions.

Similarly, chemically based control of disease vectors and nuisance organisms over large geographic areas is now considered necessary for health and comfort, both physical and psychological. As a consequence, alternative pest management schemes proposed for public health and nuisance pests must be at least as effective and economical as the ones for which they substitute in order to win ready acceptance. Value judgments that equate "quality" and "freedom from pests" are very strong and not subject to easy reversal.

REFERENCES

Ames, B. N., W. E. Durston, E. Yamasaki, and F. D. Lee (1973) Proc. Nat. Acad. Sci. 70:2281-2285.

Benedict, Murray R. (1953) Farm Policies of the United States, 1790-1950. New York: The Twentieth Century Fund, p. 83.

Blus, L. J., R. G. Heath, C. O. Gish, A. A. Beslile, and R. M. Prouty (1971) Eggshell thinning in the brown pelican. Bioscience 21(24):1213-1215.

Brown, A. W. A. (1971) Pest resistance to pesticides. *In* Pesticides in the Environment, ed. by R. White-Stevens VI, pp. 457-522. New York: Dekker.

Burbank, F. (1971) Patterns in Cancer Mortality in the United States, 1950-1967. National Cancer Institute Monograph 33. Washington, D.C.

Burnett, R. (1971) DDT residues: Distribution of concentrations in *Emerita analoga* (Stimpson) along coastal California. Science 174:606-608.

Cochrane, W. W. (1957) Farm Prices. University of Minnesota Press.

Cooke, A. S. (1973) Shell thinning in Asian eggs by environmental pollutants. Environ. Pollut. 4:85-152.

Craig, P., and G. Miller (1974) Carcinogen Dose-Response Relationships. National Cancer Institute Contract No. N01-CP-43302.

Crocker, J. F. F., R. L. Ozere, K. R. Rozee, S. C. Digout, and O. Hutzinger (1974) Insecticide and viral interaction as a cause of fatty visceral changes and encephalopathy in the moth. Lancet (July 6):22-24.

Crump, K. S., D. G. Hoel, C. H. Langley, and R. Peto (In press) Cancer Res.

Cutler, S. J., J. Scotto, S. S. Devesa, and R. R. Connelly J. Nat. Cancer Inst. Third National Cancer Survey: An Overview of Available Information 53(6):1565-1575. Dec. 1974.

Doll, R. (1967) Prevention of Cancer, Pointers from Epidemiology. London: Nuffield Trust.

Doll, R., and A. B. Hill (1964) Brit. Md. J. 1:1460.

Federal Register (1974) Vol. 39, No. 203, pp. 37246-37272, Oct. 18.

Georghiou, G. P. (1972) The evolution of resistance to pesticides. Ann. Rev. Syst. Ecol. 3:133-167.

Haenszel, W. (1961) J. Nat. Cancer Inst. 26:37-132.

Hayes, W. J. (In press) Mortality in 1969 from pesticides including aerosols. Arch. Environ. Health.

Headley, J. C. (1968) Estimating the productivity of agricultural pesticides. Am. J. Agric. Econ. 50:13.

Hooker, A. L. (1972) Genetics of disaster. J. Environ. Qual. 1:212-215.

Hurlbert, S. H., M. S. Mulla, and H. R. Wilson (1972) Effects of an organophosphate insecticide on the phytoplankton, zooplankton, and insect populations of freshwater ponds. Ecol. Monogr. 42:269-299.

Innes, J. R. M., B. M. Ulland, M. G. Valerio, L. Petrucelli, L. Fishbein, E. R. Hart, and A. J. Pallotta (1969) J. Nat. Cancer Inst. 42:1101-1114.

International Agency for Research on Cancer (1974) Monographs on the Evaluation of Carcinogenic Risk of Chemicals to Man, Vols. 1-6. Lyon, France.

Jablon, S. (1975) Environmental Factors in Cancer Induction: Appraisal of Epidemiologic Evidence. Leukemia, Lymphoma, and Radiation: *In* Proceedings of the XI Int. Cancer Congr., Florence, 1974, Excerpta Medica International Congress Series No. 351, Vol. 3, Cancer Epidemiology, Environmental Factors. Amsterdam.

Kahn, H. A. (1966) *In* Epidemiological Approaches to the Study of Cancer and Other Chronic Diseases, W. Haenszel, ed. National Cancer Institute Monograph No. 19. Washington, D.C.

Kmet, J. (1970) J. Chron. Dis. 23:305-324.

Lee, D. J., J. H. Wales, and R. O. Sinnhuber (1971) Promotion of aflatoxin-induced hepatoma growth in trout by methyl malvalate and sterculate. Cancer Res. 31:960-963.

Lee, P. N., and J. A. O'Neill (1971) Brit. J. Cancer 25:759.

Lichtenstein, E. P., T. T. Liang, and B. N. Anderson (1973) Synergism of insecticides by herbicides. Science 181: 847-849.

Lilienfeld, Levin and Kessler, eds. (1972) Cancer in the United States. Cambridge: Harvard University Press.

Mantel, N., and W. R. Bryan (1961) J. Nat. Cancer Inst. 27:445-470.

Mason, T. J., and F. W. McKay (1974) U.S. Cancer Mortality by County, 1950-1969. USDHEW Publ. No. (NIH) 74-615. Bethesda, Md.

McCann, J., N. Spingarn, J. Kobori, and B. Ames (1975) Proc. Nat. Acad. Sci. 72(3):979-983.

Miller, J. A., and E. C. Miller (1971) Chemical carcinogenesis: Mechanisms and approaches to its control. J. Nat. Cancer Inst. 47:v-xiv.

National Academy of Sciences (1972a) Genetic Vulnerability of Major Crops. Committee on Vulnerability of Major Crops, Agricultural Board, Division of Biology and Agriculture. Washington, D.C.

National Academy of Sciences (1972b) The Effects on Populations of Exposure to Low Levels of Ionizing Radiation, Report of the Advisory Committee on the Biological Effects of Ionizing Radiations. Washington, D.C.: U.S. Government Printing Office Publ. No. 0-489-797.

National Center for Health Statistics (1952-1972) Monthly Vital Statistics Report. Annual Summaries. Washington, D.C.: USDHEW.

National Pesticide Monitoring Program (1974) Washington, D.C.: National Agricultural Chemicals Association.

Pimentel, D. (1971) Ecological Effects of Pesticides on Non-Target Species. Washington, D.C.: U.S. Government Printing Office. 220 pp.

Pinthus, M. J., Y. Eschel, and Y. Schori (1972) Field and vegetable crop mutants with increased resistance to herbicides. Science 177:715-716.

Rall, D. (1974) J. Wash. Acad. Sci. 64(2):63-68.

Ryan, R. G. (1970) Resistance of common groundsel to simazine and atrazine. Weed Sci. 18:614-615.

Snyder, Noel F. R., H. A. Snyder, J. L. Lincer, and R. T. Reynolds (1973) Organochlorines, heavy metals, and the biology of North American accipiters. BioScience 23(5):300-305.

Stanley, C. W., J. E. Barney, M. R. Helton, and A. R. Yobs (1971) Environ. Sci. Tech. 5:430-435.

U.S. Congress (1972) Federal Environmental Pesticide Control Act of 1972 (Public Law 92-516), 92nd Congress, House of Representatives. Washington, D.C.: U.S. Government Printing Office.

U.S. Department of Agriculture (1965) Losses in Agriculture Agricultural Handbook No. 291. Washington, D.C.: U.S. Government Printing Office.

U. S. Department of Health, Education, and Welfare (1969) Report of the Secretary's Commission on Pesticides and Their Relationship to Environmental Health. Washington, D.C.: U.S. Government Printing Office. 470 pp.

U.S. Department of Health, Education, and Welfare (1971) Leading Components of Upturn in Mortality for Men, U.S. 1952-1967. Publ. No. (HSM) 72-10008. Rockville, Md.

U.S. Department of Health, Education, and Welfare (1972) End Results in Cancer, Report No. 4. Publ. No. (NIH) 73-272. Bethesda, Md.

U.S. Department of Health, Education, and Welfare (1973) Mortality Trends: Age, Color, and Sex, U.S. 1950-1969. Publ. No. (HRA) 74-1852. Rockville, Md.

U.S. Department of Health, Education, and Welfare (1974) Mortality Trends for Leading Causes of Death, U.S. 1950-1969. Publ. No. (HRA) 74-1853. Rockville, Md.

U.S. Public Health Service (1964) The Change in Mortality Trend in the U.S. Publ. No. 1000, Series 3, No. 1. Washington, D.C.

U.S. Public Health Service (1975) Vital Statistics of the U.S., 1970, Vol. II, Mortality, Part A, Tables 1-25. Rockville, Md.: USDHEW Health Resources Administration, National Center for Health Statistics.

van den Bosch, R., and K. S. Hagen (1966) Predaceous and Parasitic Arthropods in California Cottonfields. Univ. Calif. Agric. Exp. Sta. Bull. 820. 32 pp.

Where Are the Markets? (1973) Farm Chem. Croplife 136(1): 26-30.

2

CONTEMPORARY PEST CONTROL PRACTICES

THEORY

Behavior of Decision Makers

Pest control is not free, and the users who apply the technology accordingly match their activities--at least approximately--with the benefits they perceive from controlling pests. At the simplest level, it is clear that some resources are managed much less intensely than others. Pest control activities are only one of many inputs available to resource managers, and the level at which they are employed generally parallels the level at which other inputs are used. The management of land for agricultural and forest production illustrates how pest control activities vary along a spectrum of management intensity. At one end of this spectrum are natural forests and grazing lands that receive little or no input. The returns per acre or per dollar invested of such lands are not expected to be high, and managers invest little or nothing in them. Further along the intensity spectrum are even-aged managed forests, improved pastures, and planted cereal grains. At the high end are various vegetable crops and, finally, very valuable crops grown in such artificial conditions as sterilized soil or greenhouses. As the value expected from the resource increases, so does the manager's desire and ability to purchase inputs--including pest control--to enhance or protect the output.

A manager of a given resource has a threshold tolerance for pest organisms; population densities below the tolerance level will evoke no action, but those above will elicit some type of control activity. The magnitude of this threshold population density will vary depending upon the resource being managed, the pest in question, the motivatic of and resources available to the manager, and the ability of the manager to substitute some input for pest control in order to meet his broad goals. Consider, for example, the following agricultural situations: The cotton grower who sprays insecticide on his field in the beginning stages of a bollworm infestation; the cash-grain corn farmer who applies a herbicide to his field; and the strawberry grower who fumigates his soil before starting his crop. The first is taking out a form of insurance against the risk that the insect pest may explode to a population density at which he will suffer heavy losses; the second is substituting capital for labor; and the third is attempting to increase yield by reducing populations of soil pathogens. In each case, however, the grower has the ultimate aim of maximizing profits, and he will apply pest control measures when he believes that the pest population density has crossed an *economic threshold*.

The economic threshold is commonly defined in three different ways, which correspond to increasing levels of sophistication on the part of the decision maker:

1. The population at which the pest is perceived to cause a reduction in yield. This is more appropriately regarded as a physical, rather than an economic, threshold.
2. The population density at which the pest is perceived to cause a reduction in the value of the crop that is greater than the cost of control treatment. At this level of density at least a minimal outlay for pest avoidance is justified by the value of the pest damage it avoids.
3. The population density at which the perceived difference between (a) the cost of the most effective treatment applied at the optimal level and (b) the value of the crop saved is greatest. Any pesticide application beyond this level adds more to the cost than to the value of the crop saved.

To understand and predict pest control practice, it is also necessary to know how the decision maker gathers the data he uses to determine this threshold. He may rely on his own experience or on the experience of the neighboring growers whom he views as most successful; he may use advice from the cooperative extension service or some other public-sector professional source; he may employ either a pest management consultant who is affiliated with a chemical firm or an independent specialist. These sources of information differ from one another in a number of ways: in direct experience with the problem, in extent of scientific training, in cost, and in objectivity. Which one the grower chooses may influence the intensiveness of his own pest management practices. Whichever he chooses, he confronts a treacherous area for decision making. The growth of a pest population is complex and hard to predict. Studies on the actual effects of various levels of pest populations on yields are expensive to perform, and we simply lack them for many important crop pests. It is not surprising that the understanding about economic thresholds is imprecise.

This concept of threshold tolerance is, in principle, applicable to decision makers managing other resources. For example, the manager of a mosquito abatement district knows that when the number of mosquitoes in his area reaches a certain concentration, his constituents will inform him that the tolerance level for these nuisance organisms has been passed and that control activities are indicated. Similarly, the director of a malaria control project will monitor the population in his district for active cases and initiate control activities if a certain number appear during a given time. The actual level of nuisance and disease-bearing organisms needed to trigger activity will vary considerably over time and space, depending on the values attached to comfort and health and the abilities of the population protected to pay for the control efforts.

The point to be emphasized about threshold tolerances is that the classification of an organism as a pest does not depend just upon the nature of the damage it does, but also on the judgment made by people about the relationship of the organism to their own welfare. Standards for health, comfort, and aesthetics all have a genuine value to people, even if a dollar value cannot be assigned to them. The designers of a pest control policy cannot simply label certain desires as frivolous or unimportant; to do so would be to substitute a normative standard of their own for judgments made by their fellow citizens. The economic and social priorities assigned to a pest problem by the victim must be dealt with responsibly regardless of the nature of the loss.

The individual pesticide user, household or firm, can be expected to utilize the most effective pest control practice available up to the point at which the marginal cost of control is equal to the marginal financial benefit (in the case of a firm) or the marginal utility of the practice. Because the firm or the individual typically considers only the private costs and private benefits of a pest control practice, costs that impinge on other individuals or firms, or on the environment generally, may be ignored. When these "spillover effects" are large, as they may be in the case of some pesticides, the optimal level of use by the firm or individual will exceed that for society as a whole. We face a problem of public policy for pesticide development and use precisely because of the frequent divergence between private and social costs. The objective of pesticide policy would be to achieve reasonable consistency between private and social costs and benefits.

Pest control activities causing damage or risk to other valued resources, however, must be examined critically and rationally. Our continued ability to survive in comfort will depend in part upon our skills at analyzing the risks and benefits of treating pests in a variety of ways, including that of accepting them under certain circumstances.

An unfortunate drawback in the determination of threshold tolerances is the interference between different management practices. Efforts to maximize returns from one input may interfere with pest control objectives, or practices to control one type of pest may lead to different pest control problems. To make matters more complex, social and legal constraints or incentives may negate efforts to foster better pest control practices. One example of such interaction is discussed in more detail in the report of the cotton study team but can be briefly mentioned here. Control of the pink bollworm in the Imperial Valley of California would be made easier if farmers harvested their cotton early in the season and destroyed the stalks. Farmers, however, prefer to let the cotton grow as long as possible, so as to maximize their yields in pounds per acre. Government support payments based on historical yields further exacerbate the situation by giving the farmer an incentive to achieve a high yield. Lower expenses to the farmer for insecticidal control methods and environmental benefits from less toxicant released would follow from early destruction, but these benefits would be accompanied by lowered income to the farmer.

A similar example is discussed in more detail in the report of the public health study team. Some countries with considerable malnutrition might benefit from higher uses of pesticides and consequent higher supplies of food, provided distribution systems are adequate. The increases in food, however, might be balanced by increased poisonings because the accident rate with pesticides is probably higher in poorer countries compared to the wealthy nations. Hence management practices designed to yield higher food supplies interfere with management designed to protect the general public health. Decision makers should expect such trade-offs to be the rule rather than the exception in pest control practices.

The Biological Setting for Pest Control Activities

Pest Organisms

The most common pests include weeds (a mixture of green plants, usually angiosperms), pathogens (a mixture of microbial organisms, including fungi, bacteria, viruses, and nematodes), insects and other arthropods (such as mites), and vertebrates. The discussion below briefly summarizes the biological characteristics of each of these classes and the strategies employed in dealing with them as pests.

Weed Control

Weeds are unwanted plants, but the classification of a particular plant as a weed depends upon the situation and upon the nature of the plant. It is inherent in human contact with vegetation that man will place some value on the relative merit of the species present and that some will therefore be preferred to others. Such appraisals occur on all sorts of lands, rural and urban, and are made even of the vegetation in lakes, streams, and canals; but most especially they are made in an agricultural setting. Persons engaged in plant industry, particularly in intensive crop production, differentiate between crop and weed, strenuously encouraging the one and suppressing the other.

In crop monocultures, all plants except the crop are considered weeds, largely because weeds often reduce crop yield by competing for the same resources. Yield is determined by the availability of water, light, nutrients, and other resources; in specific situations yield may be determined almost exclusively by the availability of a single limiting resource. On a worldwide basis, water most frequently acts in this way; but in some instances competition for elemental nutrients or light occurs. When water is the limiting factor, crop yield can be expressed in units of commodity per unit of water; each unit of water removed by weeds is directly reflected in a reduction of crop yield. In humid regions, for example, the limiting resource may be light. Weeds that overgrow and shade the crop can destroy the latter entirely. In contrast, low-growing weeds in an orchard, for example, may have little effect on the crop.

Competition among plants is not entirely for external resources. Some weeds are known to release chemicals toxic to other plants. This mode of competition was once thought to be exceptional, but new instances continue to emerge. Other plants, such as dodder, witchweeds, mistletoe, and broomrape, are considered weeds because they are parasitic on crops. Still other weeds interfere with the planting, transplanting, pruning, thinning, harvesting, and processing of crops. For example, foliage can clog grain harvests or foul the spindles on cotton pickers, thus interfering with harvesting and causing staining of the lint; and thorny vegetation can limit access to orchards. The quality of agricultural products may also be affected: Unpalatable, thorny, or irritating plant material in hay can greatly reduce its value; weed seeds in cereal grains and particularly in planting seeds reduce the value of these commodities; and weeds in feed may cause off-flavors in milk. Indeed, in some crops (e.g., hay) the quality is primarily determined by the extent of weed contamination.

Unlike monocultures, ranges, forests, and watersheds are made up of mixed vegetation; and the boundary between crop and weed is less sharply drawn. Such lands may be occupied by several preferred species, accompanied by various plants of negative or ambiguous value. Management is less intensive than that of monocultures, often involving only such harvesting operations as the selective removal of timber species or the grazing of livestock. Both practices suppress beneficial plants and shift the competitive advantages to low-value species. Excessive utilization pressure tends to convert certain marginal forests and ranges to brushfields populated by shrubby species of little value for either forage or forest products. Large areas of forests and grasslands throughout the world have been converted to brush as a result of selective utilization pressure.

Range weeds include such poisonous plants as larkspur, lupine, hemlock, and loco weed, in addition to woody species. Other herbaceous plants are viewed as weeds because they have no forage value and reduce productivity by displacing plants of higher worth. Some range plants bear thorns or spiny fruit and have seeds that may injure livestock by causing lameness or blindness.

In commercial forests differences exist not only between the value of trees and shrubs but also in the relative value of dominant tree species. Major stands of timber may be classed as weedy in terms of original growth or potential managed growth. Parasitic plants and species that are alternate hosts and sources of infestation by diseases, insects, and other pests also act as forest weeds. For different but obvious reasons, poison oak and ivy are similarly classified.

Aquatic weeds are a major problem in surface waters, where they find adequate nutrients and sunlight. For example, the recreational value of lakes and ponds is reduced by heavy algal blooms and associated decomposition; and floating meadows of water hyacinth and alligator weed destroy the usefulness of canals and inland waterways. In irrigation and drainage canals, water flow can be restricted by submerged or emergent weeds, necessitating expensive dredging or other cleaning procedures. Weed parts in irrigation systems can clog pipes, water meters, and sprinklers. Thick vegetative algal growth can be detrimental to fish and waterfowl and prevent bathing and boating, or interfere with the operation of waterworks or impart unpleasant tastes to domestic water. In arid areas, rank vegetation along irrigation canals and around reservoirs wastes large amounts of water by transpiration.

These problems posed by water vegetation are exceptionally difficult because of the many uses of water and the multitude of needs and interests affected. Our knowledge of life cycles and of the conditions required for growth and survival of water plants is often not adequate to the task of managing aquatic vegetation; and our capacity for assessing and reconciling conflicting values in such management is equally deficient.

The most uncompromising confrontation between man and plants occurs in urban and industrial environments. In the central districts of cities virtually all land is paved or built upon and maintained free of all vegetation, except for occasional specimen plantings in curbed enclosures and boxes or scrupulously planned parks. In residential areas there is also a preoccupation with the species composition and arrangement of vegetation, in which very local monocultures are created and all but the prescribed plants are looked upon as weeds. Nevertheless, weed control by the city dweller is typically more energetic than thorough; close inspection shows that cities continue to support a remarkably diverse volunteer flora. Weeds thrive on every hand--in window boxes, flowerpots, cracks in pavement, strips along foundations, on vacant lots, and in parks and gardens. The attitude toward vegetation on industrial lands is largely one of keeping plant growth in check enough to prevent excessive interference with operations of such facilities as storage yards, airfields, and transportation and utility rights-of-way. The objective is usually control of all vegetation, but near highways and utility rights-of-way the need for visibility, accessibility, and fire protection is balanced against requirements for erosion control and aesthetic considerations. These needs may be met by plantings of turf or other low-growing vegetation.

Man-made and man-managed habitats have the common characteristic of disturbance, usually in the form of bare areas of simplified and unstructured plant communities. The values assigned to plants and plant communities are determined by management objectives for the site as affected by the nature of plant species present. Some plants are particularly adapted to prosper in close association with man and are thus more often a nuisance than others.

The most common agricultural and urban weeds are pioneer or ruderal species adapted to invade and survive in frequently disturbed conditions. Such common plants as crabgrass, lamb's-quarters, pigweed, shepherd's purse, bindweed, purslane, cheatgrass and a few score others are pandemic weeds of cultivation that are found worldwide in populated areas. They spring up in newly disturbed habitats, but in the absence of further disturbance they are displaced by more stable plant communities. Such plants are essentially obligatory associates of man and his ceaseless manipulation of the land.

The survival capacity of ruderal plants in seemingly hostile environments results from a high degree of specialization. They are usually wind pollinated and are profuse seeders. Their seeds remain viable in the soil for 25 years or more and often have dormancy adaptations that phase germination over a long period of time. Successive crops of annual weeds in cultivated lands build up enormous reservoirs of seeds in the soil. This potential, rather than the vegetative stand of weeds at any moment, is the measure of a weed population. Annuals do not resprout from their roots following destruction of their tops, making them seem easy to control; but even a lifetime of preventing seeding may not cleanse the soil of weed seed to a practical extent.

Perennial weeds are usually less-prolific seeders, and their survival mechanisms depend upon survival of the vegetative phase of the life cycle rather than upon seed adaptations. These mechanisms include resistant roots, rootstocks, or other underground parts capable of resprouting following destruction of the aerial parts. Underground organs store sufficient food to survive repeated tillage for months or years. In fact, moderate tillage, far from being harmful to these plants, may break up overcrowded clusters of tubers, thus rejuvenating and spreading them to form new colonies. Such evolutionary strategies have made perennial weeds remarkably persistent.

Weed control, or weed management, is a part of the more general field of vegetation management, which attempts to create conditions most favorable to the crop or other preferred vegetation and least favorable to weeds. Measures taken to improve growing conditions for the crop seek to improve its competitive position in relation to weeds. Vigorous, well-adapted crop varieties properly managed for maximum growth can accomplish much toward the competitive suppression of weeds. Yet the weeds may be equally vigorous and well adapted and respond equally well to the same management regimes. Such generally beneficial measures as preparing a seedbed, proper irrigation, and supplying plant nutrients do not benefit the crop exclusively, nor even preferentially. They must be combined with direct, often drastic, measures aimed at suppressing weeds. Obviously, the measures applied must not be indiscriminate, or the crop may suffer equally with the weeds.

The classical approach has been to destroy each offending plant by hand. The hoe has historically been so much the preferred implement that it has become the very symbol of agriculture and likewise of human toil. The hoe and other hand methods have everything to recommend them except labor efficiency. For this reason, animal-drawn and, in modern times, motorized cultivation have been substituted for handwork wherever possible. Machine tillage is a highly efficient means of weed control; however, its critical limitation is that it cannot remove weeds immediately adjacent to crop plants. This problem was met by combining machine tillage between crop rows and handwork within the narrow band not reached by the machine.

Mechanical energy can be applied to cutting, uprooting, crushing, and shredding weeds in an unlimited number of ways, including plows, disks, sweeps, rotary tillers, blades, giant crushing rollers for brushfields, harrows and drags, draglines, and so on. The advantages and limitations of each are usually obvious from the characteristics of the machine and the nature of the vegetation. Mowing, for example, is only effective against tall vegetation and is the recognized weed control procedure for establishing and maintaining turf.

Burning, selective flaming, flooding, smothering with nonliving materials and the application of steam are employed in weed control where conditions are appropriate. Burning is employed in land clearing and long ditch banks and waste places; it is also used extensively in forest and range weed control and combined with chemicals and other procedures in converting brushlands to pasture. Flooding is employed against unusually tenacious perennial weeds; the opposite, drainage of canals and reservoirs, is used to destroy aquatic weeds by drying or by cultivating the exposed banks and bottoms.

Other methods include preventive, biological, chemical, and managerial methods. Preventive methods include quarantine and sanitary methods aimed at forestalling the introduction and spread of weeds to new areas. Biological control is the employment of insects, plant diseases, and other natural enemies, except domesticated plants and animals, in weed control. This method focuses on target weeds that are objectionable wherever they occur; the control of *Hypericum* in western rangelands by an introduced insect is a striking example of its utility.

The first herbicides to be used were brines, natural ores, arsenicals, and industrial wastes; over the past quarter-century these older materials have been replaced by chemically and functionally diverse, often highly selective, organic chemicals. Some of these agents select between plant families or larger groups, while others discriminate between genera or species. They must be toxic to the common weeds and not to the crop at the rates used. Herbicides that meet these requirements and are reasonably free of health and environmental hazards are now available for all major crops and most minor ones. These materials are widely used in cropland and moderately on forests and rangelands in the developed nations. A number of relatively nonselective herbicides are extensively used on railroads, utility rights-of-way, and other industrial properties. Use of herbicides on rangelands, brushlands, and forests is important, but not extensive in terms of the total areas of these lands.

The modern devices, machines, and chemicals employed in weed control accomplish nothing that was not possible by hand methods 10 centuries ago. Unlike smaller organisms, terrestrial weeds are large enough to be seen, grasped, and destroyed individually. This provides a basis for selective weed control that is as discriminating as a person's ability to distinguish one plant from another. The limitation, of course, is cost. Chemicals and machines substitute for labor--by doing nearly as good a job as could be done by hand much more cheaply.

Managerial methods are concerned with such decisions as the choice of a crop, planting times, or the intensity and timing of grazing. The objective is to devise optimum systems of mutually supporting methods. Individual actions taken without regard to one another may cause problems for surrounding lands or crops in later years. Sound practice involves thorough analysis and resolution of such conflicts to the extent possible.

An important principle in weed control is rotation--of crops, herbicides, and grazing and other management practices. Continued production of a crop on the same land using the same herbicide leads to the development of a weed flora that is tolerant to the particular conditions. This result can be ameliorated by timely rotation of crops, or rotation of herbicides having different modes of action. Such procedures are but modern expressions of a time-honored principle in land and vegetation management.

Plant Pathogens

Plant diseases caused by biotic agents--as opposed to those resulting from chemical or physical environmental factors--result from the pathogenic effects of a parasite on a particular host. The severity of the disease may be markedly influenced by environmental factors, such as temperature, moisture, and light. Effective control of diseases requires knowledge of the life cycle of the causal agent, an understanding of the biological interactions between the pathogen and the host, and information about the biology of the host plant.

Most plant pathogens are either fungi, bacteria, viruses, mycoplasmas, nematodes, or parasitic seed plants. By far the greatest amount of chemicals applied for disease control is for diseases caused by fungi.

Fungi All higher plants are subject to infection by one or more fungus pathogens, and most agricultural plants are susceptible to attack by many different fungi. Potatoes, beans, and cotton, for example, are attacked by more than 30 fungus pathogens. Fungi alone cause losses amounting to hundreds of millions of dollars per year on a worldwide basis. Fungi may attack any of a plant's organs: roots, stems, leaves, flowers, fruits, and seeds are all susceptible. Both the vitality of the plants and the germinability of seeds may be reduced. Fungi also attack plant products such as fruits in transit and storage, tubers, corms, bulbs, rhizomes, fiber, and wood.

Fungus pathogen control is complicated by the diversity of fungi. Many have life cycles that permit them to live on more than one host plant. Some can live for a period as saprophytes, and only become infectious when susceptible plants are available. Others have spore stages that facilitate their dissemination in air, water, or soil. Still others may survive for periods of many years in soil or on plant parts because they produce resistant structures or resting spores.

Fungus spores are produced in great numbers and may be spread by air or water to susceptible hosts. Chemical control of fungus-caused diseases usually involves protecting plants against infection by germinating spores.

Bacteria Though bacteria cause relatively fewer diseases than fungi, some bacterial diseases are highly destructive--including fire blight of apples and pears and certain bacterial blight diseases of beans, as well as the soft rots of fruits and vegetables.

Most fungi that infect leaves require a spore germination period of several hours. Bacteria, however, are small enough to enter foliage rapidly through natural openings, such as stomata or hydathodes. Bacteria may also enter through wounds caused by insects or by cultivation and, in this way, escape the toxic effects of chemical protectants applied as sprays or dusts. Most of the chemicals used in disease control are far more effective against fungi than against bacteria.

Viruses Viruses also attack almost all species of higher plants. Virus diseases have caused serious losses to a variety of vegetable and tree fruit crops throughout the world.

When a plant has become infected with a virus, it is usually not economically feasible to restore the plant to normal health. Control practices, therefore, for the most part are directed toward preventing infection by eliminating or decreasing the source of inoculum, preventing spread of the virus, controlling insect vectors, or developing resistant varieties. Use of insecticides to reduce the population of insect vectors has given control of some virus diseases. It often has been impossible, however, to reduce insect populations adequately to obtain satisfactory results. Disease resistance has been used effectively for control of virus diseases, for example, control of the curly-top virus disease in sugar beets.

Nematodes Nematodes cause several severe diseases on plants, alone and in combination with fungi or bacteria. Plant pathogenic nematodes are known to transmit several viruses that infect plants and also are associated with fungi and bacteria in certain root disease complexes. Some weak fungal and bacterial root pathogens that alone are not highly infectious may cause severe damage once they gain entry into a root as a result of nematode feeding. Nematode diseases are controlled principally by treatment of the soil with chemicals or by use of resistant varieties.

Parasitic Seed Plants Parasitic seed plants, such as dodder, broomrape, and mistletoe cause diseases that are important on field crops, ornamentals, and forest trees. Control principally involves prevention of distribution of seeds and removing parasitic plants from a crop or crop site by physical or chemical treatments.

Control of Plant Diseases

Plant pathologists and others involved in plant disease control have used many different types of plant disease control measures that can be placed under three broad groupings: cultural and biological control, breeding for disease resistance, and chemical control.

Control measures may also be grouped to emphasize certain principles of disease control.

1. Avoidance--not allowing conditions favorable to a disease.
2. Exclusion--preventing contact with the inoculum.
3. Eradication--eliminating the inoculum at the source.
4. Protection--preventing infection by rendering inoculum ineffective in the infection court.
5. Disease resistance--altering the genetic makeup of the host plant to render the causal agent less effective.
6. Therapy--treatment of infected plants to reduce severity of disease or eliminate the causal agent.

Plant disease control efforts usually are directed at prevention of disease or reduction in the amount or severity of diseases. Control programs are usually concerned with populations rather than individuals, although with some perennial tree crops, in particular, control may be on an individual basis. Effective control of most diseases usually involves an integrated program in which environmental, biological, and chemical measures are applied.

Development of effective control programs requires a knowledge of the cause of disease, of the life cycle of the causal organism, of the interaction of the causal organism and the host plant, and of the effect of environmental factors on disease dynamics. Temperature, moisture, and light intensity may markedly affect disease severity, as well as the response of inoculum to control measures. It is important in developing a control program that one have as complete an understanding as possible of the pathogen, of the host and of the cultural and climatic conditions under which the crop is grown, as well as a knowledge of cultural, genetic, and chemical approaches to disease control that might be applicable to the disease under consideration.

Examples of Disease Control Procedures under Each Principle

1. Avoidance of the pathogen. Plant pathogens may be avoided by selecting locations within an area that is unfavorable for disease development, a geographical area in which certain diseases will not develop because of climatic conditions, and planting dates that are more favorable for the host plant than for the organism; by the use of disease-free planting stock; and by modification of cultural practices to avoid conditions favorable for disease development.

2. Exclusion. Pathogens may be excluded from an area by treatment of propagules of plants, inspection and quarantine procedures that prevent the introduction of an organism on plants or plant materials, and elimination of insect vectors of virus diseases. Treatment of seed or other plant propagating parts would normally involve the use of chemicals or heat, and elimination of insect vectors would normally involve use of insecticides.

3. Eradication of the pathogen. Efforts are made to reduce the population of a pathogen or eradicate pathogens through crop rotation, removal and destruction of susceptible plants or diseased plant parts, heat and chemical treatments applied to plants, and soil treatment with chemicals or by flooding or fallowing. Biological control of plant pathogens has also been attempted. This means of control is based on efforts to reduce the activity of a pathogen by increasing the population of another organism that would be antagonistic or parasitic to the pathogen. Action would be either biocidal, in which one

organism kills the other, or biostatic, in which one organism inhibits another. This method of control has not been successful for the most part, but it is assumed that significant biological control occurs naturally in the soil and is a part of the effect obtained by using different cropping sequences. It has been difficult to develop methods that are effective in agricultural practice.

4. Protection. Protection of plants is usually attempted by spraying or dusting foliage and fruit with chemicals and by treatment of propagules to protect against infection. Other protective measures include control of insect vectors of pathogens, modification of the environment to make conditions less favorable for disease development, inoculation with a benign virus to protect against a more virulent form, and modification of nutrition to avoid conditions of high susceptibility.

5. Disease resistance. Development of resistant hosts includes selection and breeding for disease resistance. It has long been a major plant disease control measure. Presently, it is estimated that more than 75 percent of the U.S. crop acreage is planted with varieties that have been developed for disease resistance.

6. Therapy applied to diseased plants. This principle involves treatments of diseased plants to reduce severity or eliminate an agent. It usually involves chemotherapy, heat treatment, or survey of diseased parts.

Grouping disease control measures on the basis of (1) cultural and biological control, (2) control through resistance, and (3) control through the use of chemicals may add to the total understanding of plant disease control.

Chemical Control of Plant Disease The development and use of satisfactory disease-resistant varieties and the use of cultural methods are normally the principal methods used to control plant diseases. Chemical control, however, is frequently more economical and effective than other measures and may be the only effective way to attack a particular disease problem. In spite of the advances in development of resistant varieties, there are still many diseases that cannot be controlled by any means other than use of chemicals. The only means of ensuring adequate quality and yield of crops in many instances is through the use of chemicals.

Chemicals may be used as eradicants, protectants, or therapeutic agents. By far the greatest majority of chemicals are used as protectants, to prevent inoculum from entering the host and initiating an infection. Protectant chemicals are used to prevent growth or sporulation of microorganisms and to kill or inactivate the inoculum at the source, in transit, or at the infection court.

While most of the emphasis in plant disease control is on the use of fungicides, chemical control of disease may also require nematocides, insecticides, herbicides, or bactericides. Nematocides are used primarily for direct control of plant pathogenic nematodes, but also may be used to control nematode vectors of pathogens or to kill nematodes that provide infection routes for fungi and bacteria. Insecticides are used in disease control principally to control insect vectors of plant viruses or to reduce injuries from insect attack that may provide infection courts, such as in the fungus-induced Dutch elm disease. Herbicides may be used in disease control to prevent the production of inoculum by eliminating an alternative host for a plant pathogen. As an example, herbicides are used to kill currants and gooseberries that are alternate hosts for the white pine blister rust fungus and are also used to kill weeds that may serve as alternative hosts for pathogens that attack specific crops.

Prior to 1940, various combinations of inorganic sulfur, copper, or mercury compounds were the principal agents used in disease control. With the discovery of the fungicidal properties of the quinones and the dithiocarbamates in the 1930s, major efforts were devoted to the development of organic fungicides that are now the principal chemicals used in plant disease control.

To be effective as protectant fungicides, these compounds must resist deterioration from weathering and degradation in the environment. They also must be highly fungitoxic but relatively nonphytotoxic to host plants and should have low mammalian toxicity. Ideally, such compounds should be compatible also with certain insecticides that may be applied simultaneously to achieve insect and plant disease control.

Since most of the compounds used in plant disease control are of relatively low mammalian toxicity, they have not been the subject of as much concern as have the insecticides, which as a class of compounds are considerably more toxic to humans.

Arthropods

To a large fraction of the general public, pest control *means* insect control. Some of the most catastrophic instances of crop destruction, as well as some of the most feared human diseases, are traceable to the activities of insects and related arthropods; and these organisms were the objects of the earliest and most intensive efforts at chemical control by man. These efforts, in turn, have generated special human concerns; they have produced a number of secondary problems for man and for other animals man wishes to protect.

Three-fourths of all the animal species known in the world are included in the phylum Arthropoda--more than 800,000 named species, of which 700,000 are insects. Of the latter number, about 10,000 are important as pests of crops, livestock, humans, or stored foodstuffs. Of the remaining arthropods, several groups--including mites and spiders--include organisms that are either significant pests themselves or figure importantly in control as predators or parasites.

A prolonged history of coevolution with terrestrial plants (and, later, with the terrestrial vertebrates that accompanied them in the colonization of the land surface) has brought the insects and other arthropods into resource competition with man. This history has produced an impressive radiation of groups that are classified in part by their feeding habits and related adaptations. Among the major groups of insects that contain important pests are orders that feed primarily on plant juices and employ sucking mouthparts (Heteroptera, Homoptera), that have plant-eating larvae (Lepidoptera, many Coleoptera), that are larval or adult parasites upon mammals (Diptera), and a rich variety of others. Fortunately, a number of groups have evolved a mode of existence that exploits other insects, either as parasites or predators; and these constitute the main sources of biological control for our natural populations of pests.

Activities of Arthropod Pests The kinds of damage done by arthropod pests span nearly the entire range of human interests. Termites and other wood-boring insects destroy man-made structures; various species of beetles and lepidopterous insects damage growing timber in the forest, causing the death of trees and also losses in incremental growth. Nearly every agricultural or ornamental plant and every domestic animal is subject to attack by a set of insect and other arthropod pests; the range of these activities is discussed in the study team reports and in the sections on various control strategies. The arthropods include the most important vectors of epidemic disease in man: fleas and plague, simuliid flies and

Chagas' disease, mosquitoes and malaria and yellow fever, ticks and Rocky Mountain spotted fever, and many others (see Volume V, *Pest Control and Public Health*). Not only are stored agricultural products attacked by various insect pests, but also a wide range of manufactured goods are vulnerable enough to require protection. A number of arthropod species, finally, constitute simple nuisances, of sufficient intensity to persuade man to expend resources in their alleviation.

It is clearly beyond the scope of this volume to attempt a comprehensive discussion of the variety of insects, the kind of damage they do, or the properties of each taxonomic group that make its members special as pests or potential control agents. But certain attributes of the biology of insects (which apply, with only minor modifications, to related terrestrial arthropods) are important to an understanding of their efficiency as pests.

Metamorphosis All of the higher insect orders undergo a dramatic postembryonic developmental change, in which a juvenile, nonreproductive form (the larva) enters a quiescent stage and later emerges as an adult, reproductive form having very different properties. Generalities are difficult to make; but in most cases the larva is adapted for feeding (usually on plant, sometimes on animal, tissues) and the adult is adapted for reproduction. It is normally the adult stage that is winged and therefore possesses the mobility necessary to find a mate. These metamorphoses are crucially important in the design of control strategies. In the case of the major lepidopteran crop pests, the larvae often feed internally and the adults emerge late in the season. In the case of mosquitoes and other insect disease vectors, the larvae are often aquatic. Control in either stage may be possible, but is apt to require totally different strategic approaches. Frequently, the most damaging stage is especially difficult to reach with chemical control techniques. Insect metamorphosis and growth is under the continuous control of hormones, which makes specific chemical intervention in the life cycle a possibility.

Population Size and Dynamics Insects possess enormous reproductive potential, and--especially when resource availability is large, as with crop monocultures--population sizes are correspondingly large. Such populations are impressive reservoirs of genetic variability and can thus respond effectively to an imposed regime of selection. A vast variety of reproductive strategies is found, ranging from those in which sexually mature adults have wide mobility to those in which the reproductive units occupy a restricted space and the flow of genes throughout the entire range is correspondingly limited. Mechanisms for controlling emergence and timing development are diverse and may be keyed to such environmental variables as the length of day. Furthermore, complex chemical communication systems often function in bringing the sexes together for reproduction or integrating the activities of members of the population. Because these features are not easy to ascertain, except by detailed studies on each species, *general* control strategies that will be applicable in a variety of pest situations are hard to design. Even more important, most insects have several generations per year or even per growing season; they have adopted reproductive strategies that emphasize speedy population responses to exploit a newly available resource. A number of control problems are traceable to this property of insect pests, including the appearance of *secondary pests* when a primary target has been reduced to low density by one control measure and the phenomenon of *pest resurgence* that often follows an apparently effective intervention.

Genetic Mobility Large populations, with substantial genetic variability and short generation time, also enable insects to respond to selective influences quickly. The genetic systems of some insects contribute to this ability, by arrangements that promote reshuffling of genes and thereby continually present them to the environment in new combinations. Their long history of coevolution with plants contributes to this ability. The plants have evolved defenses (including chemical ones) that deter herbivore attack; the insects, in turn, have evolved resistance mechanisms to overcome the defenses. In some cases this history has given the insects a built-in advantage in confronting defenses of human devising, either through the presence of resistance mechanisms that preadapt them to human control measures or through adaptations of the genetic systems themselves that make quick responses possible.

Biological Features Several other properties of insects as a group are relevant to their success as pests. Small size and mobility obviously make control difficult. Less obvious, but equally important, are some features evolved as basic protections against the vicissitudes of terrestrial life. Insects have chitinous, wax-covered exoskeletons that function in water conservation but are also effective in restricting damage from contact toxins. A similar protection, though perhaps even more impressive, is enjoyed by the various inactive developmental stages in which insects may overwinter or pass through other difficult times; eggs and pupae, during such periods of diapause, are highly resistant to temperature change, oxygen lack, or chemical attack.

A wide repertoire of stereotyped, "instinctive" behaviors is also characteristic of insects in general; and behavioral evolution, like that of structural features, is rapid. Thus, for example, a new behavioral trait that restricts mate recognition and thus encourages reproductive isolation can be rapidly fixed in the population; or behaviors that promote resistance to chemical attack by taking the insect away from areas of concentration can evolve, just as can biochemical mechanisms for dealing with the compound.

All these features have contributed to the problems mankind has had with insects over the centuries of human agricultural existence. Insects, well adapted to terrestrial life, have enjoyed a success that must be viewed, in objective biological terms, as at least equal to that of man and the other higher vertebrates. It is hardly surprising that the insects and their arthropod relatives, being harvesters of long standing of those resources in which man is interested, prove vexing human competitors.

Vertebrate Pests

Vertebrate animals interfere with human activities, thereby achieving pest status, in a remarkable variety of ways; yet, in no single instance has a vertebrate pest problem been pervasive enough to warrant the kind of control effort mounted, for example, against arthropod pests. The diversity of vertebrate pest problems is well illustrated by the following summary.

Lower Vertebrates Minor control activities are mounted against sharks or other fishes that threaten human safety. Poisonous reptiles, though occasionally causing fatality, are quantitatively unimportant as a health hazard and attract little or no control activity. Economically "undesirable" fish species frequently displace more desirable ones; where this occurs in contained bodies of water, reclamation following the use of toxicants (rotenone, toxaphene, and a few other compounds) has been common practice. The special situation of the sea lamprey in the Great Lakes, which virtually destroyed a valuable commercial fishery, is discussed elsewhere in this volume.

Birds In local situations, granivorous finches, blackbirds and starlings, and a few other bird species cause economically significant crop losses and may be important feedlot pests. Other examples of agricultural damage include grainfield losses in the northern prairie states from waterfowl, sheep kills attributable to raptorial birds, and the damage done to orchards by fruit-eating birds. Nonagricultural damage includes a number of relatively minor cases, including the impact of woodpeckers on utility poles, of a variety of roosting birds on human constructions, and of forest birds on conifer reseeding operations. A more serious matter is the hazard presented by birds to aircraft operations, a high-priority pest control issue since the well-publicized crash in 1960 of a commercial airliner at Logan Airport, in Boston. This disaster, which involved the loss of 62 lives, was caused by a flock of starlings and marked the beginning of extensive tallying of air-strike information and the initiation of research into bird-dispersal techniques. In addition to control using poisoned grain and other chemical methods, a variety of behavioral strategies have been devised and have met with mixed success. These include noise-making "scare" devices that produce recurring explosions; chemicals that produce illness and distress behavior in a small number of individuals in the flock, resulting in the dispersal of the rest; various traps; and control by introduced predators.

Mammals Small mammals, including rodents (mice and rats, ground squirrels, gophers, etc.) and members of a few other groups, give rise to agricultural, public health, and "aesthetic" pest problems. Small mammals do inconsiderable damage to field crops in some regions and are a really major source of loss of forest seed (see Volume IV, *Forest Pest Control*). Various kinds of industrial damage (to power and communication cables, etc.) are done by gnawing rodents. Of major direct economic importance is the role of rodents (and a few other small mammals) as consumers and contaminators of stored foods, especially cereal grains. Finally, small mammals are vectors of a number of diseases (see Volume V, *Pest Control and Public Health*).

A combination of the last two properties, together with some aesthetic feelings, leads householders to take private action against rodents, and rodent control is a significant part of the commercial and suburban structural pest control operator's business in some areas. The special problem of urban rats has received substantial recent political attention. In U.S. cities, neither plague nor typhus--the two diseases especially associated with commensal rodents--constitutes a public health problem at this time; nor is the incidence of rat bite high enough to justify viewing urban rodents as a public health problem. But powerful aesthetic considerations are involved, and these demand serious consideration of the problem.

Perhaps the most dramatic and politically sensitive set of vertebrate pest control problems involves predation by carnivores upon livestock. Substantial losses, especially to sheep ranchers in the West, are caused by the depredations of coyotes and a few other predators. In the middle 1960s, federal expenditures for predator control exceeded $6 million; and two states paid out a total of more than half a million dollars in bounties. Poisoned baits, airplane gunning, and other techniques were used until recently in such operations; but in 1972, following a recommendation by an advisory committee to the Secretary of the Interior, the EPA cancelled the registration of sodium monofluoride (1080), thallium, and sodium cyanide. Current research efforts in this area are discussed elsewhere in this volume. At the present time, various ranchers' groups are pressing for a relaxation of predator control restrictions, while a number of conservationists and wildlife preservation groups are in favor of maintaining or increasing them.

If there is a single property that characterizes vertebrate pests and vertebrate pest problems, it is diversity. With the possible exception of predator control in the 1960s, control of unwanted fish, and control of commensal rodents, no single operation has involved a substantial use of the kind of chemical technology that we find in the control of arthropods, weeds, or pathogens. Instead, there is a host of special situations--each characterized by a unique set of behavioral and ecological properties of the pest organism and each in a particular social and economic setting. It is small wonder that, where possible, such applications have gotten along primarily on "borrowed" technology. For example, until the development of a compound especially designed as a lampricide in the 1960s, all chemical control of fish in aquatic ecosystems was accomplished with compounds devised as insecticides. In many situations, however, the control problem is so unique that special approaches are absolutely necessary: The control of flocking or roosting birds is an example. These situations present policy problems of two kinds. First, where they employ materials that find their major uses elsewhere, they are liable to become the inadvertent victims of demand changes in a distant sector. Second, the size of market involved may not provide substantial incentives for private-sector research on effective control techniques, and public concern for the seriousness of the problem may be insufficient to produce activity unless a dramatic signal--like the Logan Airport crash of 1960--underscores it.

Thus, there are no general rules for approaches to the control of vertebrates. Instead, there are a series of pest problems that--for the most part--are not large in extent or total dollar value, but may produce locally serious consequences. Each control problem demands a specific solution based on special features of the biology of the pest organism; often, since the group includes many animals in which man is interested for other reasons, the spillover costs of control procedures are high. On both accounts, vertebrate pest control demands a heavy research input; the results raise questions about its present adequacy.

Agroecosystems

Agricultural ecosystems (AES) differ from forests and other natural communities in several important respects, including the following:

1. They are more frequently and fundamentally disrupted, particularly when the crop is short-lived and the ground is cultivated between successive crops. The shorter-lived the crop, the more frequent is such disruption.

The disruption of an AES clearly poses problems for all species in the system; they require mechanisms to achieve continuity in the face of the disruption. Species that are pests of the crop are generally well adapted to it and survive disruption either by passing into a resistant stage (e.g., seed, spore, egg, pupa, diapausing adult) or by moving to alternative hosts. It is important for pests with the second strategy that disruption does not occur in all crops and other appropriate vegetation in a given area at the same time and that the pest is mobile. Weeds and fungal pests tend to fall into the first class, but the majority of insects are in the second class and depend on alternative hosts.

2. AES are simpler than most natural communities. They have fewer species (including enemy species of the pests), less physical structure, are more homogenous in space, and the crop itself has little genetic or demographic diversity. The simplicity of its genetic and age structure means that, if and when the crop is susceptible to certain pests or diseases, it is uniformly so; conversely, it may be uniformly resistant to others. When uniformly susceptible crops cover large areas as a pure stand, with little interspersed structural diversity, the spread of the pest or disease organism is likely to be enhanced. These potential dangers need to be balanced against the greater ease of management that follows from such uniformity.

3. There has been less coevolution among the species in AES. Existing natural communities are, by definition, those whose constituent species have survived, interacting together, over long periods. Thus, the plants have evolved mechanisms that prevent the herbivores from exploiting them to the point of extinction; a broad range of such chemical and physical defenses have been described. The herbivores similarly have avoided being wiped out by their enemies (predators and parasites) while the enemies have nonetheless evolved mechanisms to keep up with the herbivores, thus ensuring themselves a harvest. This coevolution probably partly explains the stability as well as the persistence of natural communities.

An AES is fundamentally different from a natural community in that its dominant plant species is more or less arbitrarily thrust into a foreign ecosystem; it then collects species from the surroundings. Where the crop is grown regularly in the area, many of the species on it may form a group that has evolved together in these circumstances during the recent past; but the shared history is normally much shorter, and the proportion of coevolved species is smaller than in a system undisturbed by man.

In addition to the arbitrariness noted above, the genetic makeup of some of the major species, and therefore their ecological characteristics, has been recently altered. Thus, even where the major crops of the world have evolved alongside their herbivores, the advantages to the plant of that shared evolutionary history have largely been lost as the crop has been selected for other properties. Furthermore, breeding material and the crop may have been protected from pest attacks. Although the problem has not been examined, herbivorous insect pests have no doubt been changing genetically in response to our crop breeding programs;

certainly plant diseases have. Presumably, the species that attack the pest also respond to its altered genetic character. Thus, crop systems not only consist of a partially haphazardly selected and reduced set of species, but the ecological characteristics of the major species ar constantly changing rapidly so that the sort of dynamic stability that arises during coevolution is not achieved. Of course, where introduced pests (alien species) are involved, these problems are more severe.

The three features of AES discussed above result, among other things, in the pest problems that concern us here. Because insect pests are usually highly mobile in at least one of their life stages, and because crops are disrupted ecosystems, pests frequently arrive early in the life of the crop and often before their enemies. In addition, the crop is a concentrated, homogeneous stand of a plant species missing most of its natural defenses against insect attack. The pest is therefore able to increase and spread rapidly. Enemies of the pest, when they do appear, comprise a smaller part of the spectrum than would be the case in a natural community. Furthermore, although they may be attacking a species that they have encountered for many generations, it may not have been their major prey throughout their recent evolutionary history; and the pest species may have recently evolved different ecological characteristics. In these circumstances, it is not surprising that pest species escape control by their enemies, even when they are not aided and abetted by poorly managed spraying programs.

Aquatic Ecosystems

Pest control measures are often carried out in natural or culturally eutrophied aquatic ecosystems; these activities may involve the control of nuisance vegetation with herbicides or the control of nuisance fish species. Because such ecosystems are particularly delicate and complex and depend heavily upon semiclosed nutrient and energy cycles, they merit separate discussion in this section.*

Unlike many of the managed agroecosystems, aquatic ecosystems--even those managed primarily for the production of one kind of organism--seldom are monocultures

---

*We acknowledge the contribution of Professor A. D. Hasler, University of Wisconsin, in supplying the Executive Committee with an early draft of this section.

even in the restrictive sense of being dominated by a single species. In addition, the properties of the aquatic environment make containment of a toxic chemical even more difficult than it is on land. As a result of these features, chemical control programs often add nutrients (from the decaying tissues of the target organisms) that spur further growth of the nuisance organism. Suggested alternatives for weed control include two main features: first, reduction of nutrients, in particular phosphorus, entering the system; second, developing uses for the so-called "nuisance" vegetation. Technology is now available for converting some kinds of lake weeds into compost or high-protein feeds. Nutrient entry can be reduced through erosion control, proper control of fertilizer application to croplands, and improved reclamation of nutrients from urban sewage.

Similar problems attend the use of fish toxins (e.g., antimycin) that are used to eliminate unwanted species from natural freshwater systems. Most of the pesticides now employed have fairly broad toxicity, and affect non-target organisms; furthermore, except where they are applied in total-poisoning operations preceding the re-planting of small lakes, their effect is to hasten the eutrophication process.

One of the largest efforts aimed at the eradication of a fish pest has been the U.S.-Canadian program against the sea lamprey (*Petromyzon marinus*) following its invasion of the Great Lakes in the 1940s. Because of its major impact on the commercial and sport fisheries for trout and whitefish, the lamprey became the focus of an effort mounted finally through an international body, the Great Lakes Fishery Commission; the control program involved first electrical wires in breeding streams and, eventually, a new chemical control agent, 3-trifluoromethyl 4-nitrophenol. Although these approaches have reduced lamprey populations in Lake Superior especially, the lamprey still causes large losses to the Great Lakes fishery. In large bodies of water, even when the target species is restricted to tributary streams for breeding, the scale of the ecosystem may simply be too large to make chemical control a practical possibility.

Ecosystem Dynamics and Pest Control

Stability and Diversity Ecologists and others have suggested on theoretical grounds that diverse communities tend to be more stable than simple ones and that agroecosystem diversity may therefore be desirable as a preventive strategy against pests. Unfortunately, the generalization that diversity leads to stability depends in substantial part on observations of artificially simplified systems, so the generality is somewhat circular. Many simple natural communities are no less stable than complex ones. It seems likely that community stability owes more to processes that occur among constituent species than to diversity per se. Similarly, while diversity sometimes helps pest control in crops (e.g., by providing sources of natural enemies), it may exacerbate pest problems by providing alternative hosts for the pest.

It is important, however, not to discard the diversity-stability idea in its entirety. Genetic diversity in the crops, and age-diversity in long-lived forest ecosystems, probably helps prevent the rapid spread of pests and diseases that attack susceptible individuals. There is also evidence from simple laboratory systems and mathematical models that *spatial* diversity--the presence of barriers or structures that subdivide the area--might be useful in preventing pest outbreaks. This latter possibility has been recognized and used in limited application and deserves further study through large-scale experiments in real crop systems. Finally, where natural predators are a part of the pest control program, they may require components of the system other than the crop to complete their life cycle. It would be poor strategy to plant too large an area to the crop monoculture in such situations, because the predators will no longer be able to survive.

On the other hand, there are practical considerations that weigh against applying this kind of control strategy. A sound economic rationale tends to concentrate the production of each crop in an area to which it is highly suited; this strategy subjects the crop to less annual variability due to weather vagaries. Mixed cropping systems also force part of the acreage into plants that are less suited to the climate or lower yielding; greater annual variability and reduced yield are likely to result.

The possible benefits of increasing crop diversity in order to reduce the need for pesticides must thus be balanced against lower yield and higher variability. This trade-off may be economically desirable or even essential, if certain pests cannot be controlled with chemicals becau of resistance in the pest of because the pesticides cause environmental damage. Benefits other than pest control, such as better distribution of the work load through the year, may also result from crop diversity.

In some cases, serial crop diversity may exacerbate pest problems. Certain insects are more prevalent in rotations than in monoculture--as, for example, wireworms following clover or white grubs following grass sod. In other situations, serial diversity merely exchanges one set of pest problems for another. For example, certain weeds are favored by monocultures of corn, soybeans, or wheat when the same herbicide is used for 6 years. But when the three crops are grown in sequence, different weed species are favored, because competition takes place in different seasons and because tillage practices change.

Even monocultural crops can vary in amount of spatial diversity. In the San Joaquin Valley, for example, a million acres of cotton are planted to a single, genetically pure cultivar. In contrast, wheat and other cereal grains are sometimes deliberately planted in mechanically or genetically mixed combinations because it has been found that this procedure improves their resistance to fungal diseases. This approach, which seeks to increase diversity by broadening the regional genetic base of a single crop plant, is being attempted with a number of species (Browning and Frey 1969).

Regulation of Populations Populations of a single species in a community tend to fluctuate about "equilibrium" densities that are characteristic over long periods for a given species but differ between species. The equilibrium density of a population can be set by many factors in the environment, e.g., the amount of food, pressure from predators, and so on. If some event occurs that significantly alters the density, the population, in interacting with various environmental factors, *tends* to return to its equilibrium density. This process occurs because the system is regulated, i.e., it has negative feedback. In the terms of population ecology, there are processes (e.g., competition for food or space, predation) that are *density-dependent*; that is, the proportion of the population killed depends upon the amount by which the population exceeds the equilibrium density.

Population ecologists are not unanimous that this process is universal in all species at all times, and a few would even deny that it is common. Furthermore, there are theoretical and practical problems in stating and testing the theory. Nevertheless, population regulation is a widely accepted first approximation to reality; and it can often be seen to operate in pest populations. The application of pesticides generally kills a large fraction of pests, independent of their population density; but a common pattern (in the absence of more applications) is for the pest to recover to roughly its previous density--that is, the pest population has compensatory reproductive abilities. Thus, subsequent density may be the same in sprayed and unsprayed areas, because higher survival in the sprayed area has compensated for the mortality caused by spraying. This latter process occurs because survival is density-dependent, tending to be higher at lower densities.

In principle, both competition and predator-prey interactions ought to be useful in the design of strategies for controlling pest populations. Competitive displacement of insects or weeds could be a valuable result if the replacing species produced a smaller effect on yield. Unfortunately, a simple rationale predicts that such strategies are unlikely to work. Whether it is an exploiter of the crop plants (most insects) or a competitor (most weeds), the pest is highly adapted to utilize the resources present in the agroecosystem. The main resource is the crop itself, or something the crop utilizes; any effective competitor, by definition, is an organism that competes for that resource. Thus, if it displaces a pest, the new organism is likely to be a pest itself, in exactly the same sense as its predecessor. The only substantial applications of the competition principle have thus involved modifications in the capacity of crop plants to compete with weeds.

Predator-Prey Interactions and Biological Control Biological control rests on the premise that many pests can be, and are, *regulated* largely by their natural enemies (predators and parasites in the case of insect pests). When biological control by enemies is consistently successful in an area, the best conclusion is that the enemies are indeed operating in a density-dependent fashion. To this extent, biological control uses the principle of population regulation, but the principle has rarely been applied in a more sophisticated way. Such applications are at least foreseeable. For example, the kind and number of natural enemies released under different circumstances could be geared to pest abundance, so as to take deliberate advantage of the principles of density dependence. In practice, however, the rule has been simply to release or foster natural enemies on the assumption that they will operate in a density-dependent fashion.

Against this background, we can now look at the properties of predators, specifically those that are insects. They are of two types: true predators that eat many prey individuals during their development and insect parasites that are free-living as adults but develop in or on a single host individual as immatures. Parasites tend to have shorter generation times than do predators, and their life histories are sometimes synchronized with those of the pest. Entomologists have generally held the belief that parasites that are highly specific to the pest are the best type of enemy, because they are synchronized with and highly adapted to the pest and have high rates of increase. This belief probably rests mainly on evidence from the control of introduced (alien) pests that do not have a complex of parasites and predators of their own, and it may be valid for such situations. By contrast, there is no good evidence that any one type of enemy is "best" for native pests. For example, in California cotton, *Lygus* is controlled by a large array of predators (such as bugs) that are *not* highly specific to the pest. No situation has been studied well enough to describe the dynamics of the enemies that are successful in control programs, though, in theory, both true predators and parasites, and both specific and general predators, could function separately or together to control pests by causing density-dependent mortality.

When natural enemies are a part of a control program, the first problem is to ensure that the enemies persist in the area--that is, the farmer needs to pursue a *continuity* strategy. The more closely tied to the pest the enemy is, the less need is there for the farmer to maintain ancillary components in the system. Thus, a specific parasite that needed no other host would be perfect from this point of view. But present pest species rarely have such parasites that are adequate. Most enemy species in fact are linked to other components of the system, i.e., to nonpest species and to species that occur on other crops or other vegetation. In some cases the links may be few and simple. For example, the parasite of a leafhopper pest in California vineyards needs only one alternative (winter) host, a bug that lives on blackberry bushes. In cotton in California, on the other hand, a wide variety of enemy species occur that attack *Lygus*, and these species maintain continuity by living in other crops such as alfalfa and sunflower and in "weeds."

The need for continuity and for ancillary components in the system are related to the question of diversity in relation to pest control. The evidence suggests that certain specific components are necessary for control, but that diversity per se offers no obvious advantages. Unfortunately the enemies' natural histories are rarely sufficiently well known for us to determine precisely which components are needed. We can, however, make one general statement: Where the enemies depend upon other components of the agroecosystem, the scale of crop monoculture can be so large that much of the crop will be too distant from sources of enemies. The actual field size at which this is a problem will depend upon the movement patterns of the enemy species.

The above discussion illustrates that one important set of properties of predators can be lumped under the heading "natural history": range of hosts, degree of movement, synchrony with the pest, and so on. There seem to be no general rules about which features are important, but together they determine what tactics the farmer needs to employ to maintain continuity in the enemy complex and constrain the disruptive tactics he can use on the pest.

The second set of important properties includes more dynamic ones, such as those that determine the attack rate of the enemy and the degree to which the attack rate is responsive to changes in prey density. Hunting efficiency, the time taken to deal with each pest individual, reproductive rate, generation time, effect of alternative foods, variation in growth rate in response to changing prey density, and others are in this group. A moderately sophisticated branch of modern ecology analyzes, by experiment and mathematical modeling, these dynamic aspects of predation.

The attack rate of a predator population at any time is clearly the product of the average number of prey killed per predator and the number of predators present. The number killed per predator depends on the predator's characteristics--whether it is a true predator or parasite its size if it is a predator, the time it takes to deal with each prey ("handling time"), etc. Models suggest that the shorter the handling time, the more likely the predator is to be a good controlling agent. Other features may also be important--for example, the presence of alternative prey, the preference of the predator for the pest, and the degree to which this preference can be altered by experience. Thus, it can be shown by modeling that a general predator can be useful in *preventing* pest outbreak when the pest is at low numbers, provided an alternative prey is available and the predator switches to the pest as it becomes abundant.

Such features determine the short-term response of the predator to change in pest density, referred to as the predator's *functional response*. In simple laboratory situations, this response tends not to be controlling (i.e., not density-dependent). But in nature, such complicating features as unevenness in prey distribution and the presence of other prey species can make the functional response density-dependent.

During development, the increase in size of the predator can alter its attack rate; changes in growth rates, which in turn are referable to pest density, can thus increase predation efficiency through a kind of *developmental response*. Predator numbers can also change ("numerical response") as pest density changes. Where this response depends upon predator reproduction or upon large-scale movement of predator populations, it is likely to operate with quite a long time lag. However, among true predators, especially those that are cannibalistic, the survival rate of the predator can change fairly quickly in response to changing prey density. Entomologists have usually considered the numerical response to be the most important one in controlling pests and in particular have favored parasites that have a short generation time. The theoretical and practical analysis of these problems is not sufficiently advanced to enable us to evaluate this claim.

The analysis of predation dynamics discussed above, while of considerable interest to ecologists, has played essentially no part in control programs. It may be that the cost of discovering the necessary information for any field situation is too high. Nonetheless, this is an area that should be pursued because of its potential for the future. At present, even in the most sophisticated programs for integrated control, the enemy complex is not understood at this level of detail. Yet a knowledge of its dynamics might allow us to use and augment this complex in quite powerful ways. Perhaps the most useful approach would be to try to simplify the theoretical and analytic framework so that a bare minimum of information is needed from the field and then design the predation models so that this information could be usefully employed. Whatever the specific solution, further investigation of theoretical research in population biology for the improvement of pest control methodology is needed.

REFERENCE

Browning, J. A., and K. J. Frey (1969) Multiline cultivars as a means of disease control. Ann. Rev. Phyto pathol. 7:355-382.

3

INSTITUTIONS AND PEST CONTROL: RESEARCH AND DEVELOPMENT CAPABILITIES FOR THE INNOVATION OF NEW PEST CONTROL TECHNOLOGIES

GENERAL THEORY

Concerns about efficacy and side effects of current pest control practices have led to the hope that new techniques or practices will emerge from innovative processes, be found successful, and then come into use. We are, accordingly, led to an examination of our society's capacity to generate new technologies for controlling unwanted organisms.

Although we lack a well-developed theoretical understanding of the factors leading to innovation, the following six points provide an analytical framework for inquiri about our research and development capacities:

1. A variety of perceived needs can motivate a society toward innovation. By far the most common motivation in the United States during the nineteenth and twentieth centuries has been the desire to improve the efficiency of a process so as to foster economic growth. Examples of innovations so arising are too numerous to list; they are found in the fields of mechanical engineering, metallurgy, chemistry, electronics, and biology, among others. In agriculture, the development of hybrid corn is one of the most familiar examples of an innovative response to the desire to increase economic productivity.

A second motivation for fostering innovation arises when society perceives that a contemporary technology has begun to fail. In such cases, there may be hope that the replacement technologies will generate new economic growth but the major criterion for success of new inventions will be their ability to maintain a threatened status quo of economic productivity. A variety of examples of this category have emerged in recent years. Pressures for research on methods of mass transportation, for instance, ma be viewed as arising from perceptions that the automobile is failing as a technology for moving people. These perceptions are, it is supposed, a result of increased air pollution, the likelihood of decreased supplies of fuel, and an increasing disenchantment with the social disruptions caused by the construction of modern highways.

As we described earlier in this chapter, a major factor pressing the search for new pest control technologies is the view that our current methods are in danger of failure. The resistance of pest organisms to toxic chemicals, the replacement of new pest organisms for old following the use of chemicals, and the socially perceived threat to human health and to the environment from the use of pesticides all combine to threaten the continued reliance on our present chemical technology. Accordingly, in our evaluation of society's R&D capacity for pest control techniques, we will emphasize our ability to replace endangered technologies rather than merely to foster economic growth. We note that this kind of capacity to innovate may differ from the first type mentioned.

A knowledge of motivations is important, because motivations reflect the desired outcome of innovative efforts. Put another way, societies do not engage in research on randomly selected subjects. Man turns to science and technology because he wants to know something about a natural phenomenon he has observed (basic research) or because he wants to do something more efficiently (applied research). It follows that in order to evaluate the adequacy of any given R&D effort, one must have some understanding of its goals. Normally, research will not lead to the solution of a question that was not asked or to a process that was not sought. The history of scientific and technological progress is, of course, marked by some significant exceptions to this generalization; for this reason, we would hardly hope for the discontinuance of serendipity. But we cannot rely on it: We must recognize the likelihood that scientific and technological advances will be forthcoming from those fields that are given priority in access to research resources and not from those areas that are starved.

2. On a short-term basis, the ability of a society to invent depends upon the resources that it allocates to research and development activities. Three major types of resource inputs exist: (a) a knowledge base built by past research and development activities in the applied and basic sciences; (b) physical, institutional, and monetary resources in the form of laboratories, universities, instruments, computing centers, and others; and (c) manpower with the proper training and adequate direction.

The long-term rate of change in a technology depends at least in part on the resources society allocates to R&D. There are two possible relationships, however, between resource allocation and the long-term prospects for technological change: (a) If the resources allocated to R&D are the only or the dominant variable in determining the rate of change, we may expect that our technological abilities will always increase as long as we allocate resources to R&D. According to some models (Starr and Radman 1973), the increase will be exponential. (b) If other factors have a potential influence on the rate of change of our technological abilities, we cannot yet predict our future capacity to innovate: It may rise or fall. We make no judgment as to which possibility is more likely. We do assume, however, that for the short run (the next 5-10 years), the resource allocation for R&D on pest control will be the most important factor in determining our ability to innovate in this field.

3. The cost (and hence the magnitude) of the resources allocated to R&D depends upon the demand for and the suppl of the resources necessary for R&D. Demand depends upon the magnitude of the "payoff" expected from any innovation arising from the R&D efforts. Payoff can come in the form of higher economic returns, prestige, or the preservation of the status quo when a decaying technology is replaced. The magnitude of the payoff depends upon the projected demand for the product of the innovation and the expected cost of making it, and capture of the payoff by the innova tor depends upon protection provided by the patent laws. The supply function depends upon the number and quality of three classes of resource inputs: (a) knowledge, (b) phys cal and institutional resources, and (c) manpower.

Managers of R&D efforts generally can channel resources into a variety of projects. The projects selected will be those judged most likely to produce products with the high est demand relative to the cost of innovation. Because R&D efforts are learning processes in which the precise output cannot be predicted, the managers responsible for the exercise are constantly using judgment, guesses, and faith to estimate the relationship between the projected returns and the current costs of performing innovative wor R&D that is judged unfeasible or useless at one point in time may quickly become quite feasible if the managers of R&D change their estimation of the projected demand or if the costs of research are changed.

4. There are potential interactions between different spheres of innovative activity. An innovation in one

process may create a demand for innovation in a separate but related function. An example of this type of interaction was given by Headley and Lewis (1967), who suggest that the introductions of the cotton-picking machine and cotton herbicides may have been complementary. The elimination of hand labor for picking by the machine created a demand for technologies to replace the other major type of hand labor in cotton production--hoeing of weeds.

5. There is generally a lag between the perfection of a new technique and the time at which it wins general acceptance. Among the factors causing the delay may be (a) the necessity for the applier of the technique to purchase new capital goods before use; (b) the necessity for him to train users; and (c) the hesitancy of a potential user to risk adopting a new and untested practice when its success is essential to his economic well-being. The length of this adoption lag is quite variable; it can range from less than 1 year to decades.

The lag between development and adoption of a technique presents yet another problem in evaluating R&D efforts. We must carefully distinguish between the efforts to modify or foster the adoption of already-developed techniques and those designed to foster the development of entirely new practices. Both types of research are important, but the institutions best suited to doing the former may be quite different from the institutions best suited to the latter. Furthermore, in order to set priorities for research, we must know which of these two kinds of activities is most in need of attention; otherwise, scarce intellectual resources will be misallocated.

6. It has been suggested that R&D on a given type of technology will improve that technology through time. In some cases, a technology's ability to perform a process appears to follow a sigmoid curve (Starr and Radman 1973). Data are not available to reconstruct the curve of improvements in pest control technology during the past 50-100 years, but it is certain that pest control technologies--particularly those directed against insects--have exhibited a tendency to "decay" after a period of use, i.e., the technology loses effectiveness after a certain amout of time has passed. The cause of this decay is, of course, the development of genetic resistance by the pests, and the disruption of control by natural enemies. If we assume that past R&D on pest control techniques has improved the efficiency of pest control technologies at a rate approximated by a sigmoid curve, the "decay" would appear as a decline in the effectiveness of the technique (Figure 1). The significance of the decay property is that pest control

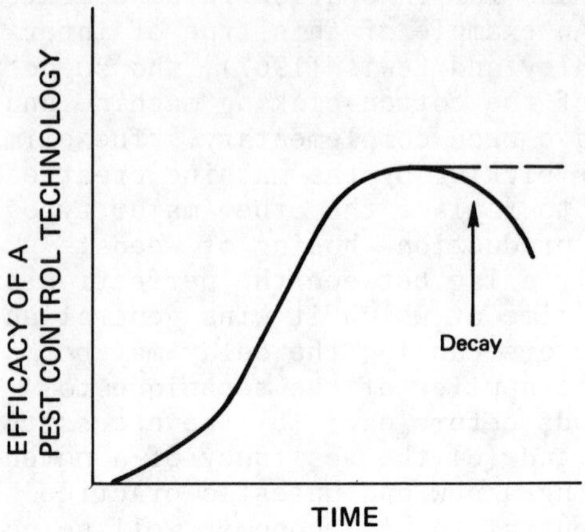

FIGURE 1 The effect of R&D on pest control technology.

R&D must be a continuous effort just to maintain the status quo. Without continuous new developments in our ability to combat pests, we will not be able to maintain our current level of control efficacy, let alone increase it.

We now turn to the specifics of the innovative processes in the private sector and in the publicly financed research institutions.

PRIVATELY FINANCED RESEARCH

The Chemical Industry

The pesticidal chemical industry is composed of about 88 companies. We do not know how many of these companies are now performing R&D activities, but it may be a majority; some companies, however, have apparently terminated their activities. Our understanding of the decision-making process in the industry is fragmentary, and the extent of hard information on the magnitude of industry's efforts is meager. The major source of information about the industry's activities is contained in a report, "1970 Industry Profile Study," prepared for the National Agricultural Chemicals Association (NACA) by the consulting firm of Ernst and Ernst (1971). We thank NACA for use of this study.

We assume that R&D managers of industrial firms allocate R&D resources so as to maximize the economic return to their companies. Such a decision maker will estimate the projected demand for the products of his division's R&D efforts. Because his research division is likely to have a variety of projects, all of which could result in some new type of product or technology, the manager must project the relative demands for the products of these alternative R&D activities and expend his resources on those he judges most likely to meet the company's economic goals. Even within profit-motivated firms, there will be differing opinions about which resource allocations will be best. Diversity may yield stability, but R&D on a wide variety of projects is likely to be more expensive than specialized efforts on a small number of projects. Managers must, therefore, weigh the stability that comes from a diversified product line against the strength that may (or may not) come from specialization in a narrow product field.

Although we can only make educated guesses about the specific goals of current industrial research efforts, we can note some specific properties of the products that have emerged during the past three to four decades. The nation's pest control technologies now depend upon uniform chemical products that can be patented, packaged, stored, and sold in a wide variety of formulations and package sizes and for a variety of pest species. Furthermore, they are inexpensive and easy to use. These properties contrast with those of almost all suggested replacement technologies. Integrated control probably cannot be patented, packaged, or stored. Only a few biological control agents can be sold in a wide variety of formulations and package sizes. Selective chemicals, by definition, cannot be used against a wide variety of pest species.

This list of contrasts could be extended, but the point is clear: The products produced by the chemical industry appear to meet admirably the economic goals that we assume dominate the decision-making processes in private industry. Unfortunately, the sole example of overlap between the properties desired by the industry and those of the most-promising alternatives would seem to be the observation that the industry favors short persistence over long by a wide margin. Conversely, however, industry favors a broad spectrum of biological activity over a narrow spectrum by almost the same margin (Von Rumker et al. 1970). We conclude, therefore, that our society may be able to rely on the conventional agricultural chemicals industry only for the development of more broad-spectrum chemicals with less persistence than earlier inventions.

There is strong evidence that the cost of developing a new chemical of the conventional type is rising. The 1971 survey by Ernst and Ernst gives some quantitative estimates of the situation in the industry. A total of 33 of the 87 companies then producing pesticides participated. In 1969, these 33 companies had sales of $692,520,000, or 81.4 percent of the industry. Between 1967 and 1970, their sales had risen from $639 million to $723 million, or 13 percent. During the same period of time, their R&D expenditures rose from $52 million to $70 million, an increase of 33.4 percent. Not only was the rise in R&D expenditures larger than the rise in sales, but the percent of sales devoted to R&D rose 18 percent, from 8.2 percent in 1967 to 9.7 percent in 1970. The total number of employees devoted to R&D efforts also rose from 2,238 in 1967 to 2,547 in 1970, a rise of 13.8 percent. The level of training represented in the R&D work force--and hence the expense in maintaining this work force--rose faster than did the total work force: The numbers of people holding doctor's and master's degrees rose by more than 20 percent, but the number of people holding bachelor's degrees or some other type of training rose by less than 12 percent. Experience of 30 of the companies with their latest two pesticides indicated that the number of R&D man-years necessary to take a compound from screening through registration increased from 49 prior to 1963 to 65 during 1967-1971. The cost from discovery of a chemical to commercialization rose from an average of $3.4 million in 1967 to $5.5 million in 1970.

There has been at least one important qualitative change in the industry's innovative efforts: a large increase in the resources allocated to maintain the registration of existing products rather than efforts to develop new products. In 1967, $7 million was spent for maintenance research, but in 1970 this type of expenditure was $16 million--an increase of 129 percent. The percent of R&D expenditures allocated to maintenance purposes compared to total R&D expenditures rose from 13.4 percent to 22.9 percent during this same time period.

The significance of this change is not altogether clear. On the one hand, resources spent for maintenance of an existing compound may represent a drain on the research capacity of a company and hence a loss of innovative activity that otherwise would have occurred. Thus, for the period 1967-1970, the industry increased its R&D expenditures by $18 million, of which $9 million went for maintenance. The industry might have spent more money on innovative R&D had they not had to perform the maintenance activity. On the other hand, it is possible that the increase in expenditures for maintenance merely reflects the fact that the pesticide industry is no longer a "growth industry." New products now must compete with established materials, instead of moving into unoccupied sales niches. R&D managers may accordingly be switching their resources into less competitive areas, within their zone of competence, rather than attempting to develop yet one more chemical that will merely compete with the company's already-established products. Under such circumstances, the $9 million spent on maintenance in 1970 that was not spent in 1967 can best be described, from the point of view of the industry, as an unwanted expense that is nevertheless justified to protect a lucrative product. From this viewpoint, the funds are not a drain from pesticide research, because the company would have spent the money in another area had they not spent it for maintenance research.

Interaction between Public Regulation and Private Research

The increase in scientific man-years and dollars necessary to obtain and maintain a registration for a pesticide is but one factor among several moving the cost of innovative activity upward. The importance of these two increases is that they stem from a change in public policy: Increasingly stringent requirements are being set by the federal government, through Congress and the EPA, for the granting and continuation of pesticide registrations. In large part, these changes reflect the growing concern about the unintended side effects of pesticides. Representatives of the chemical industry have expressed dismay, because they claim the new policies are sharply decreasing their abilities to invent new chemicals.*

---

*This Committee interviewed representatives of a number of companies, and the transcript of the discussion is on file at the National Research Council. Opinion among different representatives is clearly not uniform, and the NACA informed us that it had no formal statement about the impact of government regulation on R&D or the best remedy of any problems. The NACA did establish an ad hoc committee chaired by E. S. Bauer of Monsanto, to meet with William D. Ruckelshaus, former Administrator of the EPA, to discuss the R&D situation. The departure of Ruckelshaus to head the FBI in 1973 caused the meeting to be cancelled. According to correspondence between Bauer and Ruckelshaus, copies of which were given to this Committee, it is not simply environmental protection legislation that the industry considers important. Inadequacies of the patent laws, government policies on patents resulting from research financed by the government, and antitrust laws are also involved.

We agree that the new regulations do increase the cost of R&D and therefore must clearly act as a disincentive for research efforts. That does not automatically lead, however, to the further conclusion that opportunity cost of the research inhibited by the regulations is too much for society to bear and that the latter ought to be made more lenient. The regulations may indeed eliminate some research by raising its cost; nevertheless, they represent a step forward in protecting the general public interest, and caution should be exercised in relaxing them. More importantly, we found no evidence (1) that a mere relaxation of environmental regulations in the registration of pesticides would result in an increased flow of new research monies into pesticide R&D, or (2) that the products of the innovative activity would be an increased flow of new and desirable chemicals into our pesticide arsenal. If the public interest would be served by the introduction of new chemicals, we may have to investigate a variety of other ways to foster the innovative process.

Research and Innovation on Markedly New Products

A further problem may affect the introduction of very new types of insect control devices: conscious or unconscious competition with private industry by government agencies. For example, one company has contended that the manufacture and distribution of pheromone traps by the Animal and Plant Health Inspection Services of the USDA disrupted potential private markets for the traps and thus constituted a strong disincentive for private companies to perform research and development on the devices. At the same time, the manufacturing activities by APHIS were not considered adequate to meet the actual need for the devices.

The charge of unfair competition with private industry by the government is likely to arise whenever both government and industry scientists are associated with the development of new devices that are characterized by high skill or knowledge content, low or simple material content per unit, and the need of many units for the technique to succeed. The situation is even more likely to develop when government scientists have played a prominent role in the development of the knowledge base. Industry is accustomed to producing devices that can be patented, packaged, and sold in large quantities. Any potential market for such items will appeal to the private sector when the government is not also involved in the production process. The new insect control technologies, such as the use of pheromone traps, are exactly such devices in that a successful design could be patented, packaged, and sold in large numbers. The successful use of the traps will ultimately depend much more on the skills and knowledge base of the user than on the actual physical makeup of the trap, and government scientists were very influential in laying the groundwork for their use.

Government, however, has had a traditional responsibility to develop skills and knowledge and to make them freely available to all citizens. This task, in fact, has been a major assignment of the USDA since its inception. As a result, it may be both logical and necessary for the USDA to perform simple manufacturing activities, especially for their field-research efforts. It is a different question, of course, whether the Department should continue to manufacture and distribute devices once they are used as a matter of common practice. The fact that devices such as pheromone traps require little in the way of sophisticated machinery to produce--their major ingredient other than the pheromone is a paper drinking cup--means the Department can produce them with relatively little effort.

The problem, then, is not "How do we insure no unfair competition for industry?" but "How do we insure that pest control devices with a predicted favorable risk-benefit ratio for the user and for the general public will be adopted?" It may well be that the most efficacious method for introduction is to assign all research, development, testing, manufacturing, and marketing responsibilities to a public agency. The situation clearly to avoid is one in which no organization--public or private--has a suitable environment in which to develop and introduce a desirable innovation. The probability of confusion, misunderstanding, and distrust will be very high in such cases. Furthermore, there is a good possibility that the introduction of the device will not occur. Because the legal framework and economic ideology of the United States generally give preference to private industry in most manufacturing tasks, it is likely that the Congress will have to consider the problem before a satisfactory solution can be developed.

The Introduction of New Chemical Entities--Where We Stand

The kind of information needed to evaluate the health of the pesticide industry's research efforts depends upon the viewpoint of the evaluator. From the perspective of the industry itself, the most important information is the number of new chemical entities per year, with some estimation of the market potential of the new products. From the viewpoint of society as a whole, the number of new materials per se is meaningless; interest falls instead on the abilities of new compounds to perform some task having social utility. For example, the introduction of one new chemical that can control some pest not previously controlled, or replace some chemical rendered ineffective by the development of resistance in the target organism, is more useful to the general public than the introduction of several new products that compete with chemicals already available.

The Committee encountered considerable confusion about this distinction, which accompanies the conclusion that the number of new products coming to market has declined in recent years. The inference drawn by many is that society is in trouble in this area and that the most likely source of the malady is the increased stringency of registration requirements for new pesticides. The conclusion is that pesticide regulations are too costly and are discouraging the development of needed new compounds.

This Committee judges that it is premature to conclude anything about the health of the research enterprise in the chemical industry. There is no evidence that the number of new products coming to market has dropped significantly in the last 8 years; in general, the number has ranged from 10 to 20 per year (Table 6). Furthermore, there is no evidence that the number of chemicals introduced for use in agriculture has changed from its traditional proportion of about half the new compounds introduced.

There was a drop in the number of registrations granted in 1971 and perhaps in 1970, but the number granted subsequent to those 2 years is very close to the number granted from 1967 to 1969. The cause of the drop in 1970-1971 is not known, but there are at least two factors that might have played a role. First, the EPA was formed in late 1970, and the transition to a new agency may have caused a delay in the granting of some registrations. Second, there was a general business decline in the years 1970-1971, and companies may have cut back their expenditures devoted to the registration process, thus causing a drop in the number of registrations granted.

Our inability to conclude whether sufficient new materials are being developed or not should not be interpreted as a lack of concern about the subject. As discussed elsewhere, we are concerned about the apparent difficulties in effecting the registration of such materials as juvenile

TABLE 6 Registration of New Pesticides, 1967-1974[a]

Year	Class[b]									Function[c]			
	A	H	F	S	R	B	GR	M	N	Agricultural	Nonagricultural	Unknown	Total
1967	4	2	2	5	2	1	0	0	0	9	7		16
1968	6	2	5	4	0	0	1	0	0	9	9		18
1969	7	4	0	2	0	0	1	0	0	9	5		14
1970	1	2	2	3	0	1	1	0	0	3	7		10
1971	0	0	1	1	1	0	1	0	0	1	3		4
1972	4	5	2	7	0	0	0	1	1	7	13		20
1973	5	1	2	4	0	0	0	0	1	7	5	1	13
1974[d]	4	4	2	1	0	0	0	0	0	6	4	1	11
TOTAL	31	20	16	27	3	2	4	1	2	51	53	2	106

[a]Environmental Protection Agency, 1974. Personal Communication.

[b]Abbreviations:
A = Arthropod control agent (includes insects, mites, barnacles)
H = Herbicide or weed control agent
F = Fungicide; some control other microbial organisms also
S = Slime control agent or other microorganism control agent
R = Rodent control agent
B = Bird control agent
GR = Growth Regulator (includes sucker control agents and desiccants)
M = Mollusk control agent
N = Nematode control agent

[c]Function: "Agricultural" means that the material is registered for use on at least one crop or animal raised by commercial farmers, except turf, ornamental plants, and plants raised in greenhouses. "Nonagricultural" means the compound is not registered for use on any commercially grown crop or animal; includes turf, ornamentals, and plants raised in greenhouses. Two compounds could not be classified.

[d]January-May 1974.

hormone insecticides and viral insecticides. The problem lies in the unavailability of criteria to tell us whether the new chemicals being registered are useful for solving the problems associated with the contemporary pest control practices. It was beyond the responsibility and resources of this Committee to develop such indicators, but we recommend that the EPA give strong consideration to their formulation. Among other items, we feel that the following types of information must be included:

- How many new materials are registered for the first time each year? (This Committee was surprised to learn that the EPA did not keep a running count of new products entering the market; the agency had to perform a special tabulation to give us the information produced in Table 6. We are further dismayed that the information on new chemical registrations given us apparently disagrees with similar information given the President's Science Advisory Committee and published in *Chemicals and Health*, September 1973, Table D-14, p. 176. We strongly recommend that the EPA immediately upgrade their efforts to monitor and analyze the introduction of new chemical entities.)
- Does the chemical perform some task previously not done by any other agent; or does it replace an agent rendered unusable by the development of resistance in the target, deregistration, or some other factor?
- What is its toxicity spectrum compared to compounds already available? Spectrum of toxicity is as difficult to quantify as it is important. Nevertheless, we can claim little improvement in our collection of pest control chemicals unless the average width of the spectrum declines.
- What is the persistence of the chemical compared to materials already available? Long persistence may be justified in some cases, such as for termite control in houses; but, in general, shortness of durability is a desired characteristic.

There will always be many pitfalls in attempting to measure the rate of flow of new chemicals into the marketplace and to understand its significance. Many of the criteria suggested above as being important to an evaluation will at best be qualitative estimates. Furthermore, the number of chemicals introduced by a group of private companies will be a complex function of their internal estimates of the projected payoffs of R&D expenditures, in addition to such external events as corporate mergers and general business conditions. The privacy with which corporate planning occurs will always make it difficult to interpret the cause of a perceived change in the flow of new chemicals. Adoption of new public policies will therefore be even more difficult. Because we depend upon private companies for new materials, there is a continuing need for monitoring their products by public agencies that have broader views of societal benefits and costs.

A problem of particular importance, which will be further considered in Part III of this volume, has to do with the impact of regulation on new developments at the innovative margin of the contemporary technology.

There is reason to believe that regulations devised for the conventional mainstream of a particular activity may work particular hardships on innovations, because by definition these occur in situations to which the regulations were not primarily designed to apply. This point has been explored for the third-generation chemical pesticides by Djerassi et al. (1974). Using "critical path" analysis, the authors conclude that regulations sometimes hinder innovation by delaying corporate commitments at specific, crucial points in the course of research and development.

PUBLICLY FINANCED RESEARCH

Coordination between Public Agencies in Pest Control Research

Public research in pest control is conducted or funded by many federal, state, and local agencies. The federal efforts include the Departments of Agriculture; Interior; Defense; and Health, Education, and Welfare. Such independent agencies as the NSF, the EPA, and others are also involved. Research at the state level is in the land grant universities; state departments of health, natural resources, agriculture, fish and game, and forestry; and in various state and local educational institutions and operating agencies. Funding and management of research are frequently shared between different agencies or levels of government. Funds are derived from federal, state, and local appropriations, as well as from foundations, industry, and individuals.

Pest control research is not treated as a unified subject, nor are research agencies organized in a uniform way. The structure of public pest control research reflects the diverse nature of pest control problems and the varying types of responsibilities mandated to different government agencies and other types of public institutions.

Some organizations, such as the Forest Service, Bureau of Reclamation, Bureau of Land Management, Corps of Engineers, and the Tennessee Valley Authority, are direct users of technology in the management of public lands and waterways. Others, such as the Environmental Protection Agency, are concerned with regulatory and enforcement functions relating to public safety and welfare. Another major category of institutions performing research--the land grant universities--are responsible for basic and applied research and for general education about pest control.

Planning and funding of public research tends to be based on the individual interests and responsibilities of the different agencies. There are, however, formal and informal mechanisms for coordinating research efforts between different organizations. Protocols for interagency consultation and the reviews provided by the Congress and the state legislatures provide a strong means of formal cooperation. Most importantly, there is a formal basis for the joint planning of pest control research between the USDA and the land grant universities, the two institutions responsible for most expenditures for research. The agricultural experiment station (AES) in each state university receives federal funds through the Cooperative State Research Service (CSRS) of the USDA under provisions of the Hatch Act. The CSRS must approve all state projects, and it is usually required that federal funds be matched by funds from state sources. Some Hatch Act funds are designated as Regional Research Funds and can be expended only on joint projects among two or more states.

Progress and expenditures on research funded by the Hatch Act are reviewed periodically by federal authorities, often with the aid of scientific review panels. The review sometimes extends to an evaluation of an entire department. The federal audit gives due respect to the autonomy of each AES and is intended largely to insure that federal funds are expended only for authorized purposes.

Pest control research in states is also funded under the provisions of the McIntyre-Stennis Act for forestry, through special grants under Public Law 89-106, and to a minor extent under other laws. The McIntyre-Stennis Act provides formula funds for research to schools of forestry, including those that are not affiliated with land grant universities. Projects funded by the McIntyre-Stennis Act are initiated, terminated, reviewed, and audited in essentially the same way as projects funded by the Hatch Act.

Public Law 89-106 originally established a program for USDA grants to states on a nonformula basis: that is, grants are made to the experiment station judged best able to perform the research. Funds were to be expended in an effort to meet national, rather than merely state, objectives. Over time, the Congress has placed increasing restrictions on the nature of the research to be supported and the institutions that are to receive the grants. Appropriations have focused heavily on corn, soybeans, and cotton. Congress has also increasingly specified that funds be granted to the Colleges of 1890--state universities established in the South for black students.

Coordination of research in the public agencies is also effected through numerous informal channels, such as exchange of plans and progress reports among administrators and informal contacts among research scientists. Interaction among individual scientists is a major part of the planning process. Scientists are acutely aware of the accomplishments and failures of other scientists and are sensitive and responsive to budgetary and managerial decisions affecting research. These interactions are both cooperative and competitive and serve to stimulate and coordinate research activities. Although personal interchanges among scientists serve a useful function in that they tend to coordinate and stimulate research, they also tend to focus along disciplinary or narrowly functional lines to the neglect of interdisciplinary exchanges and the translation of disciplinary knowledge into technology.

Thus there is a basis in law and tradition for cooperation in the management of pest control research, especially between the USDA and the states. The cooperative effort involves:

1. Consultation and cooperative planning between the USDA and the states on research policies and priorities.
2. Consultation in the allocation of certain federal funds.
3. Operation of a nationwide inventory system for agricultural research projects with computerized cataloging, indexing, and data access.
4. Operation of a national agricultural research planning system involving joint state and federal committees and the development and annual revision of a 5-year moving plan.

These procedures cause the research personnel and administrators of the Agricultural Research Service (ARS), the Economic Research Service (ERS), the U.S. Forest Service (USFS), and others to have frequent contact with counterpart agencies and programs in the states. The relationships throughout the system remain, however, largely voluntary; and the autonomous nature of the participating organizations is clearly recognized.

Cooperation is further strengthened by direct interchanges between scientists. The USDA and land grant university personnel often participate in joint projects and share laboratories and field facilities. USDA scientists are frequently stationed in university departments, and university scientists not uncommonly occupy federal facilities.

Research in the USDA/Land-Grant-University Complex

The agricultural research establishment in the public sector is undoubtedly the largest and the most significant factor in the development of new technologies and practices in American agriculture. In fiscal year 1972, some 10,500 scientific man-years (SMY) and more than $627,000,000 were expended by the federal and state governments for the support of research on agricultural problems. This outlay of resources is conducted largely in a complex of organizations known as the land grant university and USDA system. A report from the NAS (1972) described this complex well:

> In the United States two complex networks of agricultural experiment stations have evolved, one administered by the State Agricultural Experiment Stations (SAES), and the other by the United States Department of Agriculture. The SAES system has in each state a central facility on the campus of the land-grant college and one to several branch stations in outlying areas of the state. The branch stations are located to service a specific geographic area, or specific commodity needs. In the 50 states, the SAES have over 300 such outlying branch stations and field laboratories. They represent a wide range of investments in buildings, land, and manpower, and vary considerably in mission and program sophistication. Some are primarily extension centers and demonstration farms. Others have substantial capital and manpower investments for research and have programs comparable to the central stations.
>
> The USDA system has its national center in the Beltsville, Maryland-Washington, D.C. area with several major national or regional laboratories scattered throughout the United States. The USDA also has a large number of smaller research

laboratories or centers in the states. Many of these have rather specific commodity orientations with little national or regional orientation. Others, however, do have national or regional significance.

The scope of the work in our agricultural research system is very large. Agronomists, entomologists, plant pathologists, weed scientists, and others conduct a wide variety of research efforts in the applied biological sciences. Chemists and engineers investigate such problems as food preservation, pesticides, commodity utilization, and mechanization; and social scientists conduct research in economics, statistics, marketing, foreign trade, and rural sociology. Approximately 42 percent of the total resources allocated are spent in the USDA portion and 58 percent by the SAES.

A significant portion of the resources in the agricultural research complex are devoted to pest control research: In FY 1972 approximately 2,268 SMY (21 percent) and $132 million dollars were devoted to research on new pest control technologies in the USDA and SAES. The primary locations of research activities on pest control are in the ARS and USFS of the USDA and in the SAES. The Schools of Forestry perform some of the work on control of forest pests. For some problems, such as the control of weeds in fruits and vegetables, most work is performed in the states; for others, such as the control of insects and other invertebrate pests of field crops and rangeland, most of the effort is in the federal laboratories. Investigations on the control of forest pests are conducted almost entirely by the USFS; the states conduct a small portion of the work, and the ARS has a minor effort directed toward the control of forest insects. A detailed presentation of the resources allocated to research on different areas of pest control is presented in Table 7.

The incentives affecting the individual researcher in the USDA/land-grant-university complex differ from those acting on scientists in the private chemical industry. Whereas the latter are rewarded for finding and recognizing a chemical that can be produced and sold at a profit, the former produce papers that are judged for quality by both their superiors and their peers. Accordingly, success for the chemical worker is tied directly to the success of his discovery in the marketplace; success for the agricultural worker is immediately tied to the number and quality of publications he can produce, even though he is ultimately beholden to the legislative bodies supporting him and to various interests making use of his results. The lack of direct market pressures on his research product has allowed him room to explore that would not be possible for the chemical industry researcher. The difference in incentive structure helps explain some important differences in the current status of research in these two sectors. The techniques and devices for pest control that have come out of the agricultural research groups could be characterized as diverse, imaginative, and--in some cases--incongruent with the economic matrix within which they must operate. Industry scientists, on the other hand, have produced results that are fairly characterized as more homogeneous: broad-spectrum, inexpensive, biocidal materials that have as much (or more) persistence as they must have to function within the economic setting in which they are sold.

The diversity of the pest control remedies emerging from the agricultural complex is at once a boon and a bane. A range of options is surely needed at the present time, yet sorting them is no trivial undertaking. The challenge is to create a public policy that will encourage the continued production of new ideas for pest control and the allocation of resources for the experimental work needed to select out the useful ones.

Clearly, a supply of monetary resources for pest control research is necessary to ensure progress in the field. It is very difficult, however, to determine just how much money is "enough." This question is especially treacherous for a group of scientists to answer, because they have a built-in bias to urge more funding. Clearly other elements of society must also play a role in the judgment of adequacy of funding for a research area.

We are compelled to point out, however, that the resources currently devoted to solving some of our pest problems are minute compared to the resources devoted to other areas of research endeavor. We estimate that, for 1970, slightly more than 5,000 SMY and $180 million dollars were devoted to research on pest control in the public and private sectors (Table 8). This was 0.9 percent of the total SMY and 0.7 percent of the total funds devoted to all R&D activities in the United States during 1970. In contrast, for example, defense research consumed some $8 billion, or 30.2 percent of all R&D activities, while research for the space program required about $3.5 billion, or 13.5 percent of the total (Table 9).

TABLE 7 Location of Research on Pest Control, Fiscal Year 1972[a]

Problem Areas	SAES, Forestry Schools and Other Cooperating Institutions			USDA ARS			USDA USFS			Total		
	SMY	$ Million	%[b]	SMY	$ Million	%[b]	SMY	$ Million	%[b]	SMY	$ Million	%[b]
Biology, culture, and management of forests and timber-related crops[c]	110.1	5.15	36	0	0	0	181.3	9.34	64	291.4	14.5	100
Control of insects affecting forests	36.9	1.80	24	1.6	0.12	2	101.5	5.59	74	140.0	7.51	100
Control of diseases in forests	25.9	1.13	24	0	0	0	73.9	3.64	76	99.8	4.77	100
Control of insects on fruits and vegetables	109.6	5.49	57	64.9	4.09	43	0	0	0	174.5	9.58	100

TABLE 7 (continued)

Problem Areas	SAES, Forestry Schools and Other Cooperating Institutions SMY	$ Million	%[b]	USDA ARS SMY	$ Million	%[b]	USDA USFS SMY	$ Million	%[b]	Total SMY	$ Million	%[b]
Control of diseases on fruits and vegetables	190.6	0.26	71	68.9	3.79	29	0	0	0	259.5	13.0	100
Control of weeds in fruits and vegetables	42.4	1.81	79	5.3	0.496	22	0	0	0	47.7	2.31	100
Control of insects on field crops and range	143.3	7.86	43	161.7	10.5	57	0	0	0	305.0	18.3	100
Control of diseases on field crops and range	184.2	8.76	53	127.4	7.67	47	0	0	0	311.6	16.4	100
Control of weeds on field crops and range	77.4	4.10	67	40.9	2.07	34	0	0	0	[illegible]	[illegible]	[illegible]

external parasites on livestock and poultry	30.0	1.41	42	28.6	1.95	58	0	0	0	58.6	3.36	100
Control of internal parasites of livestock and poultry[d]	33.8	1.71	36	37.9	3.03	64	0	0	0	71.7	4.73	100
Control of insects affecting man	35.5	1.55	46	28.0	1.81	54	0	0	0	63.5	3.36	100
TOTAL	1,232.6	61.9	47	678.8	51.5	39	356.7	18.6	14	2,268.1	132	100

[a]Cooperative State Research Service, U.S. Department of Agriculture (1973) Inventory of Agricultural Research FY 1972, vols. I and II. Washington, D.C.: U.S. Department of Agriculture.
[b]% = Percent of total dollars in research problem area.
[c]It is thought that this category includes research closest to control of "weed" species in forests. Undoubtedly this overestimates the resources allocated to forest weed control research, but the error does not significantly affect the generalizations made here.
[d]This category is not strictly within the responsibility of this Committee.

TABLE 8 Resources Allocated to Research on Pest Control and Pesticides by Federal and State Governments and by the Chemical Industry (1970)

Problem Area	SMY	$ Million
Public[a]		
Biology, culture and management of forests	277.5	12.310
Control of forest insects	123.7	5.590
Control of forest diseases	93.1	3.752
Control of insect pests on fruits and vegetables	169.6	8.171
Control of diseases on fruits and vegetables	260.0	11.700
Control of weeds on fruits and vegetables	46.9	1.810
Control of invertebrates on field crops and range	289.7	13.850
Control of diseases on field crops and range	322.2	13.660
Control of weeds on field crops and range	115.3	5.262
Control of insects and external parasites on animals	56.8	2.654
Control of diseases of animals	376.1	24.670
Control of internal parasites of animals	87.3	4.595
Control of insects on man	52.9	2.514
SUBTOTAL	2,271.1	110.538
Private[b]		
Chemical pesticides	2,768.0	69.870
TOTAL	5,039.1	180.408

[a]Figures are from Inventory of Agricultural Research FY 1970 (Current Research Information System 1972) and hence include only that research performed at USDA, State Agricultural Experiment Stations, Schools of Forestry, and a limited number of other types of cooperating institutions. Public expenditures are reported by Fiscal Year.

[b]Figures are for 33 chemical companies holding 81 percent of pesticide sales (Ernst and Ernst 1971). Expenditures may be reported by either fiscal or calendar year. Additional research on pest control may be performed by other private firms such as seed companies. We have no estimates of research expenditures by such companies, but we do not think their omission seriously affects the overall estimates made here.

TABLE 9 Resources Allocated to Pest Control Research Compared to Total National Research Resources (1970)[a]

Research Area	SMY	$ Million	% Gross National Product	% of National Total SMY	% of National Total $
National total	549,700[b]	26,700	2.73[c]	100.0	100.0
Defense[d]		8,067.1	0.83		30.2
Space[d]		3,596.7	0.37		13.5
Agriculture[e]	10,539	536.6	0.05	1.9	2.0
Research in total chemical industry[f]		1,624	0.17		6.1
Research for pest control[g]	5,039	180.4	0.02	0.9	0.7

[a]Except where otherwise specified, all figures are from National Science Board (1972).
[b]Scientists and engineers employed in research and development rather than SMY.
[c]Gross National Product was $976,400,000,000.
[d]Federal expenditures only.
[e]Federal and state funds form the vast majority of these expenditures. "Industry" and "other" sources given to public laboratories account for only $21.6 million. (Current Research Information System, 1972)
[f]Industry funds only.
[g]Expenditures in the agricultural research establishment and in the pesticide industry.

Reorganization of the Agricultural Research Service

The productivity of research in the public sector, compared to the private sector, makes it essential that public research efforts be maintained in a healthy state. The ARS, the largest research agency in the USDA, adopted a new organizational arrangement on July 1, 1972. Discipline-oriented divisions were replaced with a combination of geographically defined divisions working in conjunction with a national program staff. Among the motivations for the reorganization were the desires (1) to foster work in multidisciplinary teams, (2) to increase response capacity to new research problem areas, (3) to produce more initiative in project managers and researchers for research problem definition and management, and (4) to increase the ability of ARS scientists to cooperate with scientists from SAES. Furthermore, the mere increase in the size of the ARS from fewer than 4,000 employees on a budget of $32 million in 1953 to some 10,000 in 205 locations and a budget of almost $200 million in 1972 suggested that geographical decentralization was desirable in order to relieve top ARS managers from intense administrative burdens.

The reorganization was accompanied by two distinct traumas that made what in any case would have been a difficult step even harder. First, the reorganization plans were conducted without consultation with ARS entomologists in high positions. The resulting low morale of the entomologists in the ARS was deemed sufficiently important by the Entomological Society of America that the Society appointed a special committee to examine the reorganization and determine its possible impact on entomology.

The conclusions of the special committee were mixed in their level of optimism. On the other hand, the committee found there was a general discontent among ARS entomologists with many aspects of the reorganization but that these may disappear with time and increasing familiarity with the new arrangement. There was genuine concern on the part of the committee that entomological representation in the bureaucracy of the ARS had been weakened because of the reorganization. No entomologist was on the ARS Board of Directors, the policymaking and managing collection of regional and national administrators; furthermore, the creation of regions in the ARS meant that regions would compete for funds at the national level. The committee was concerned that the regional administrators might "sell" their entomological programs with insufficient vigor. The committee also worried about the flexibility of the new ARS in coping with emergency developments. On the other hand, the committee found that entomologists high in the ARS were optimistic that the reorganization would benefit agricultural research in the long run (Maxwell et al. 1974).

The second shock to the public research establishment was the report of the Committee on Research Advisory to the USDA, otherwise known as the Pound Committee after its chairman, Dr. G. S. Pound, of the University of Wisconsin

(NAS 1972). The Pound Committee issued its report to the USDA shortly before the ARS reorganization became effective. The report was critical of the quality of public agricultural research in general and devoted considerable attention to pest control research. We make no judgment as to the correctness of the Pound Committee's report, nor have we examined the actual operations of the ARS in the detail of the Pound Committee. We note, however, that the tone of the report created some defensiveness among some ARS and SAES scientists. Furthermore, because the reorganization was effected after the Pound Committee completed its report, it is difficult to counter the inevitable argument that the reorganization corrected any deficiencies that may have been present before.

It is still too early to judge the health of the reorganized ARS and SAES system. Within the next 1-3 years, however, it would be most appropriate for a balanced committee to examine the state of pest control research in the public sector. Ideally the committee would look not only at the ARS and SAES, but also the Cooperative State Research Service, the Forest Service, the Economic Research Service, and the Animal and Plant Health Inspection Service. The major questions to be answered include:

- Is the research atmosphere one that encourages innovation?
- Has interdisciplinary research between entomologists, weed scientists, plant pathologists, plant breeders, ecologists, environmentalists, agricultural engineers, economists, sociologists, and others been fostered?
- Has unnecessary duplication between national and state laboratories been eliminated?
- Are research resources allocated according to reasonable schemes?
- Are sufficient resources devoted to large-scale pilot tests of new and untried control technologies?
- Do research agencies have the capacity to handle pest emergencies adequately?

USDA Pest Control Research by Target Area

Estimations of R&D efforts must be made not only by the general area of research (e.g., Table 7), but also by the resources allocated to specific goals of research. The USDA has attempted to quantify the resources they put into R&D by type of goal. All of its activities relating to pests are assigned to one of the six targets listed in Table 10.

TABLE 10 USDA Pest Control Research by Target[a]

Target[b]	FY 1971[c] $ Million	FY 1971[c] SMY	FY 1972[d] $ Million	FY 1972[d] SMY	FY 1973[d] $ Million	FY 1973[d] SMY
I	22.1	418	25.7	428	24.8	421
II	30.0	535	34.5	546	35.4	551
III	14.9	266	16.3	264	15.9	264
IV	6.0	93	7.4	93	7.4	102
V	0.4	8	0.4	8	0.4	10
VI	<0.1	1	<0.1	1	0.1	1
Other	1.8	40	1.9	37	1.8	36
TOTAL[e]	72.2	1,361	86.2	1,376	85.8	1,384

[a] U.S. Department of Agriculture (1974) personal communication. Includes expenditures of the Agricultural Research Service, the Cooperative State Research Service, the Forest Service, and the Economic Research Service. Excludes expenditures of the Animal and Plant Health Inspection Service, and expenditures made by land grant universities.

[b] Target I: To gain knowledge of the taxonomy, biology, ecology, physiology, pathology, metabolism, and nutrition of pests and host plants and animals.

Target II: To improve and develop means of controlling pests by nonpesticidal methods.

Target III: To develop safer and more effective pesticide use patterns, formulations, and methods of application and improved methods for detecting, measuring, and eliminating or minimizing pesticide residues in plants, animals, and their products and in other parts of the environment.

Target IV: To study the toxicity, pathology, and metabolism of pesticides and investigate levels, effect, and fate of their residues in plants, animals, and their products and in other parts of the environment.

Target V: To study economic aspects of pest control and its impact on the environment; determine the supply, requirements, and use of pesticides; and give assistance to control agencies and industries in emergencies.

Target VI: To control pests and protect the environment during and after control operations.

[c] Actual expenditures.

[d] Estimated expenditures.

[e] Totals may not be exact due to errors in rounding.

Estimates of the numbers (SMY) and dollars allocated at the federal level to each target for the fiscal years 1971 through 1973 are given in Table 10. The figures clearly indicate that basic studies and nonchemical control receive the heaviest funding (over $50 million per year). Studies on pesticidal control and on pesticide toxicology receive substantial funds (somewhat over $20 million per year) but less than the first two targets. Research on economic aspects receives comparatively little support (less than $0.5 million per year).

Chemicals for Specialty and Small Acreage Crops and Other Minor Uses

FIFRA as amended requires that any pesticide placed on the market be registered for *each* use: That is, a registrant must submit data on the efficacy and safety of the chemical for every intended use of the material. It is approximately as expensive to register a chemical for use on a major crop, such as corn, as for use on a minor crop, such as quinces, or for some such special application as weed control in irrigation canals. For understandable economic reasons, private firms measure the market potential in a given use before they develop the data to obtain registration for that use. The rising costs of registration, caused in substantial part by the increased stringency of the safety testing, have created a problem in that new chemicals being registered for the first time and older materials being reregistered are often being registered for major uses only.

This absence of registrations for minor uses, coupled with the prohibition in FEPCA against using materials for unregistered purposes, has created a number of problems. First, growers of specialty crops may soon face a serious lack of adequate pest control materials. Second, government agencies using pesticides in small amounts but for crucial purposes--such as quarantine programs--may soon be forced to use the pesticides in illegal ways or curtail their programs. The effect of these limitations could range from increases in the price of specialty food items to possible breakdowns in pest control operations in the public sector.

The problem of registration of pesticides for minor uses has been recognized since at least 1964, when the "IR-4 Project" was formed. This project is a publicly sponsored vehicle for assisting in obtaining pesticide registrations for minor uses. Organizationally it is an interregional project among different state agricultural experiment stations; its purpose is to coordinate the efforts of all state experiment stations to obtain registrations for minor uses. More recently, in 1972, a University EPA-USDA Coordinating Committee for Environmental Quality Research, Monitoring and Extension Education was formed. This committee appointed the Specialty Crop Pesticide Subcommittee, which issued a thorough review of all minor uses except those with public health purposes (University-EPA-USDA Ad Hoc Subcommittee 1973).

The major findings of the Subcommittee can be summarized as follows:

1. The problem of minor use registrations has increased since the passage of FEPCA in 1972.

2. The chemical industry has lost interest in minor uses because of the increasing costs of registration compared to the market for the use and because of the liability dangers inherent in a small market. Furthermore, uses formerly registered but now expired include many compounds on which patent protection has lapsed. In sum, the incentive for data development has disappeared.

3. Public agencies have not yet responded adequately to the problems posed either for specialty crop growers or for the fate of their own programs.

4. The IR-4 Project is a help but it is only a coordinating effort. Research for actual registration is not being conducted at an adequate level.

5. The IR-4 Project coordinates efforts only within the state agricultural experiment stations. There is insufficient coordination in registration efforts among other public agencies.

It was beyond the charge of this Committee to review thoroughly the problems of registrations for minor uses, but we approve the formation of the Subcommittee and recommend that the concerned public agencies continue the analysis begun in its report. We also recommend that future work on the registrations for minor uses include the problems of registering materials for use in control of disease vectors, actual or potential.

The Importation of Organisms for Biological Control by the Agricultural Research Service

It is estimated that a third or more of all pest organisms--especially insects and weeds--are exotic to the place in which they achieve pest status. Some common examples in the United States include the boll weevil *(Anthonomus grandis)*, the European corn borer *(Ostrinia nubialis)*, alligator weed *(Alternathera philoxeroides)*, and Eurasian water milfoil *(Myriophyllum spicatum)*. An exotic fungal disease that has caused considerable consternation is *Ceratocystis ulmi*, the Dutch elm disease, which kills ornamental elm trees.

Exotic organisms become pests because they are imported without the predators, parasites and pathogens that generally keep them in check in their native lands. A potential control mechanism, therefore, is to seek out the origin of the pest, identify its enemies, and import these into the United States for release. This method is not without hazards: It is essential to know the habits of the newly imported organism before it is released, so as to avoid damage to valued organisms. Furthermore, it is not always possible to establish a controlling organism in a new place. Nevertheless, the importation of organisms for biological control is an important activity, because a successful introduction can have an enormous payoff in terms of damage prevented.

The importation of new organisms for biological control is inherently a research activity, even though successful introductions result in control operations. The ARS's responsibility for importation at the federal level reflects the research nature of the activity. The resources expended for controlling organisms by the ARS are summarized in Table 11. The search for biological control organisms has involved the expenditure of ARS funds for their own staff members plus extramural grants to the SAES and expenditures of funds received from the P.L.-480 program in foreign countries. An estimated 140 SMY of ARS staff and slightly over $2 million dollars in extramural domestic and foreign programs has been expended during the last 10-20 years. Searches for controlling organisms for a few pests, such as alfalfa weevil, cereal leaf beetle, imported cabbageworm, alligator weed, and halogeton, have taken about 50 percent of the man-years expended by the ARS.

It was beyond this Committee's tasks to evaluate thoroughly the allocation of resources to biological control importation. We are concerned, however, that the effort by the ARS appears so small. An average of no more than 14 SMY per year--compared to 679 SMY in ARS devoted to pest control in 1972 (Table 7)--may represent an insufficient allocation of resources to such an important control methodology. Furthermore, a rationale for resource allocation for different pests does not emerge clearly. Research

TABLE 11 Efforts Made by the Agricultural Research Service (USDA) to Import Organisms for Biological Control of Insects and Weeds

Pest		Cumulative Effort		
Common Name	Scientific Name	SMY (ARS only)	Funding (Extramural)	Period[a]
Insects				
Alfalfa weevil	*Hypera postica*	28.0	$ 121,000[b]	SMY: 1957-1972; Funding: 1966-1975
Gypsy moth	*Porthetria dispar*	5.0	125,000[c]	SMY: 1963-1972; Funding: 1972-1975
Cereal leaf beetle	*Oulema melanopus*	14.0	146,000[b] 176,000[c]	1963-1972
European pine shoot moth	*Rhyacionia buollana*	1.5		1959-1964
Lygus bugs	*Lygus* sp.	3.5	22,000[c]	1961-1972
Sugarcane borer	*Diatraea saccharalis*	0.5 (Sugarcane borer and Rice stalk borer)	218,000[c] (Sugarcane borer and Rice stalk borer)	SMY: 1962-1971; Funding: 1961-1971 (Sugarcane borer and Rice stalk borer)
Rice stalk borer	*Chilo plajadellus*			
Citrus snow scale	*Unaspis citri*	0.5		1971-1972
Alfalfa blotch leafminer	*Agromyza frontella*	0.5		1971-1972
Mexican bean beetle	*Epilachna varivestis*	0.5		1966-1972
Smaller European elm bark beetle	*Scolytus multistriatus*	2.0		1963-1969
Green peach aphid	*Myzus persicae*	2.0	207,000[c]	1966-1972
Egyptian alfalfa				

Imported cabbage-worm	*Peiris rapae*	7.0		1966-1971
Grasshoppers	*Schistocerca* sp. and *Melanoplus* sp.	1.5	120,000	SMY: 1963-1972; Funding: 1969-1974
Bollworm-corn earworm	*Heliothis zea*	0.5	121,000[c]	1962-1968
Tobacco budworm	*Heliothis virescens*			
European corn borer	*Ostrinia nubialis*			
Brown soft scale	*Coccus hesperidum*	0.5	15,500[c]	1962-1972
Face fly	*Musca autumnalis*	3.0		1963-1972
Balsam wooly aphid	*Adelges piceae*	2.5		1959-1965
Pea aphid	*Acryrthosiphon pisum*	6.0		1957-1972
Green bug	*Schizaphis graminum*	1.0	26,000[c]	1968-1972
Corn leaf aphid	*Rhopalosiphum maidis*			
Comstock mealybug	*Pseudococcus comstocki*	0.5		1967-1968
Linden aphid	*Myzocallis tilliae*	0.5		1969-1970
Alfalfa snout beetle	*Otiorhynchus ligustici*	0.5		1970-1972
Elm leaf beetle	*Pyrrhalta luteola*	0.5		1971-1972
SUBTOTAL: Extramural Domestic Funding			267,000[b]	
SUBTOTAL: Foreign Fund Expenditures			1,046,000[c]	
TOTAL: Insects		84.5	$1,313,000	

TABLE 11 (continued)

Pest		Cumulative Effort		
Common Name	Scientific Name	SMY (ARS only)	Funding (Extramural)	Period[a]
Weeds				
Scotch broom	*Cytisus scoparius*	2.5		1951-1972
Tansy ragwort	*Senecio jacobaca*	5.5		1956-1972
Mediterranean sage	*Salvia aethiopis*	4.5		1956-1972
Composite thistles	*Cirsium*, *Carduus*, *Centaurea*, and *Silybum* sp.	5.0	33,000[b] 78,500[c]	1956-1972
Halogeton	*Halogeton glomeratus*	9.0	3,000[b] 59,000[c]	1956-1972
Puncture vine	*Tribulus terrestris*	3.0		1957-1972
Dalmation } Yellow toad flax }	*Linaria dalmatica* } *Linaria vulgaris* }	1.5	3,000[b] } 31,000[c] }	1958-1972
Alligator weed	*Alternathera philoxeroides*	13.0		1959-1972
Witchweed	*Striga lutea*		62,000[c]	1960-1972
Purple nutsedge } Yellow nutsedge }	*Cyperus rotundus* } *Cyperus esculentus* }	1.0	62,000[c]	SMY: 1971-1972; Funding: 1960-1972
Water hyacinth	*Eichhornia crassipes*	6.5	26,000[b] 160,500[c]	1968-1972
Russian thistle	*Salsola iberica*	2.0	3,000[b] 81,000[c]	1967-1972
Eurasian water milfoil	*Myriophyllum spicatum*	0.5	135,000[c]	SMY: 1970-1972; Funding: 1965-1972

Milkweed vine	*Morrenia odorata*	0.5		1972
SUBTOTAL: Extramural Domestic Funding			68,000[b]	
SUBTOTAL: Foreign Fund Expenditures			669,000[c]	
TOTAL: Weeds		55.0	737,000	
TOTAL: Weeds and Insects		139.5	$2,050,000	

[a]Unless otherwise noted, period includes both SMY and Funding.
[b]Domestic Extramural Funding.
[c]Foreign Fund Expenditures, usually P.L.-480 funds.

administrators may skew the distribution of resources from one problem to another for good reasons independent of the economic damage done by the pest, but on the surface it appears appropriate for ARS to reevaluate their current allocation system.

The USDA, in cooperation with the states, performs the vast majority of innovative work in the field of pest control. Nevertheless, important contributions are also made by other federal agencies. The following briefly describes the roles of other parts of the government and estimates the resources they allocate.*

The Environmental Protection Agency

A variety of research, monitoring, standard-setting, and enforcement activities were given to the EPA when it was established in 1970. As will be discussed later, most of EPA's research efforts in the area of pest control are centered on the effects of pesticides on nontarget organisms, including man. The Agency has, however, ventured into some research designed to create new methodologies for controlling pests. Until fiscal year 1973, these efforts were quite small: $65,000 for FY 1971 and $790,000 for FY 1972. The major cause for the rapid increase in expenditures in FY 1972 was the Agency's participation, with the USDA and the NSF, in funding a comprehensive research project on "The Principles, Strategies and Tactics of Pest Population Regulation and Control in Major Crop Ecosystems" under the direction of Prof. C. G. Huffaker. A total of $700,000 was allotted this project by EPA from FY 1972 funds. The Agency's participation in pest control R&D continued to grow, and a total of almost $2.4 million was expended in FY 1973; of this amount, $1.3 million was given to the Huffaker project. The Congress has made it clear that it desires a continuing involvement of EPA in R&D on pest control technologies. A special appropriation of $5 million and 50 positions was made for FY 1974 through P.L. 135 (The "Agricultural, Environmental, and Consumer Protection Bill") for assessing the potential and evaluating the development of "safer" chemical and other alternative pest control strategies.

Is it proper for the EPA--a regulatory agency--to be involved in basic research for the innovation of new

---

*Estimates of research expenditures were provided to the Committee by the General Accounting Office.

technologies? It might be argued, for instance, that basic research in the pest control field is best left to the USDA and other agencies. These agencies have a clear mandate to help develop new technologies. This Committee thinks EPA's role in technological innovation is justified, but not without qualification. One of the most important functions of EPA personnel in the regulation of pesticides is the estimation of costs and benefits associated with the uses of specific compounds. Such estimations will always be made in a context of alternatives, and Agency personnel will be more aware of innovations if they are continuously involved in the development of alternative technologies.

We would point out, however, that there are pitfalls should the Agency become involved in a major way (e.g., through intramural research) in the innovative processes. First, such activity may seriously overlap and duplicate research efforts located in other parts of the government. Second, and more important, participation in research might tend to make Agency personnel advocates of particular technologies and damage their ability to make appropriate regulatory judgments or destroy the credibility of such judgments even where sound. We recommend that the EPA consider this pitfall when extending its support of basic R&D.

The National Science Foundation

The NSF is funding basic and applied research in pest management. These funds are small compared to the funds expended by the USDA and land grant universities. In FY 1969, the NSF supported pest management research by $391,000. This sum grew over the next 5 years to $2.73 million in FY 1973. In addition, some basic research supported by the NSF in insect physiology and other fields yields indirect benefits to pest control technology.

The Department of the Interior

The Department of the Interior has wide-ranging responsibilities in the areas of conservation and resource management, including jurisdiction over timberland, wildlife, fisheries, and water located in many different geographic areas of the United States. In its efforts to manage these resources, the Department has turned to pesticides and other methods of controlling those organisms that interfere with the

Department's objectives. As a result, the Department maintains research efforts both to develop new control technologies and to determine the effects of using them.

Innovative efforts to develop pest control technologies include testing chemicals for specific uses and developing "integrated" methods of pest control that rely on a variety of management techniques. In certain cases, the Department gathers the data necessary to register a specific chemical for a particular use. In FY 1973, the Department spent a total of $1.9 million for research on pest control technologies. Most of this amount ($1.5 million) was spent by the Division of Wildlife, Bureau of Sport Fisheries and Wildlife. The Division of Fishery Research of the same bureau spent the next largest amount, $0.2 million. The remaining funds were expended by other divisions of the same bureau, as well as the Bureaus of Land Management and Reclamation, the U.S. Geological Survey, and the Office of Water Resources Research.

The Department of Health, Education, and Welfare

The USDHEW has major responsibilities for assessing the effects of pesticides and other environmental toxicants on human health. The Department also has maintained a program of research designed to produce new methodologies for controlling pest organisms. A total of $2.0 million, $2.2 million, and $1.8 million was spent during FYs 1971, 1972, and 1973, respectively, on these programs. More than three fourths of these funds were expended by the National Institute of Allergy and Infectious Disease (NIAID), a component of the National Institutes of Health. Very small expenditures have been made in the pest control area by the National Institute of Environmental Health Sciences, the National Institute of General Medical Sciences, the National Institute of Child Health and Human Development, and the Division of Research Resources, all in the National Institutes of Health, and by the Center for Disease Control, a part of the Health Services and Mental Health Administration.

Some of the Department's research efforts are directed against organisms directly pathogenic to man, a class of pests that we have excluded from our study. The efforts of NIAID directed against disease vectors are described in more detail in Volume V, *Pest Control and Public Health*.

The Department of Defense

The armed forces of the United States have often been at war in areas with major pest problems: A variety of organisms attack man directly as parasites or disease vectors, and equipment such as clothing is subject to rapid deterioration from microbial organisms. The Department of Defense accordingly maintains both in-house and extramural research efforts designed to produce control techniques for organisms that interfere with the Department's military capability. In the past few years, the Department's funding of research has fluctuated: $1.8 million in FY 1971, $2.1 million in FY 1972, and $0.8 million in FY 1973. As might be expected, the Department of the Army has been more active in pest control research than has the Air Force or the Navy. Volume V, *Pest Control and Public Health*, describes in more detail the Department's research efforts in the field of vector control.

THE ALLOCATION OF RESEARCH AND DEVELOPMENT RESOURCES

The allocation of R&D resources in the private sector is subject only to indirect influence from the general public. Industry allocates its resources according to market signals, not according to a more general consideration of the public good. If the latter corresponds well to the signals, no fundamental social problem results; but if there is a disharmony between them, society may wish to intervene.

Information on the allocation of resources to R&D in the private sector would allow an assessment on the adequacy of this match between market signals and the public good. Unfortunately, the information is lacking, and no such assessment can be made. There would be clear public benefits from a more candid and detailed disclosure from the pesticide industry of their R&D operations, so that the entire research effort in the pest control area could be judged. Potential costs from such disclosure might include alterations of resources allocated to R&D by the industry.

The allocation of R&D resources in the public sector, in contrast, is theoretically subject to direct control by the body politic. The Congress can shape the research goals in ways perceived to be in the public interest; general research priorities are subject to criticism, debate, and change.

The openness of public R&D efforts, however, does not mean it is easy to set priorities for them. The following basic questions should be asked before allocations of R&D resources in the public sector are made:

1. What is the maximum potential payoff from the R&D effort if total control of the pest is achieved? The maximum direct economic benefit cannot be larger than the dollar value of the material currently destroyed by the pest(s). If this value is low, there may be a reason to shift resources into other areas with a larger estimated payoff.

It is more difficult to estimate the payoff of R&D on a pest that causes only "aesthetic" damage. For example, gypsy moths destroy no food, and only a little fiber. Indeed, they seldom kill trees; but they do defoliate them, and people attach considerable aesthetic value to fully leaved trees. It is difficult to measure such values, but the principle that damage is limited should be adhered to by research administrators. (See Volume IV, *Forest Pest Control*, for further discussion of the gypsy moth.)

The payoff from development of a control technology for a public health pest is extremely large, because we hold increased health and welfare of humans to be of great value. It is doubtful that any sum of money spent on such R&D would ever be considered too large compared to the potential payoff, especially if the latter involved specific "saved lives." Nevertheless, the principle that payoff has an upper limit should be kept in mind by research administrators in the public health field.

2. What hazards are posed by the currently used control technologies? A crop that is currently treated heavily with pesticides, for example, insecticides on cotton, would warrant a higher R&D budget for the development of nonchemical alternatives than would be suggested by the amount of loss actually incurred due to the pest. To carry the example further, the insecticides currently used on cotton are (1) vulnerable to decay in their effectiveness through resistance, resurgence, and the development of secondary pests; and (2) a hazard to man and other organisms. Both of these factors would argue for increased funding for R&D in order to develop alternative technologies.

3. What is the potential for serious failure of pest control if research is not done? In some cases, for example, the breeding of grains for resistance to fungi, R&D can be viewed as a "holding action." The pest may already be controlled in a satisfactory way by current technologies, but there is a distinct possibility that the current methods may eventually fail. In the case of resistant grains, for example, new viral strains may evolve. If R&D does not

continually produce new cultivars of the crop plant, the emergence of a new pathogen could result in disastrous losses. Administrators, therefore, continue to fund R&D that is essentially preventive.

4. What is the likelihood that R&D will be able to contribute to the development of a solution? No administrator will fund work on which there is a consensus that no payoff will ever be attained; but it is often difficult to decide whether a given research effort is hopeless. Research efforts, after all, do sometimes turn up useful results that were totally unpredicted. The only general solutions to the problem are methodological: (1) make the review process of funding R&D a group process and insure the participation of the most competent evaluators, so that individual judgments are not allowed to dominate the allocation process; and (2) reconsider discarded suggestions periodically.

5. What is the probability that R&D in a given area will create technologies that are useful elsewhere? This issue is fundamental to any type of biological research. Some organisms are considerably easier to study than others, and it may be possible to achieve significant breakthroughs in one organism or system that could not be achieved in others. For example, the first development of the sterile-male technique for insect control was made using the screwworm fly. The screwworm fly, in addition to being a pest of economic importance on cattle and other livestock, has several advantages as a subject for sterile-male research compared, for example, to the boll weevil. Had the boll weevil been selected as the research organism, the promise of the method might never have been demonstrated. As it was, the success of the sterile-male program with the screwworm stimulated research on a variety of other organisms that are less tractable as research material.

6. Are there enough skilled researchers to spend research funds wisely? There is little point in expending resources for research if they cannot be used properly because of a shortage of trained personnel; in such cases, decision makers should allocate resources to training programs first, and only later to research efforts.

Examples of Contemporary Allocation Patterns in Agricultural Research

In the absence of reasons to the contrary, R&D resources ought to be allocated to research on crop pests in proportion to the damage done by the pest. In other words,

Question 1 above is the most important piece of information needed before resource allocations are made.

The dollar value of an agricultural crop lost may be calculated by multiplying the estimated physical loss by the unit price of the commodity. If, however, the increase in yield is very large, the fall in price could, in theory, be so large as to make the net benefit to the individual farmer of a pest control "success" very small, or even negative. In such a case, the benefits are realized by the consumer in the form of lower prices for food and fiber.

The relationship between price and supply in agricultural commodities makes it difficult to calculate the potential dollar benefits from R&D efforts, and, hence, difficult to allocate R&D resources among commodities in proportion to payoff. The results of resource allocation to, say, weed control in a given crop can be compared with that to insect control in the same crop, however, because the effect on price of a loss from weeds is the same as the effect on price of a loss from insects. Thus, unless the answers to Questions 2 through 6 argue to the contrary, R&D resources aimed at control of one type of pest in a crop should be proportional to the losses caused by that pest.

This provides a simple method for approximating the extent to which R&D resources in the public sector are allocated in proportion to the estimated losses. The USDA' *Losses in Agriculture* (ARS, 1965) gives estimates of losses due to insects, weeds, and diseases and nematodes in a variety of crops for the period 1951-1960. The *Inventory of Agricultural Research* (Cooperative State Research Service, published annually) estimates R&D resources devoted to certain pest problems on certain crops. The crops reported differ slightly in the two publications, but they overlap for corn, cotton, grain sorghum, peanuts, rice, tobacco, and wheat.* Hence, for insects, weeds, and diseases and nematodes, estimates of research allocations come directly from the *Inventory* and estimates of losses come directly from the *Losses*. Table 12 presents the ratios

*Soybeans also overlap, but the geographic location of soybean production and the consequent pest problems have changed markedly since 1951-1960--the years for which loss estimates are available. Accordingly, it was concluded tha the estimates of soybean losses in Agricultural Handbook 291 may be in serious error and soybeans ought not to be included in this discussion.

TABLE 12 Comparison of Losses Due to Classes of Pests and Research Resources Devoted to Controlling Pests[a]

	Insects/Weeds		Insects/Diseases and Nematodes[b]		Weeds/Diseases and Nematodes[b]	
	Losses	Research	Losses	Research	Losses	Research
Corn	1.2	2.8	0.8	1.3	0.7	0.5
Cotton	2.4	7.6	1.4	2.8	0.6	0.4
Grain sorghum	0.7	2.2	1.0	2.2	1.4	1.0
Peanuts	0.2	1.8	0.1	0.3	0.5	0.2
Rice	0.2	1.0	0.6	0.7	2.4	0.9
Tobacco[c]	--	--	0.8	1.0	--	--
Wheat	0.5	3.4	0.4	0.4	0.9	0.1

[a]Estimates of proportions of crop lost each year (1951-1960) were taken from *Losses in Agriculture* (ARS 1965). For example, corn losses due to insects are 12 percent per year, and losses due to weeds in corn are 10 percent per year. The loss ratio for Insects/Weeds in corn is thus 0.12 ÷ 0.10 = 1.2.

Estimates of R&D resources devoted to controlling classes of pests in crops were taken from *Inventory of Agricultural Research* (Cooperative State Research Service, published annually). The estimates of total funds spent by the USDA, State Agricultural Experiment Stations, and other cooperating institutions were utilized. An average of the funds spent in FYs 1970-1972 was used to calculate the ratio recorded in this table. For example, an average of $2.264 million per year was spent in FY 1970-1972 to perform research on the control of corn insects, and an average of $0.811 million per year was spent for research on weed control in corn. The research ratio for Insects/Weeds in corn is thus $2.264 million ÷ $0.811 million = 2.8.

If the loss ratio and the research ratios are equal, then research expenditures are being allocated in proportion to the payoff that can be expected from elimination of damage from the class of pests against which the research is directed. Inequality of the ratios implies that research allocations are not being made on the basis of the payoff that can be expected should the research be successful. The text describes conditions in which it is reasonable to skew research resource allocations away from proportionality to expected payoffs.

[b]Loss estimates are given separately for diseases and nematodes, but estimates of research on control of diseases and nematodes are given as an aggregated sum. Loss estimates for diseases and nematodes were added together to prepare this table. No loss estimates are given for nematodes on grain sorghum, rice, or wheat; and the damage was assumed to be zero for these calculations.

[c]No estimates are given for losses in tobacco from weeds, and, hence, only the Insects/Diseases and Nematodes ratios are presented. The research expenditures on weed control in tobacco are very small, about $53,000 per year in FY 1970-1972.

of losses and research resources for insects compared to weeds, diseases, and nematodes and weeds compared to diseases and nematodes for the crops mentioned above.

A few rather simple observations can be made on the basis of the numbers shown in Table 12. First, in many cases, e.g., losses in wheat from insects and from diseases and nematodes, research allocations apparently are closely proportional to estimated losses. Second, insect damage elicits heavier research allocations than does weed damage. Every research ratio in the Insects/Weeds column is considerably larger than the loss ratios. For example, 7.6 times as much is spent for research on cotton insects as for cotton weeds, even though cotton insects are estimated to cause only 2.4 times as much damage as cotton weeds.

Third, damage caused by weeds, compared with that resulting from diseases and nematodes, has generally elicited slightly less research funding than would be expected were research allocations based only on estimates of the losses caused. In most cases (corn, cotton, grain sorghum, and peanuts), the differences do not appear large, and, given the roughness of the measuring technique, should not be considered significant. In two cases--rice and wheat--the damage caused by weeds has elicited a disproportionately small response from the research enterprise. For example, weeds caused 2.4 times as much damage in rice as did diseases and nematodes but only 0.9 times as much research effort was devoted to controlling them.

Fourth, losses and research efforts on insects compared to diseases and nematodes appear to be roughly proportional except for cotton and grain sorghum. Research efforts on insects in these two crops are about twice as intensive as would be expected from the relative damage caused by insects.

These measurements should be interpreted with the caution that loss estimates are for the period 1951-1960, but research expenditures are for the period fiscal years 1970-1972. If it is assumed that loss estimates have not changed substantially since the 1950s, then the comparison is legitimate. Properly interpreted, the results shown in Table 12 stand only as a reminder that some rationality can be achieved in setting research priorities and that the primary principle should be the magnitude of the payoff expected. Explanations for the apparently disproportionate expenditure on insect control likely include: (a) the increasing awareness of the development of resistance in target species; (b) greater concern for unwanted environmental effects from insecticides than from herbicides and fungicides and (c) perhaps some subtle psychological factors based on the visibility of insect predation.

REFERENCES

Agricultural Research Service (USDA) (1965) Losses in Agriculture. Agricultural Handbook No. 291. Washington, D.C.: U.S. Government Printing Office. 120 pp.

Cooperative State Research Service (USDA) (Annual) Inventory of Agricultural Research.

Djerassi, C., C. Shih-Coleman, and J. Deikuran (1974) Insect control of the future: Operational and policy aspects. Science 186:596-606.

Ernst and Ernst (1971) 1970 Industry Profile Study. Washington, D.C.: National Agricultural Chemicals Association.

Headley, J. C., and J. N. Lewis (1967) The Pesticide Problem: An Economic Approach to Public Policy. Baltimore: Johns Hopkins University Press. p. 75.

Maxwell, F. G., H. T. Reynolds, and C. E. Palm (1974) Report of the Committee on Impact of ARS-USDA reorganization on the profession of entomology. Bull. Entomol. Soc. Am. 20(1):50-53.

National Academy of Sciences-National Research Council (1972) Report of the Committee on Research Advisory to the U.S. Department of Agriculture. Springfield, Virginia: National Technical Information Service. 464 pp.

National Science Board (1972) Science Indicators. Washington, D.C.: U.S. Government Printing Office.

Starr, C., and R. Radman (1973) Parameters of technological growth. Science 182:358-364.

University-EPA-USDA Ad Hoc Subcommittee (1973) The Development of Data Required for Registration of Pesticides for Specialty and Small Acreage Crops and Other Minor Uses.

Von Rumker, R., H. R. Guest, and W. M. Upholt (1970) The search for safer, more selective, and less persistent pesticides. BioScience 20:1004-1007.

4

THE PRODUCTION OF MATERIALS FOR PEST CONTROL

Pesticides constitute an input item for agricultural and industrial operations, a consumer item for the homeowner, and a tool for the person charged with controlling pests of nuisance and public health importance. A substantial industry has grown up--largely in the post-World War II years--to supply the chemical and biological products that form the basis of pest control materials. Most of the production of these items is located in the private sector, but the USDA is involved as a producer of certain types of materials.

The private producers can be divided into two main categories: basic producers and formulators. The basic producers manufacture the active ingredients of a pesticide, such as the chemicals toxaphene, parathion, 2,4-D, atrazine, etc. The formulators combine the active ingredients with one another and with such inert ingredients as oil, talc, and other materials. In most cases, the ultimate users of pesticides do not buy a material until after it has been formulated. There are some cases in which the basic producer and the formulator may be the same company, but the general practice is to separate the two operations.

BASIC PRODUCTION INDUSTRY

An estimated 80-100 companies are engaged in the production of active ingredients for pest control materials. The plants producing these materials are widely scattered across the nation (Figure 2).

Several types of companies are active in the production of pesticides. General chemical companies like Monsanto (methyl parathion, alachlor, propachlor) and Du Pont (bromacil, methoxychlor) are of course involved.

FIGURE 2 Locations of pesticide production plants (Lawless et al. 1972).

Pesticide manufacturing represents only a portion of the activities of such corporations. Petroleum companies such as Tenneco (Strobane) and Chevron [Bux (R)] are also active producers of pesticides. Organizations generally known as pharmaceutical houses have recently entered the field: Abbott Laboratories has developed a preparation of *Bacillus thuringiensis* (a microbial agent for use against lepidopteran insect pests); trifluralin, a major herbicide used in cotton and soybean production, is produced by Elanco Products Co., a division of Eli Lilly and Company. Finally, there are examples of corporate ownership that on the surface might be surprising: Montrose Chemical Corporation of California, the sole remaining producer of DDT in the United States, is a subsidiary of the Stauffer Chemical Company and Chris Craft Industries, Inc. The latter is a diversified company involved in producing pleasure boats and aircraft, broadcasting, and other activities.

The significance of this diversity is not clear, but there are situations in which the diversified ownership patterns might be important. For example, if the federal government were to attempt to alter the incentive structure for R&D in the pesticide industry by offering some sort of subsidy, would we expect general chemical companies, petroleum companies, pharmaceutical companies, and the other diverse types of corporations involved in pesticide R&D to react uniformly to the changed incentives? We do not know the answer to this question, but we consider it important to analyze before attempting any major changes in policies affecting the innovative activities of private industry. Otherwise, the subsidy might not accomplish its intended purpose and result in unfairness to the taxpayers bearing its burden.

The quantities of materials produced by the pesticide industry are, as expenditures on R&D, considered proprietary information by the industry. As a result, our knowledge of how much is produced is incomplete and of uncertain accuracy. The U.S. Tariff Commission collects and publishes data on the production and sales of virtually all synthetic organic chemicals, including pesticides. Data are submitted voluntarily by the companies, and the Commission must by law hold them confidential unless a given compound is produced by three or more companies in approximately equal amounts. This provision severely limits the usefulness of the Commission's data, because so many pesticides are under patent protection and, hence, have only one producer. The result is that the published data have a residual, "all-other-compounds" category into which a majority of the poundage of production frequently falls. The Bureau of Mines collects some data on the production of inorganic pesticides. Until recently, these were the only government sources of information on the production of pesticides.

The Federal Environmental Pesticide Control Act of 1972 (FEPCA) requires the EPA to register all establishments engaged in the production of pesticides. All operators of registered establishments must then submit to the EPA an annual report of the types and amounts of pesticides they are currently producing and that were produced in the immediately preceding year. The information submitted under this provision of the law is considered confidential and cannot be made public (86 Stat. 987).

These provisions of FEPCA will remedy several major defects in the information available to the government: Manufacturers now must by law report their production activities, and the reporting is to that agency in the government responsible for regulation of pesticides. We applaud this provision of the law. A major defect remains: The information is to be protected from public disclosure.

In principle, the amounts of pesticide produced can be estimated from use data. Surveys on the use of pesticides are conducted by the USDA and several states. They are also conducted by at least one private company, the Doane Agricultural Service of St. Louis, Missouri. There are differences in the methodologies of survey and the scope of coverage, but all approaches attempt to estimate the amounts of specific chemicals used on specific crops in specific geographic areas at a given time. In the case of the USDA and Doane surveys, data are also collected on the use made of pesticides by different economic classes of farmers. The public surveys are of course published, but the Doane survey is sold only to the manufacturers of pesticides for a fee.*

It seems likely that pesticide manufacturers can make estimates of their competitors' production from the use surveys with sufficient accuracy to plan corporate strategy. We conclude that the only group left uninformed about

---

*In an interview with representatives of the Doane Agricultural Service, this Committee's staff was informed that this Committee and the Environmental Protection Agency would not be permitted to purchase any current survey. Doane, however, did let us examine some past surveys so that we could see the types of information they collected. We thank Doane for this cooperation.

the amounts of pesticide being produced is the general public. We find this an unsatisfactory state of affairs, and we recommend that FEPCA be amended to allow the release of data on the amounts of specific chemicals that are manufactured each year in this country.

The production of synthetic organic pesticides in 1973, the latest year for which data are available, was approximately 1.289 billion pounds, up 11 percent from 1972. Sales from manufacturers were 1.199 billion pounds, valued at $1.344 billion (U.S. Tariff Commission 1974).* Approximately 1,500 basic chemicals are registered as pesticides (President's Science Advisory Committee 1973), but a total of 24 products or closely related product groups are responsible for almost two-thirds of the volume. These major chemicals include the herbicides atrazine and 2,4-D; the insecticides toxaphene, carbaryl, and methyl parathion; and the wood preservative and herbicide pentachlorophenol (Lawless et al. 1972).

Production and sales of synthetic organic pesticides increased each year from 1957 to 1968. Production declined in 1969 and 1970, and then rose again in 1971-1973 to levels higher than in 1968. Production of synthetic organic herbicides in 1973 was 495 million pounds or 39 percent of the total. Sales of synthetic organic herbicides, however, amounted to $764.4 million, 57 percent of the total. The production of synthetic organic insecticides was 639.2 million pounds, or 49 percent of the total, but sales of these materials were $471.1 million, or only about 35 percent of the total value of all sales of pesticides. The production of synthetic organic fungicides amount to 154 million pounds, or about 12 percent of the total pesticide production of 1971. Sales of these materials were $108.1 million, or 8 percent of the total (U.S. Tariff Commission 1974).

The EPA recently commissioned a study on the pollution potential from pesticide manufacturing in which production estimates for the major chemicals were made. We have been informed by the National Agricultural Chemicals Association that the estimates prepared for this study are ". . . the most reliable figures currently available." We have reproduced these figures in detail as Table 13. A summary is shown in Table 14 (Lawless et al. 1972).

---

*Inorganic pesticides are not included in this figure. Also not included are petroleum oils used in formulation (>1.0 billion pounds), creosote wood preservative (1.0 billion pounds) and other solvents and diluents used extensively with pesticides.

Herbicides		Insecticides		Fungicides		Fumigants/Rodenticides		Special Category	
Product	Quantity	Product	Quantity	Product	Quantity	Product	Quantity	Product	Quantity
Atrazine	90	Toxaphene[b]	50	Sulfur (inorganic)	150	Methyl bromide	22	Petroleum oil[g]	>1,000
2,4-D	45	DDT	45	Pentachlorophenol and salts[c]	46	Dibromochloropropane	10	Creosote	1,000
MSMA-DSMA	35	Carbaryl	45						
Sodium chlorate	30	Methyl parathion	45	Dithiocarbamate group[d]	40	Other halogenated C_1-C_3 organics[e]	50	Coal tar	160
								Aromatic solvents	N.A.
Trifluralin	25	Malathion	35	Trichlorophenol and salts	20	Dichlorobenzene[f]	60	Dry carriers and diluents	N.A.
Propachlor	23	Chlordane	25	Captan	18	Warfarin	12		
Chloramben	20	Parathion	15						
Alachlor	20	Aldrin	10						
CDAA	10	Methoxychlor	10						
		Diazinon	10						
SUBTOTAL	298		290		274		154		
TOTAL (All Pesticides	408		406		297		200		>2,000

[a]Quantity produced in million pounds per year for direct pesticidal use, excludes production for other uses except as indicated.
[b]Includes Strobane-T.
[c]Includes use for termite control and as herbicide.
[d]Includes 11 products.
[e]Includes eight or more compounds.
[f]Includes moth-proofing and lavatory-space deodorant usage.
[g]Includes synthetic and refined spray oils, diluents, and carriers in formulated products.

TABLE 14A Estimated U.S. Pesticide Production Volume, 1971 (From Lawless et al. 1972) Herbicides, Defoliants, and Plant Growth Regulators[a]

Common Name	Trade Name	Principal Manufacturer	Production (10^6 lb A.I.)
Atrazine	Aatrex	Ciba-Geigy	90
2,4-D	Several	Several	45
MSMA-DSMA	Several	Several	35
Sodium chlorate	Several	Several	30
Trifluralin	Treflan	Elanco	25
Propachlor	Ramrod	Monsanto	23
Chloramben	Amiben	Amchem	20
Alachlor	Lasso	Monsanto	20
CDAA	Randox	Monsanto	10
Bromacil	Hyvar	Du Pont	8
Nitralin	Planavin	Shell	8
Dicamba	Banvel-D	Velsicol	6
Diuron	Karmex	Du Pont	6
Propanil	Rogue; Stam F34	Monsanto; Rohm & Haas	6
Butylate	Sutan	Stauffer	6
2,4,5-T	Several	Several	6
N.A.	DEF	Chemagro	5
EPTC	Eptam	Stauffer	5
Simazine	Princep	Ciba-Geigy	5
Dalapon	Dowpon	Dow	5
Fluometuron	Cotoran	Ciba-Geigy	4
Propazine	Milogard	Ciba-Geigy	4
Maleic hydrazide	Several	Several	3
Dinoseb	Premerge	Dow	3
Diphenamid	Dymid, Enide	Elanco, Upjohn	3
N.A.	Folex	Mobil	3
Silvex	Several	Several	3
Picloram	Tordon	Dow	3
Cacodylic acid	Several	Ansul	2
Vernolate	Vernam	Stauffer	2
Naptalam	Alanap	Uniroyal	2
Linuron	Lorox	Du Pont	2
Norea	Herban	Hercules	2
TBA	Trysben	Du Pont	2
DCPA	Dacthal	Diamond	2
Endothall	Aquathol	Pennwalt	2

TABLE 14A (continued)

Common Name	Trade Name	Principal Manufacturer	Production $(10^6$ lb A.I.)[b]
Chlorpropham	Several	PPG	2
SUBTOTAL (37 herbicides of >1 × 10^6 lb production volume each)			408
Acrolein	Aqualin	Shell	<1
N.A.	Ammate	Du Pont	<1
Amitrole	Several	American Cyanamid	<1
N.A.	Bandane	Velsicol	<1
Barban	Carbyne	Gulf	<1
Benefin	Balan	Elanco	<1
Bensulide	Betasan, Prefar	Stauffer	<1
CDEC	Vegadex	Monsanto	<1
Chlorbromuron	Maloran	Ciba-Geigy	<1
Cloroxuron	Tenoran	Ciba-Geigy	<1
Cycloate	RoNeet	Stauffer	<1
Diallate	Avadex	Monsanto	<1
Fenuron	Dybar	Du Pont	<1
Fluorodifen	Preforan	Ciba-Geigy	<1
Molinate	Ordram	Stauffer	<1
Monuron	Telvar	Du Pont	<1
Pebulate	Tillam	Stauffer	<1
Siduron	Tupersan	Du Pont	<1
TCA	Several	Dow	<1
Terbacil	Sinbar	Du Pon	<1
Terbutol	Azak	Hercules	<1
Triallate	Avadex BW, Far-Go	Monsanto	<1
SUBTOTAL (22 listed herbicides of <1 × 10^6 lb production volume each)			15
All other herbicides			5
TOTAL, ALL HERBICIDES			428

[a] See Appendix A, Properties of Selected Pesticides.
[b] A.I. = Active Ingredient.

TABLE 14B Estimated U.S. Pesticide Production Volume, 1971 (From Lawless et al. 1972) Synthetic Organic Insecticides, Miticides, Nematocides

Common Name	Trade Name	Chemical Group[a]	Principal Manufacturer	Production (10^6 lb A.I.)[b]
Toxaphene	Several (incl. Strobane-T)	CH	Hercules, Tenneco	50
DDT	Several	CH	Montrose	45
Carbaryl	Sevin	CA	Union Carbide	45
Methyl parathion	Several	OP	Monsanto	45
Malathion	Cythion	OP	American Cyanamid	35
Chlordane	Several	CH	Velsicol	25
Parathion	Several	OP	Monsanto	15
Aldrin	Several	CH	Shell	10
Methoxychlor	Several	CH	Du Pont	10
Diazinon	Spectracide	OP	Ciba-Geigy	10
Carbofuran	Furadan	CA	FMC	8
Disulfoton	Di-Syston	OP	Chemagro	8
Phorate	Thimet	OP	American Cyanamid	8
Heptachlor	Several	CH	Velsicol	6
N.A.	Bux	CA	Ortho	6
N.A.	Dursban	OP	Dow	5
Dicofol	Kelthane	CH	Rohm & Haas	4
Azinphos-methyl	Guthion	OP	Chemagro	4
Fensulfothion	Dasanit	OP	Chemagro	4
Methomyl	Lannate	CA	Du Pont	3
N.A.	Dyfonate	OP	Stauffer	3
Ethion	Nialate	OP	FMC	3
Ronnel	Korlan	OP	Dow	3
Carbophenothion	Trithion	OP	Stauffer	2
Naled	Dibrom	OP	Ortho	2
Dimethoate	Cygon	OP	American Cyanamid	2
Aldicarb	Temik	CA	Union Carbide	2
Endosulfan	Thiodan	CH	FMC	2
Chlorobenzilate	Several	CH	Ciba-Geigy	2
Crufomate	Ruelene	OP	Dow	2

TABLE 14B (continued)

Common Name	Trade Name	Chemical Group[a]	Principal Manufacturer	Production (10^6 lb A.I.)[b]
SUBTOTAL (30 insecticides of >1 × 10^6 lb production volume each)				368
N.A.	Abate	OP	American Cyanamid	<1
Coumaphos	Co-Ral	OP	Chemagro	<1
Fenthion	Baytex	OP	Chemagro	<1
Dioxathion	Delnav	OP	Hercules	<1
Dichlofenthion	VC-13	OP	Mobil	<1
Thionazin	Zinophos	OP	American Cyanamid	<1
Dichlorvos	DDVP	OP	American Cyanamid	<1
Oxydemeton-me	Meta-Systox-R	OP	Chemagro	<1
Mevinphos	Phosdrin	OP	Shell	<1
Phosphamidon	Dimecron	OP	Ciba-Geigy	<1
Dicrotophos	Bidrin	OP	Shell	<1
Monocrotophos	Azodrin	OP	Shell	<1
N.A.	Aspon	OP	Stauffer	<1
TEPP	Several	OP	Several	<1
N.A.	Mocap	OP	Mobil	<1
Dieldrin	Several	CH	Shell	<1
Endrin	Several	CH	Shell	<1
Lindane	Several	CH	Hooker	<1
N.A.	Mirex	CH	Allied Chemical	
Chloropropylate	Acaralate	CH	Ciba-Geigy	<1
N.A.	Omite	MI	Uniroyal	<1
Binapacryl	Morocide	MI	FMC	<1
SUBTOTAL (22 listed insecticides of <1 × 10^6 lb production volume each)				15
All other synthetic organic insecticides, miticides, nematocides				10
TOTAL, ALL SYNTHETIC ORGANIC INSECTICIDES, MITICIDES, NEMATOCIDES				393

[a]CH = chlorinated hydrocarbon, OP = organophosphate, CA = carbamate.
[b]A.I. = Active Ingredient.

TABLE 14C Estimated U.S. Pesticide Production Volume, 1971 (From Lawless et al. 1972) Botanical, Biological, and Inorganic Insecticides; Attractants, Repellents, and Synergists

Common Name	Trade Name	Product Type	Principal Manufacturer	Production (1 lb A.I.
Botanical and Biological Insecticides				
B. thuringiensis	Dipel		Abbott, Nutrilite	1
Polyhedrosis virus	Viron H		IMC	<1
Pyrethrins	Several		FMC, MGK	<1
Rotenone	Several		Several	<1
Nicotine	Black Leaf		Chem. Form.	<1
Inorganic Insecticides				
Pb-arsenate	Several		Several	6
Ca-arsenate	Several		Several	2
Attractants, Repellents, Synergists				
DEET	Several	Rep.	Several	1
Piperonyl butoxide	Several	Syn.	FMC	1
Ethyl hexandiol	6-12	Rep.	Union Carbide	<1
N.A.	MGK-326	Rep.	MGK	<1
Trimedlure	N.A.	Att.	Universal Oil	<1
N.A.	MGK-264	Syn.	MGK	<1
Heliotropin acetal	Tropital	Syn.	MGK	<1
Other Attractants, Repellents, Synergists				
TOTAL, ALL PRODUCTS				13
Special Category				
Petroleum oils including synthetic spray oils, deodorized kerosene, and other petroleum distillates used as diluents and carriers[b]				

[a]A.I. = Active Ingredient.
[b]No data available; possibly 1 billion pounds or more.

TABLE 14D Estimated U.S. Pesticide Production Volume, 1971 (From Lawless et al. 1972) Fungicides

Common Name	Trade Name	Principal Manufacturer	Production (10^6 lb A.I.)[a]
PCP and salts	Several	Several	46[b]
Dithio-carbamates	Several	Several	40[c]
TCP and salts	Several	Dow	20
Captan	Orthocide	Stauffer	18
PCNB	Several	Olin	3
Dodine	Cyprex	American Cyanamid	3
Captafol	Difolatan	Ortho	2
Folpet	Phaltan	Stauffer	2
Cu-naphtenates	Several	Several	2
SUBTOTAL (9 listed fungicides >1 × 10^6 lb production volume each)			136
Aminobutane	Several	Elanco	<1
Benomyl	Benlate	Du Pont	<1
Chloroneb	Demosan	Du Pont	<1
DCNA	Botran	Upjohn	<1
Dinocap	Karathane	Rohm & Haas	<1
DMTT	Dazomet, Mylone	Several	<1
Glyodin	Glyodin	Union Carbide	<1
Tetra-CP	Several	Dow	<1
Hg-salts	Several	Mallinckrodt	<1
SUBTOTAL (9 listed fungicides <1 × 10^6 lb production volume each)			7
All other synthetic organic fungicides			2
TOTAL, ALL ABOVE PRODUCTS			145
Inorganic Sulfur and Sulfides			150
Inorganic Copper Salts			2[d]
TOTAL, ALL FUNGICIDES			297

[a] A.I. = Active Ingredient.
[b] Includes use as herbicide, desiccant, molluscicide and for termite control.
[c] Includes CDEC, Dithane M-45®, Dithane S-31®, Ferbam, Maneb, Metham, Nabam, Niacide®, Polyram®, Thiram, Zineb, and Ziram.
[d] Published data are extremely unreliable.

TABLE 14E Estimated U.S. Pesticide Production Volume, 1971 (From Lawless et al. 1972) Rodenticides, Molluscicides, Fumigants, Soil Conditioners, and Others

Common Name	Trade Name	Principal Manufacturer	Production (10^6 lb A.I.)[a]
Rodenticides			
Warfarin	Several	Several	12
Dipacinone	Diphacin	Nease	<1
Norbormide	Raticate	Pitman-Moore	<1
Pindone	Pival	Several	<1
ANTU	Several	Penick	<1
Na-fluoracetate	Several	Aceto	<1
Molluscicides			
Metaldehyde	Several	Comm. Solvents	<1
Fumigants (Soil, Stored Product, Structural, and Household)			
Methyl bromide	Several	Several	22
DBCP	Several	Several	10
Other acyclic organics			
dichloro-propene-propane mixtures	D-D Nemex, Vidden D, Telone	Dow, Shell	75[b]
carbon disulfide, carbon tetrachloride, ethylene dichloride-dibromide, chloropicrin, formaldehyde and others	Several	Several	
Dichlorobenzene	Several	Several	60[c]
Naphthalene	Several	Several	2[b]
Inorganic fumigants	Several	Several	5
Soil Conditioners			
Polyacrylonitrile	N.A.	American Cyanamid	10
Other soil conditioners			
TOTAL, ALL RODENTICIDES, MOLLUSCICIDES, FUMIGANTS, SOIL CONDITIONERS			200

[a]A.I. = Active Ingredient.
[d]Excludes nonfumigant use.
[c]Includes moth-proofing and lavatory-space deodorant use.

There is considerable international trade in pesticides, and the U.S. exports about 10 times as much material as it imports (dollar value). In 1971, pesticide imports amounted to $26.2 million, $19.6 million of which were synthetic organic pesticides. Botanical insecticides, such as pyrethrins and rotenone, constitute another major item in U.S. imports. Most imports come from the United Kingdom, West Germany, Japan, and France. U.S. pesticide exports were valued at $251.6 million in 1971. Switzerland, the Netherlands, Belgium-Luxembourg, Canada, Japan, Mexico, and Brazil were the biggest importers in 1971, with each country purchasing more than $10 million worth of materials (Fowler and Mahan 1973). The value of exports must be of considerable value to the American pesticide industry, as exports constitute approximately one-fourth of the value of the total production. For some companies, exports are of special importance: Montrose Chemical Corporation of California, for example, exports almost all of the DDT it produces because the use of the chemical is almost totally prohibited within the United States.

THE FORMULATING INDUSTRY

The technical grade pesticidal chemicals produced by the basic manufacturers must be further processed and packaged before they are ready to be used by the farmer, industrialist, public health specialist, or home owner. The basic chemicals must be diluted for safety to the applicator and rendered into a physical form providing the greatest efficiency in reaching the target organism. These two goals are reached by the general process of *formulating* the pesticide; the formulated product must then be packaged in such a way that it can be transported to a distribution outlet and sold to the consumer or applicator. The latter will usually transport the formulated pesticide from the distribution center to his own storage facilities.

The most common types of formulations are dusts, wettable powders, granules, emulsifiable concentrates, and aerosols. Dusts are mixtures of the pesticide and a carrier such as flour, sulfur, lime, gypsum, talc, volcanic ash, or some other mineral preparation. Wettable powders are mixtures of the pesticide, carriers, adjuvants, and surface-active agents. For application, the operator merely mixes the formulation with water and sprays; the surface-active agent makes it possible to mix the pesticide and keep it suspended in the water. Emulsifiable concentrates are mixtures of pesticide and surface-active agents in an organic solvent such as kerosene. For application, the concentrate is mixed with water and sprayed. Granules are mixtures in which the pesticide is impregnated on granule-sized particles of clay, vermiculite, or other mineral. The granules can be applied either by aircraft or ground machinery. Aerosols are mixtures of pesticides, solvent, a variety of auxiliary material, and a propellant. Application usually requires only a simple aim-and-spray operation. Aerosols are very popular ways of formulating and packaging pesticides for use around the home (Lawless et al. 1972).

As many as 1,800 companies may be involved in formulating and distributing pesticides (Lawless et al. 1972). The Census Bureau estimated that for 1967 344 companies were engaged primarily in the formulation of pesticides into ready-to-use packages for agriculture. These companies employed 11,500 people, 6,900 of whom were production workers, with a total payroll of $80.7 million. The total value of the products shipped by this industry was $817.0 million (U.S. Bureau of the Census 1970). In many cases, especially for formulators of agricultural pesticides, the formulating plant is located in the midst of the market for the product. These plants are thus located in many parts of the country. A variety of ownership patterns and operating arrangements are found in the formulating industry. Sometimes the basic producer of the chemical not only manufactures the pesticide but also formulates it into the appropriate mixture. There are also a wide variety of independent formulators. Some independents work under contract with the basic producer and market a product with the producer's own label. Other independents buy the basic pesticide from the producer and then prepare a formulation under their own label. Farm cooperatives are an important type of independent formulator, because a significant portion of farmers' expenditures for pesticides are made at co-op stores as compared with other privately owned outlets (Lawless et al. 1972). The preparation of aerosol containers of pesticides is highly specialized. One study estimated that some 90 percent of the 100 million aerosol pesticide containers produced annually in the United States were prepared by only six companies. Pesticide aerosols account for about 8 percent of the total number of aerosol products manufactured annually. The remainder of pesticide formulation is performed by small packagers who prepare a variety of items for sale. The volume of pesticide processed by these operations is reported to be very small (Lawless et al. 1972).

THE U.S. DEPARTMENT OF AGRICULTURE

Although the USDA is not involved in the production of chemical pesticides on a commercial scale, it is a major producer of biological materials that are used in a number of large-scale control programs. These programs may be of considerable importance in the production of certain commodities in some areas. Furthermore, they are of importance in that they demonstrate the government's capacity to conduct a large-scale production program on a continuing basis.* Production programs operated by the USDA include the rearing of sterile males, the production of parasitic insects for release, and the production of pheromones and pheromone traps.

Sterile Males

Sterilized males of a particular type of insect may be released into the environment where wild populations of the insect occur. When the sterile male mates with a fertile female, no offspring result from the union. Thus the release of sterile males is a way to control the insect population by preventing the population from reproducing itself. In order to be successful, the number of sterilized males has to be very large compared to the number of wild, fertile males; furthermore, the sterilized males must be able to compete sexually with their wild counterparts. When these conditions are met, almost all sexual unions will be between the sterilized male and wild, fertile females. Few or no progeny will result, and the population will die out or drop to very low numbers. This method of control can be used only when the target pest population is already at a very low density.

The USDA is currently involved in major efforts to control populations of three insect species by the release of sterile males: the screwworm (*Cochliomyia hominivorax*), a pest on livestock, especially in the Southwest; the pink bollworm (*Pectinophora gossypiella*), a pest of cotton in the Southwest; and the Mexican fruitfly (*Anastrepha ludens*), a pest of citrus and deciduous fruits in subtropical areas of the United States. Pilot pest control projects utilizing production and release of sterile males

*These production programs are closely tied to operation of control programs. Other control programs conducted by government will be described later.

have been initiated against the boll weevil (*Anthonomus grandis*), a pest of cotton in the Southeast and parts of the Southwest; and the Caribbean fruitfly (*Anastrepha suspensa*), a pest of deciduous fruits in Florida.

Control of the screwworm using the sterile male technique was a significant technological innovation in the mid-1950s. The first demonstration of the effectiveness of the method was carried out on the island of Curaçao in the Caribbean by Baumhover and his associates (1955). The pest was then controlled in Florida, and in 1962 a major project was launched to control it in the Southwest along the Mexican border. The USDA produced and released approximately 8.8 billion sterile male flies in 1972 in connection with this latter program, which is conducted by 580 Animal and Plant Health Inspection (APHIS) personnel and involves an annual budget of $10 million (FY 1973) The production facilities are located in Mission, Texas. Plans have been made to release the sterile flies throughout northern Mexico and thus eradicate the fly as far south as the Isthmus of Tehauntepec in southern Mexico (Fowler and Mahan 1973).

The screwworm control program suffered a severe setback in 1972 when the number of laboratory-confirmed cases rose to 95,642 compared to 473 in 1971. In 1973, the number dropped back to 14,976 (USDA 1974).

The pink bollworm first appeared in the United States in 1917. Control based on mandatory destruction of cotton stalks at the end of the growing season led to its elimination from some areas of eastern Texas, western Louisiana, northern Florida, and southern Georgia by the early 1920s. In 1952, the insect again became established in Texas, and it has now extended its range westward to California. During 1966, extensive acreages--primarily in the Imperial Valley of Southern California--were treated for pink bollworm infestations (Newsom and Brazzel 1968). Most of the cotton grown in California is located in the San Joaquin Valley, in the central part of the state. The Department of Agriculture, through APHIS, is now producing sterile male pink bollworms for release in the San Joaquin Valley in order to prevent the insect from establishing itself in this fertile area. During 1972, the APHIS production facilities at Phoenix, Arizona, raised over 282 million moths; over 99 million were sterilized and released in the San Joaquin Valley (Fowler and Mahan 1973). Costs of the program are split on a 50/50 basis between the federal government and cooperators in California; the latter contribution is raised by a bale tax on cotton. In FY 1973, $672,000 was spent, and 52 persons were employed (USDA, personal communication, 1974).

Detection surveys for the Mexican fruitfly are conducted in Arizona, California, Florida, Louisiana, Texas, and northwest Mexico by APHIS. Sterile male flies, produced in facilities at Nuevo Leon, Mexico, are released on an as-needed basis when a local infestation is detected. In 1971, a total of 16.5 million flies were released; the facility now raises about 1.5 million flies per week between June and November. The cost of this program is about $110,000 per year and it employs eight people (USDA 1972).

The boll weevil first entered the United States in 1892, and cotton production has not been the same since in areas where it became established. Before its entry, cotton had no serious pests; afterwards, the boll weevil (and a variety of insects that became pests because of the treatment for the weevil) has posed a major control problem. In 1970, the USDA, in cooperation with the state of Mississippi and Mississippi State University, launched an eradication experiment against the boll weevil in southeastern Mississippi. One part of the eradication attempt is the production and release of sterile male weevils, which are reared and sterilized in facilities at Mississippi State University, in State College.

The USDA, the Florida Department of Agriculture, and the University of Florida are currently conducting tests to determine the feasibility of controlling or eradicating the Caribbean fruitfly using the release of sterile male insects.

Production of Parasitic Insects

The second production activity engaged in by the USDA is the rearing of parasites for release. The rationale is that if the parasite of an insect pest is released into areas in which the parasite is not found naturally, it may be able to establish itself and serve as an effective control of the pest.

A variety of egg and larval parasites of the cereal leaf beetle (*Oulema melanopus*) are now being raised at 23 field insectaries in 9 states for release in a 12-state area. Slightly more than $1 million was spent on this project in FY 1973, and a total of 60 man-years was allocated to it (USDA 1974b).* Three of the parasites (*Anaphes flavipes*, *Tetrastichus julis*, and *Diaparis carinifer*) have become established (USDA 1972).

A pilot program involving the production and release of parasites is currently under way against the gypsy moth (*Porthetria dispar*), a forest pest of the Northeast. The federal government, through APHIS, has signed cooperative agreements with the departments of agriculture in New Jersey and Maryland to produce about 15 parasites of the moth. A total of $86,000 per year is currently allocated to this effort from federal sources [USDA (APHIS), personal communications, 1974].

Pheromone Traps

The USDA has also engaged in the production of pheromone traps as a part of research efforts to incorporate pheromones into practical control schemes. Experiments of this sort require that large areas be treated with the pheromone and that the chemicals and/or traps be purchased or manufactured and then distributed to growers as part of the experimental design. Such programs have been mounted by USDA-ARS, sometimes with state cooperation, against the gypsy moth (in 14 states), the pink bollworm (California), and the boll weevil (Texas and Mississippi). Some of the distributed materials were manufactured at ARS in Beltsville, Maryland; some were purchased; others were manufactured in state facilities. Questions have arisen as to whether the USDA is suited to the role of primary producer and supplier of these materials. On the one hand, the USDA argues that their enterprise requires the flexibility to change specifications for materials rapidly during the course of experimentation and that they can produce their materials more cheaply than can private contractors. Some private firms, on the other hand, have complained that such government-sponsored production acts as a disincentive for the private sector to produce equipment for new types of pest control technologies and assert that incomplete distribution of the materials prevents the development of a stable private market without substituting equivalent coverage.

---

*Five parasites are being produced, one on the eggs (*Anaphes flavipes*) and four on larvae of the beetle (*Tetrastichus julis, Diaparsis carinifer, Diaparsis* sp. and *Lemophagus curtus*).

The Committee found no evidence that a serious problem exists insofar as the USDA policy is concerned. There is a legitimate distinction between "research" efforts and "control and operations," and it is appropriate for the USDA to produce needed research items in their own facilities. No evidence of which the Committee is aware indicates that the leadership of the Department is unaware of the problems faced by the private corporations seeking to enter the field of new pest control technologies. It should, of course, give continuing attention to the question of whether distribution programs are acting as disincentives for good control practices.

A more serious problem, as yet unresolved by the Congress and the USDA, centers on the extent to which the public desires to or should subsidize or control the production of pest control materials used in actual control programs. There is a problem of equity to the taxpayer when a control program is run by the government for the benefit of a readily identifiable group of farmers-businessmen. The problem of whether and to what extent the private sector will engage in producing materials for new types of pest control methods is a serious one, however. The government should not inadvertently hinder the entry of private groups into the production of new pest control techniques. Just as important, the government will want to ensure that new technologies of general benefit to society are developed and introduced, even if the government itself must assume a major responsibility for the production of needed materials.

REFERENCES

Baumhover, A. H., A. J. Graham, B. A. Better, D. E. Hopkins, W. D. New, F. H. Dudley, and R. C. Bushland (1955) Screw-worm control through release of sterilized flies. J. Econ. Entomol. 48:462.

Fowler, D. L., and J. N. Mahan (1973) The Pesticide Review 1972. Washington, D.C.: U.S. Department of Agriculture. 58 pp.

Lawless, E. W., R. Von Rumker, and T. L. Ferguson (1972) The Pollution Potential in Pesticide Manufacturing. Technical Studies Report No. 15-00-72-04. Washington, D.C.: Environmental Protection Agency. 249 pp.

Newsom, L. D., and J. R. Brazzell (1968) Pests and their control. *In* Elliot, F. C., M. Hoover, and W. K. Porter, Jr., Advances in Production and Utilization of Quality Cotton: Principles and Practices. Ames: Iowa State University Press. 532 pp.

President's Science Advisory Committee (1973) Chemicals and Health. Washington, D.C.: U.S. Government Printin Office. 211 pp.

U.S. Bureau of the Census (1970) Census of the Manufacturers, 1967. Industry Series Agricultural Chemicals, MC67(2)-28F. Washington, D.C.: U.S. Government Printing Office.

U.S. Department of Agriculture (1972) 1971 Progress Report, Plant Protection and Quarantine Programs. Washington, D.C.: USDA (APHIS). 31 pp.

U.S. Department of Agriculture (1974) Screwworm Progress Report for 1973. APHIS, Mimeographed. 2 pp.

U.S. Tariff Commission (1974) United States Production and Sales of Pesticides and Related Products, 1973 Preliminary. Washington, D.C.: U.S. Tariff Commission 11 pp.

5

TRANSPORTATION, STORAGE, AND ACCIDENTS

TRANSPORTATION AND STORAGE OF PESTICIDES

Pesticides are frequently manufactured and formulated in places and at times remote from their use. Hence, transportation and storage of the chemicals is an integral part of our pest control system. We have no knowledge of the amount of pesticide in storage, but the Commodity Transportation Survey of the Census Bureau makes estimates at 5-year intervals of the magnitude of shipping involved. In 1967 (the latest year available), approximately 1.3 million tons of formulated pesticidal materials were transported 517 million ton-miles. Rail transportation accounted for over 42 percent of the tonnage shipped, and public and private trucking for about 58 percent. Water transport was responsible for less than 1 percent of the tonnage. More than 85 percent of the material was shipped in batches of 5 tons or more (Bureau of the Census 1970).

ACCIDENTS DURING THE TRANSPORTATION AND STORAGE OF PESTICIDES

Accidents during transportation and storage are to be expected. People may be injured at the accident site by the pesticide, or clothing and foodstuffs may be contaminated and later cause injury; thus, damage to people or property from accidents must be counted as a category of cost or risk arising from the use of the contemporary chemical technology. A variety of activities may help lower or minimize this risk, but it is unlikely that it can ever be reduced to zero (Wolfe 1972).

A variety of laws allow the federal government to make regulations about the packaging and labeling required for pesticides.* In particular, the Hazardous Materials Regulations Board of the Department of Transportation (DOT) is empowered to make rules governing the packaging of these compounds for shipment. All incidents in which a pesticide is unintentionally released--during transit or during temporary storage related to transit--must be reported by the carrier to the Office of Hazardous Materials DOT.† Accidents involving death, injury, or property damage above $50,000 must be reported immediately by telephone. All episodes must be reported in writing within 15 days.

Most pesticides are classified as "Class B" poisons by the DOT. In 1971-72, approximately 400 incidents of the accidental release of Class B poisons were reported. Some 200 of these accidents involved pesticides, and 35 injuries were noted, all of which were due "certainly" or "probably" to the pesticide. No deaths were reported. Property damage was tabulated at $91,418.‡

The hazard in shipping pesticides does not, on the surface, appear alarming. Somewhat more than 200 accidents, 35 injuries, no deaths, and about $90,000 worth of damage from the shipment of toxic materials for 2 years do not seem an exorbitant risk to bear. Certainly we would rank hazards of transportation as one of the lesser hazards derived from our contemporary pest control technologies. Our complacency is marred, however, by the presence of unknown factors:

- How serious is underreporting of accidents?
- Property damage estimates are not listed for about

*For example, the Federal Environmental Pesticide Control Act (86 Stat. 973), Poison Prevention Packaging Act (84 Stat. 1670), and various acts administered by the Department of Transportation.

†The requirement for reporting accidents originates in the Hazardous Materials Transportation Control Act of 1970 (84 Stat. 977).

‡Hazardous Materials Regulations Board, DOT, 1973, personal communication. The hazardous material involved in some incidents was not reported by name. We estimate 212 incidents involving pesticides and include in this figure all materials reported as "organic phosphate." Materials listed as "Poisonous liquid" (50 incidents), "Poisonous solid" (10), "Ques" (2), and "Unknown" (20) were not included in the estimate even though some of them probably were pesticides.

one-half of the accidents reported to the Department of Transportation.

- Was there really zero damage in these incidents, or did the filer merely not know how to estimate it?
- Are external damages (i.e., damages for which the transporter does not have to pay) adequately reported, or are the damage estimates merely estimates of damages for which the carrier is liable?
- In cases of injury, should follow-up reporting be required in order to appraise more fully the magnitude of the injury and its economic impact?
- Are there unrecognized, delayed effects of any exposures that do occur?

Our Committee cannot answer these and other important questions, but we feel that their consideration is important. The Secretary of the DOT has now had 3 years of experience in the reporting of accidents involving pesticides and thus a data base on which to evaluate more completely the reporting system and the hazard created by the necessity to ship these materials. We recommend that the Secretary give priority attention to such an evaluation.

ACCIDENTS DURING AERIAL APPLICATION OF PESTICIDES

In the United States, charges for aerial application are generally based on the number of gallons applied per acre, i.e., rates are established as if the operation were one of transportation--though it has not been included in that category in the foregoing account. (Accidents can result in death or injury to the pilot and other people, damaged or demolished aircraft, spilled and wasted pesticide, and damage to structures or vegetation on the ground.) The immediate costs are borne by the aerial applicators themselves or by their insurers. We could not locate centralized statistics on the magnitude of costs associated with aerial accidents and hence made no effort to estimate them carefully. The National Transportation Safety Board (NTSB), however, does collect information about each aircraft that is involved in some type of mishap. For the latest years in which statistics are available from the NTSB, 1968-1970, there were in each year approximately 275 accidents and between 29 and 33 fatalities in accidents involving application of pesticides or defoliants. There were also between 28 and 36 serious injuries for each of these years (NTSB 1968, 1969, 1970). There are severe methodological and ethical problems in placing a value on a human life, but we do note in passing that if human lives are valued at $250,000 each (the Warsaw Convention liability limit) and if there are 30 deaths each year, then the cost of fatalities alone from aerial spraying is $7.5 million per year. The cost of damaged aircraft and other damaged materials would make the total costs of aerial application even higher.

PESTICIDE ACCIDENT SURVEILLANCE SYSTEM (PASS) OF THE ENVIRONMENTAL PROTECTION AGENCY

The EPA has central responsibility for the regulation of pesticides, and it is therefore appropriate that they have maintained the Accident Investigation Branch (AIB), which was initiated when the USDA had regulatory authority over pesticides. PASS is the framework by which AIB will develop estimates of the extent and nature of all accidents involving pesticides and make recommendations for new regulations necessary to alleviate the damage incurred. The 10 regional offices of the EPA are each responsible for encouraging the voluntary submission of data on accidents by state agencies in their area. The AIB, located in Washington, D.C., then collates and analyzes this information from the states with information received from other federal agencies, such as the DOT.

PASS in its present form is new and developing. It would be premature to examine its workings in any detail. We recommend, however, that accident investigation be given a thorough review within the next 1-2 years. The costs of accidents represent a totally negative drain on the benefits of pesticides, and they should be reduced to the lowest level practicable. The major questions a review should address include the following:

- What estimates of the nature and costs of accidents emerge from PASS?
- Do these estimates agree in a reasonable way with other estimates?
- Has the Registration Division of the Office of Pesticide Programs utilized the PASS information fully in reviewing registrations?
- Is the voluntary submission of data on accidents adequate to protect the public interest or should some system of mandatory reporting be required?
- Is coordination adequate between the EPA and other federal agencies, especially the DOT, DOL, DI, and DHEW?

REFERENCES

Bureau of the Census (1970) 1967 Census of Transportation. Vol. III. Commodity Transportation Survey.

National Transportation Safety Board (1968, 1969, and 1970) Briefs of Aircraft Accidents Involving Aerial Application Operations. Washington, D.C.

Wolfe, H. R. (1972) Safety problems related to transportation and storage of toxic pesticides. *In* Industrial Production and Formulation of Pesticides in Developing Countries. Vol. I. General Principles and Formulation of Pesticides. United Nations Industrial Development Organization.

6

INFORMATION ABOUT PESTICIDES AND PEST CONTROL

INFORMATION FOR THE FARMER

The Sociology of Information Transfer

The independent grower always confronts a choice in applying a pest control program: What shall I use? Once he has made this decision, the operation he will conduct is virtually fixed. He has only to go to a source of supply, buy the material, and make or arrange for the application. The decision of what to use is thus of considerable importance, and we need to know the informational basis on which the farmer makes his decision. Professor George M. Beal and his associates (1966a) at Iowa State University have published the most extensive sociological studies available on how farmers gather the information they use in making their decisions. We quote from these studies in the following several paragraphs, but the reader should bear in mind that the data are all from Iowa; important regional variations in information gathering may exist.

Farmers in Iowa have a wide variety of outlets at which they can buy pesticides and other materials for pest control operations. The three most commonly used sources were feed and seed stores (62 percent), co-op elevators (61 percent), and veterinarians (53 percent) (used only for purchase of livestock insecticides and animal health products that were farmer-administered). The other sources mentioned by 10 percent or more of the growers were general farm supply stores (28 percent), drug stores (24 percent) (for livestock products), Farm Bureau outlets (22 percent), privately owned elevators (20 percent), on-the-farm chemical salesmen (16 percent) (mostly for livestock products), petroleum dealers (14 percent), and wholesalers of agricultural chemicals (12 percent). A variety of other sources accounted for the remainder of sales.

The sources from which farmers learn about agricultural chemicals are diverse. In the Iowa survey, the 10 most commonly mentioned sources of information were farm magazines and newspapers (94 percent), labels on the product (90 percent), other farmers (68 percent), local agricultural chemical dealers (61 percent), radio (49 percent), county extension personnel (48 percent), veterinarians (44 percent), television (42 percent), agricultural chemical company publications (42 percent), and general newspapers (36 percent).

Although farmers received information from a wide range of sources, their perceptions of the quality of the information they were getting indicated that only a few different sources were regarded as "most useful" for specific types of questions. In choosing which chemical to do the job properly, farmers relied most on local agricultural chemical dealers (22 percent), farm magazines and newspapers (21 percent), county extension personnel (11 percent), and other farmers (6 percent). Understandably, labels on the pesticide were the most useful source of information for methods and rates of application (69 percent) and for handling and safety precautions (62 percent). The most useful sources of information about hazards and possible harmful consequences of misuse were more diverse and included the label (36 percent), farm magazines and newspapers (23 percent), chemical company sources (13 percent), USDA-extension sources (11 percent), and other farmers (6 percent). Learning of a new chemical for the first time is clearly a very important aspect of the adoption of new technologies, and farmers indicated three sources as being most useful to bring them word of new chemicals: farm magazines and farm papers (52 percent), USDA-extension (20 percent), and chemical company sources (16 percent).

To the extent that the results of the Iowa survey are generally applicable, they suggest a few conclusions. First, the chemical industry and local chemical dealers are regarded as a very useful source of information for pest control operations. The grower relies heavily on the local dealer and the label in selecting a chemical. He also relies heavily on farm publications in the selection process; and in so doing, he may be obtaining additional material from the chemical salesmen, because the chemical industry advertises heavily in farm journals. (See Chapter 8 for discussion of recent actions by the Federal Trade Commission on pesticide advertising.)

The potential impact of this advertising is suggested by the circulation figures for these magazines. Exclusive of corporate "house organs" and newsletters, etc., that carry no advertising, about 110 different publications in magazine format are distributed on a monthly basis to U.S. farmers. Most of these specialize in some crop or region, and most are distributed free or at trivial cost. The total national circulation of these publications in 1973 was more than 120 million. The advertising density on behalf of pesticides is high: During the season of heavy application, 30-40 percent of the total space in a journal devoted to a crop like cotton may contain pesticide advertising.

The farmer also relies on these same types of industry sources to learn of new chemicals. The label prepared by the chemical industry, and approved by the federal government through registration, is clearly the most useful source of information on the details of handling, application, and possible harm from misuse. Second, the farmer does rely significantly on public sources of information--the USDA-extension publications and personnel--but these sources apparently are viewed as less useful than private sources.

It is not clear why public sources are not listed as "most useful" more often, and the Iowa study did not attempt to determine the answer. It should be noted, however, that there are between 10 and 20 times more dealers than county extension agents; the resources available to farmers from the public sector are thus very low in terms of manpower and funding. Furthermore, the public sources spend considerable time in dealer education and in supplying information to farm magazines; thus the public sector may have an importance in terms of quality of information that is not reflected in whether public personnel are considered "most useful" in advising farmers.

Third, dependence on other farmers, while it exists, is slight; the utility of information from other farmers (cited as "most useful" by no more than 6 percent of respondents) contrasts sharply with the rate of contact (68 percent of respondents say they ask other farmers). It appears that most farmers want an "outside expert" to help them with their pest control problems, rather than one of their peers.

Because the local agricultural chemical dealer is seen by the farmer as an important source of information in pest control decisions, it is important to know about the dealer perception of his own role, sources of information, and factual knowledge. As in the case of farmers, the only studies we have on chemical dealers were performed in 1965 by Beal and his associates (1969a) on a sample of 126 dealers in the state of Iowa.

Most dealers claim a responsibility to be a source of information: 70 percent of them said that "The dealer has a responsibility to keep himself well informed and to make recommendations to his customers about agricultural chemicals and their uses." Some 90 percent believed that the dealer either probably or certainly should make recommendations about the methods of application of chemicals. In part, the dealer's perception of his own role was a reflection of what he thought the farmer wanted from him. Forty percent stated that farmers expected them to be highly qualified sources of information on all aspects of agricultural chemicals and their uses. Thirty-two percent maintained that farmers expected them to be at least a source of information on some aspects of chemicals and their uses. A substantial majority of dealers indicated that they should assume major responsibility for providing information about hazards or harmful consequences that might result from the use of chemicals and about safety precautions that should be taken in handling and applying pesticides.

In practice, however, dealers question certain points only infrequently. Although 70 percent of them reported that more than one-half of their customers asked them about chemicals and their use, only about 15 percent of the dealers said that a majority of their customers asked about safety precautions in handling and applying pesticides or about hazards and harmful consequences from the use of the chemicals. More than half the dealers reported that fewer than 25 percent of their customers asked about safety precautions, hazards, and possible harmful consequences.

Dealers give out a wide variety of information to their customers. Well over 80 percent advised on methods, rates, and timing of applications. Ninety percent told customers what chemical is best for a specific purpose and 61 percent named a particular brand as being the best quality. Seventy-one percent told customers of the economic returns that would result from the use of chemicals, but only 50 percent gave information on the research results on the effectiveness of pesticides. Many dealers have given information on some of the possible dangers that can result from the use of pest control chemicals:

- hazards in application (86 percent)
- danger of residues in the marketable product (70 percent)
- safety measures to be taken during application (82 percent)
- hazards from excessive application (59 percent)

On the other hand, only a minority of dealers mentioned ever having given advice about certain hazards of pesticide use.

- unintended and possible harmful effects on humans (42 percent)
- proper disposal of empty containers (38 percent)

A majority of dealers thought the information they gave about the uses of chemicals, the safety precautions to be taken, and the hazards to be avoided were fairly adequate; approximately an additional one-third of the dealers thought the information they gave was completely adequate. Only about 10 percent of the respondents felt their advice on use, safety, and hazards was less than adequate or very inadequate.

Dealers, like farmers, rely on a variety of sources for information. The most commonly mentioned source (90 percent) was the label. Some 60 percent of the respondents said they utilized manufacturers, wholesalers, and distributors of the chemicals; or farm magazines, papers, and publications. About 50 percent mentioned extension publications and staff as a source. Estimates of the quality of information received, however, indicated that relatively fewer sources were considered "most useful." When making recommendations for the best chemical to use, dealers relied most heavily on manufacturers, wholesalers, distributors, the label, and short courses sponsored by Iowa State University and the agricultural extension service. As with farmers, the label is the single most important source of information about methods and rates of applications, procedures for handling and applying, and hazards. Publications and personnel from the Extension Service, USDA, and Iowa State University were considered most useful sources by fewer than 20 percent of the dealers.

Because of the importance of the local dealer in providing information on chemicals to the farmer, Beal and his associates examined chemical dealers in Iowa to determine their level of knowledge about the chemicals. Participants were given statements about specific insecticides and herbicides and asked to say whether they agreed, disagreed, or had no opinion. In almost all cases, dealers who sold the chemical in question gave a higher percentage of correct answers than did dealers who did not. The questions were categorized into eight classes according to the uses of the chemicals in question. The average percentage of correct answers in the different classes varied from 80-42 percent. The percentage of incorrect answers varied from 6-51 percent. A sample of the test questions are shown in Table 15. It is not possible to conclude from this study whether or not the incorrect answers are resulting in damage either to the farmer directly or indirectly to the larger

TABLE 15 Sample of Knowledge Questions Asked of Agricultural-Chemical Dealers from Iowa[a]

Statement[b]	Correct		Incorrect		No Opinion		Total[c]
	Number	%	Number	%	Number	%	
Grass killers							
Randox should not be applied to corn until it is 6 inches tall. (Correct answer is Disagree)	66	79.5	5	6.0	12	14.5	83
Weed killers							
2,4-D is an effective insecticide for mosquito control. (Correct answer is Disagree)	99	89.2	5	4.5	7	6.3	111
The recommended dosage for spraying 2,4-D on corn at "lay-by" time using a drop-extension nozzle is 1/2 lb or one pint of ester per acre. (Correct answer is Disagree)	45	40.6	54	48.6	12	10.8	111
Grass or weed killers							
Dosage recommendations of weed control chemicals are more accurate when given in terms of pounds of active ingredients per acre. (Correct answer is Agree)	89	80.2	18	16.2	4	3.6	111
Amiben is an effective perennial weed killer in soybeans. (Correct answer is Disagree)	40	36.0	55	49.5	16	14.5	111

TABLE 15 (continued)

Statement[b]	Correct		Incorrect		No Opinion		Total[c]
	Number	%	Number	%	Number	%	
Grass, weed, or brush killers							
Amino-triazole will kill only broad-leaf weeds. (Correct answer is Disagree)	87	78.4	11	9.9	13	11.7	111
When amino-triazole is applied to thistle patches in a pasture, it is recommended that livestock not be allowed on the treated area for 8 months. (Correct answer is Agree)	34	30.6	58	52.3	19	17.1	111
Livestock insecticides							
For a fast knockdown of pests attacking livestock, one should use a "combination" spray made up of all recommended livestock insecticides. (Correct answer is Disagree)	92	96.8	3	3.2	0	0.0	95
Methoxychlor is recommended for lice control on feeder cattle and can be used up to time of slaughter. (Correct answer is Agree)	33	34.7	26	27.4	36	37.9	95

Two lb per acre of actual aldrin or heptachlor, which is broadcast and disked in, will control all major soil insects attacking corn on sod ground. (Correct answer is Agree)	38	42.2	46	51.1	6	6.7	90
Crop or livestock insecticides							
DDT is a recommended chemical to be sprayed on milk cows for fly control. (Correct answer is Disagree)	103	93.7	4	3.6	3	2.7	110
Toxaphene is a recommended mange-control treatment for hogs if it is applied no later than 28 days before marketing. (Correct answer is Agree)	35	31.8	37	33.6	38	34.6	110
Soil, crop, or livestock insecticides							
Alfalfa that has been sprayed with diazinon should not be cut for hay until at least 7 days after treatment. (Correct answer is Agree)	61	51.7	14	11.9	43	36.4	118

[a] Adapted from Tables 22-29 of Beal et al. (1969a).

[b] Statements were put into eight classes by the function of the chemical in question. There was only one question in three of the classes. In each of the other five classes, the table shows the two statements eliciting (1) the highest percentage of correct responses and (2) the highest percentage of incorrect responses.

[c] The only dealers included in this table are those who sold materials for the function in question. For example, only dealers who sold "grass killers" are shown for the question about "grass killers."

society, but it is not comforting to realize than an information source that is very important to farmers may have a high error factor.

One feature common to farmers and dealers is the reliance placed on information sources in the private sector (Table 16). The Iowa survey indicates conclusively that both groups of decision makers rely most on people who manufacture, distribute, or sell the chemicals, and on publications that carry much advertising of chemicals. The label, a quasi-public channel of communications, is important for details of handling and application but less so for selection.

Public sector information in the form of personnel and publications from the Extension Service-State University-USDA community are mentioned as most useful sources but only by a minority of farmers and dealers (Table 16). The significance of this is not entirely clear. On the one hand, it may merely reflect the conclusions of farmers and dealers that the information they get from the private sector is both adequate and conveniently available; there is no need or incentive to go further for help. On the other hand, the importance of private sector information may reflect suspicion of public sources among some farmers. In either case, there is the implication that the most efficient channels for transmitting information on pest control with pesticides lie in the private sector. The figures and our conclusions based thereon, however, cannot take into accunt the indirect effect of the experiment stations and extension service on the advice ultimately conveyed to the farmer by the local dealer, farm magazine, or other private source. Furthermore, the programs of the extension service, described below, show some initial promise as a way of bringing alternative methods of control to farmers. It is not clear, however, whether the channels in the private sector are capable of transmitting, without public subsidy, the information necessary to conduct the more sophisticated methods of pest control.

The Cooperative Extension Service

The Cooperative Extension Service (CES), created in 1914 by the Smith-Lever Act, is the established institution in the public sector for transfering new technologies from centers of research and development to users. Financing, planning, and conducting the educational programs of the CES are responsibilities shared jointly by the USDA, the state land grant universities, and the Extension personnel.

TABLE 16 Sources of Information Judged "Most Useful" by Farmers and Agricultural Chemical Dealers in Iowa

	Farmers[a] (%)					Dealers[b] (%)				
Source	Which Chemical to Use?	Methods and Rates of Application	Handling and Safety Precautions	Hazards and Possible Harmful Consequences	New Chemicals	Which Chemical to Use?	Methods and Rates of Application	Handling and Applying	Hazards and Possible Harmful Consequences	New Chemicals
Industry[c] sources	28.9	11.8	10.9	13.1	16.2	50.8	41.2	37.3	38.9	57.8
Advertising media[d]	22.7	6.1	11.4	29.6	56.3	8.0	3.2	3.2	4.8	20.7
SUBTOTAL	51.6	17.9	22.3	42.7	72.9	58.8	44.4	40.5	43.7	78.5
Label[e]	14.4	69.0	62.4	35.5	0.4	13.5	43.7	42.8	34.1	0.0
Public[f]	22.2	6.0	8.3	11.3	21.3	24.5	7.9	15.1	20.6	19.1
Other private[g]	8.7	5.6	5.2	7.8	3.4	2.4	3.2	0.8	0.8	1.6
Miscellaneous or no answer	3.1	1.3	1.7	2.6	2.2	0.8	0.8	0.08	0.8	0.8
TOTAL[h]	100.0	99.8	99.9	99.9	99.8	100.0	100.0	100.0	100.0	100.0

[a] Taken with rearrangement from Table 32 in Beal et al. (1966). A population of 229 farmers selected to be representative of the "general farmer" completed the interview.

[b] Taken with rearrangement from Tables 78-82 in Beal et al. (1969b). A population of 126 dealers completed the interview.

[c] Industry sources include (a) clinics and short courses sponsored by agricultural chemical companies, (b) clinics and short courses sponsored by local agricultural chemical dealers, (c) agricultural chemical company publications, (d) local agricultural chemical dealers, and (e) salesmen for agricultural chemical manufacturers or wholesalers.

[d] Advertising media include (a) farm magazines, farm papers, and farm publications; (b) television; (c) radio; and (d) other private publications.

[e] The label must be regarded as a hybrid public-private means of communication. The label is first prepared by the person or corporation that wants to register a pesticide; it is then approved and registered. At the time this survey was taken, registration was the responsibility of the U.S. Department of Agriculture. The Environmental Protection Agency now registers pesticides.

[f] Public sources include (a) clinics and short courses sponsored by Iowa State University and the Agricultural Extension Service, (b) agricultural extension publications, (c) U.S. Department of Agriculture publications, (d) county extension personnel, (e) Iowa State University specialists, (f) vocational agricultural teachers, and (g) *Iowa Farm Science*.

[g] Other private sources include (a) Farm Bureau, (b) community organizations, (c) family members, (d) other farmers, (e) landlords, (f) bankers, (g) farm managers, (h) veterinarians, and (i) Hybrid Seed Corn Corn Co.

[h] Total may not equal 100 percent because of rounding errors.

These groups are located in each county of the United States, on the campuses of the land grant universities, and in Washington, D.C.

The CES conducts a wide variety of programs in the areas of agricultural and natural resource management, home economics, community improvement, and youth development (the "4-H" program). Much of the advice on agricultural management is, of course, intended for farmers and consists of information on how to grow, market, process, and use farm products. Information is also provided to farmer cooperatives, wholesalers and transporters of farm goods, and professionals such as veterinarians, vo-ag teachers, and bankers.

The professional staff of CES numbers about 17,000 people. Approximately 40 percent of their time is devoted to programs on agricultural and natural resource management. The overall budget is about $385 million (FY 1973, excluding $1.5 million for the District of Columbia). About 42 percent of the total budget comes from federal sources, and the remainder comes largely from state and county tax funds. A small portion comes from nontax sources (<2 percent).

The Secretary of Agriculture formalized the USDA's policy on pest management in 1973: "Non-chemical methods of pest control, biological or cultural, will be used and recommended whenever such methods are economically feasible and effective for control or elimination of pests. When non-chemical control methods are not tenable, integrated control systems utilizing both chemical and non-chemical techniques will be used and recommended in the interest of maximum effectiveness and safety (USDA 1973). Prior to this formal statement, the CES had begun the development of new pest management programs designed in accordance with the Secretary's policy. The Director of these programs for the federal government has stated that the goal is the establishment of "multiple and alternate choice systems of pest control that are effective, economical, and environmentally sound" (Good 1973).

Tobacco in North Carolina and cotton in Arizona formed the bases for the first two pilot pest management programs in 1971. The effort has subsequently been expanded and now includes 39 projects in 29 states on 19 crops. Target areas for the programs vary from 10-acre blocks in orchards involving 20 apple growers in Pennsylvania to 224,900 acres of cotton on 901 farms in Arkansas. A summary of the programs planned for 1973 is shown in Table 17. No major changes were planned for 1974.

When a program is initiated, it is simple: Pest and beneficial insects are monitored in each field and

TABLE 17 Status of Pest Management Projects of the Cooperative Extension Service (FY 1973)[a]

State	Crop(s)	Pest(s)	Federal Funds ($)
Alabama	Cotton	Insects	75,510
Arizona	Cotton	Insects	63,750
Arkansas	Cotton	Insects	50,400
California	Cotton	Insects	31,250
	Pears	Insects and diseases	80,000
Delaware	Peppers, potatoes	Insects and diseases	36,000
Florida	Citrus	Insects, diseases, weeds, nematodes	75,000
Georgia	Cotton	Insects	54,959
Idaho	Potatoes	Insects	60,000
Illinois	Corn	Insects, weeds	70,000
Indiana	Alfalfa	Insects	50,000
	Corn	Insects, weeds	25,000
Iowa	Corn	Insects, weeds	35,000
Kansas	Grain sorghum	Insects, weeds	39,250
Louisiana	Cotton	Insects	50,000
Maryland	Sweetcorn, beans	Insects	60,000
Michigan	Deciduous fruits	Insects, diseases	100,000
Mississippi	Cotton	Insects	63,295
Missouri	Corn	Insects	25,000
	Cotton	Insects	22,500
Nebraska	Corn	Insects and weeds	35,000
	Grain sorghum	Insects and weeds	33,350
New Jersey	Sweetcorn and lettuce	Insects	50,000
New Mexico	Cotton	Insects	20,000

TABLE 17 (continued)

State	Crop(s)	Pest(s)	Federal Funds ($)
New York	Apples	Insects and diseases	62,500
North Carolina	Cotton	Insects	54,900
	Tobacco	Insects	148,000
Ohio	Corn	Insects and weeds	25,000
Oklahoma	Cotton	Insects	59,900
	Grain sorghum	Insects	33,000
	Peanuts	Diseases, nematodes, insects and weeds	12,400
Pennsylvania	Apples	Insects and diseases	62,500
South Carolina	Cotton	Insects	70,000
Tennessee	Cotton	Insects	28,333
Texas	Cotton	Insects	119,303
	Grain sorghum	Insects and weeds	44,400
	Peanuts	Insects	38,400
Washington	Alfalfa (seed)	Insects	30,000
	Apples, pears, and peaches	Insects	90,800
SUBTOTAL COTTON			764,100
SUBTOTAL OTHER CROPS			1,320,600
TOTAL FUNDING			2,084,700

[a] Figures supplied by Dr. J. M. Good, Director of Pest Management Programs, Extension Service, U.S. Department of Agriculture.

recommendations for control measures are given to the grower. The CES hopes to enlarge the scope of the program by including surveillance of weeds, diseases, nematodes, and other pests. Eventually the CES hopes to be able to help growers formulate an optimal cropping system that will take into account pest problems, cropping history, soil and weather conditions, and agronomic practices.

Expert advice in the pest management programs comes from CES professionals working in cooperation with research specialists in the land grant universities and the USDA. Many paraprofessionals in the form of trained scouts conduct the actual surveillance in the fields. Farmers, grower organizations, and other government agencies cooperate with the CES in the establishment of a program in a local area. Expenses are split between the private participants in the program and the public CES. The CES anticipates that a progressively higher fraction of the expenses of the program will be borne by the participating farmers. Federal expenditures for these programs have been relatively small, approximately $2 million for FY 1973.

Perhaps the most interesting implication of the development of the pest management programs is that information delivered through a public channel can lead to a significant improvement in pest control practices. The conclusions that we drew from the Beal study discussed in the previous section--that the public sector was not the major channel for information on pest control for Iowa's farmers--may thus be changed. What is still not clear, however, is the extent to which the pest management programs can bring about real changes in the behavior of pest controllers in the absence of a public subsidy.

INFORMATION FOR URBAN AND SUBURBAN CHEMICAL DEALERS

Urban and suburban people use pesticides for lawns, gardens, ornamental plants, household, and other buildings. Usually they procure their materials from a local store. The amount of pesticide used on lawns and gardens and in households is unknown, but it is estimated that the retail sales of this type of product are now above $300 million per year (What's ahead for fertilizers and pesticides? 1972). Because this is an economically significant portion of yearly pesticide sales and because pesticides marketed this way have a high potential for coming in close contact with many people, we would like to know how much information on the use of these chemicals is available to the buyer. Unfortunately, we have no systematic data on the use of chemicals by households, but we do have some information about the behavior of urban pesticide dealers. Professor Beal and his associates of Iowa State University have performed the only formal study of the urban pesticide dealer, and we quote from their work below (Beal et al. 1969b). The reader should bear in mind that the study area was confined to Iowa.

Researchers interviewed 145 people who were in charge of establishments that sold at least one of the following types of chemical: lawn pesticides, garden pesticides, household pesticides, ornamental pesticides, or rodenticides. This sample included approximately 10 percent of the urban chemical dealers in Iowa selected to be representative of such dealers in the state. Most of the dealers were grocery stores, supermarkets, hardware stores, and drugstores. Sixty percent had gross sales of pesticides of less than $600 per year. Seventy-one percent sold no chemicals for agricultural purposes.

The dealers were asked a series of questions designed to test their knowledge of pesticides and their special knowledge about the particular types of chemicals that were the focus of the study. Approximately 85 percent of the dealers answered seven out of eight general safety questions correctly. The number of correct responses about general government regulations of pesticide use and about pesticide terminology was not quite as high as for the safety questions. Dealers responded correctly to questions of specific knowledge approximately 50 percent of the time. Incorrect answers and no-opinion responses were each about 25 percent. The average percentage of incorrect answers was over 30 percent for questions about household pesticides and rodenticides.

Dealers were about evenly divided as to their perceptions of what the customer expected of them. Forty-eight percent believed that the customer expected them to be either a highly qualified source of information on all uses of chemicals or at least a source of information on some aspects of pesticides and their uses. Forty-five percent indicated that their own role included a responsibility to keep well informed and to make recommendations to customers. Over 50 percent of the dealers, however, said that fewer than 30 percent of their customers ever asked for information or recommendations.

The label was the most important source of information relied upon by a substantial majority of dealers in recommending the best chemical for a job, the methods and rates of application, and the precautions to be observed. Other sources of information, such as newspapers, magazines, chemical company and extension service literature, and so forth, were relied on by only a few dealers. For the majority of dealers, information about new chemicals came from manufacturers or distributors. More than 80 percent of the dealers had attended no educational programs on pesticides in the 18 months preceding the interview. Approximately 6 percent had spent from 8 to 50 hours in such programs.

One cannot draw conclusions applicable to the entire United States from the results of a survey conducted entirely within the boundaries of one state. Nevertheless, we conclude that, at least for one section of the country, the process of selling toxic chemicals to many consumers is conducted in a very unsophisticated way. Most consumers do not ask for information; if they did, half the time they would receive an incorrect answer or an "I don't know." What the dealer does know appears to come from the label of the pesticide container. There is no evidence that the average dealer goes to much effort to keep up his knowledge by attending brush-up courses on the uses of the chemicals he sells.

From these observations, we would conclude that two principles ought to be kept in mind in the approval of registrations of pesticides for the urban consumers. First, the chemicals approved for household use should have a minimum of human toxicity and of persistence, because the skills in application and storage are likely to be very low. Second, the label should be extra clear and easy to follow, because it is likely to be the *only* source of information for both the dealer and the ultimate consumer.

INFORMATION FOR INDUSTRIAL AND GOVERNMENTAL USERS OF PESTICIDES

Pesticides are widely used in nonagricultural industries and government, as is discussed in Chapter 8. This Committee was unable to locate any data on the process of information transfer in industry. The large volume of pesticide consumed by industry makes the lack of data regretful, and we recommend that the EPA initiate studies to remedy the problem.

The federal agencies using pest control technologies coordinate their activities through the Federal Working Group on Pest Management. Information on the proper use of pesticides and other methods is presumably available to all agencies from the USDA and the EPA. Similarly, state and local agencies presumably have ready access to information from county agents, knowledgeable state personnel, and land grant universities. The effectiveness of these channels, however, especially at the local level, is unknown. We recommend that the EPA initiate studies of information flow to government agencies conducting pest control operations.

REFERENCES

Beal, G. M., J. M. Bohlen, G. W. Edwards, and W. A. Fleischman (1969a) Behavior Studies Related to Pesticides: Urban Chemical Pesticides and Iowa Urban Chemical Pesticide Dealers. Ames: Iowa State University, Agricultural and Home Economic Experiment Station. P-Bulletin 140. 65 pp.

Beal, G. M., J. M. Bohlen, and W. A. Fleischman (1969b) Behavior Studies Related to Pesticides: Agricultural Chemicals and Iowa Agricultural Chemical Dealers. Ames: Iowa State University, Agricultural and Home Economics Experiment Station. P-Bulletin 139. 80 pp.

Beal, G. M., J. M. Bohlen, and H. G. Lingren (1966) Behavior Studies Related to Pesticides: Agricultural Chemicals and Iowa Farmers. Ames: Iowa State University, Agricultural and Home Economics Experiment Station. Special Report No. 49. 24 pp.

Good, J. M. (1973) Pilot programs for integrated pest management in the United States. Paper presented at US-USSR Pest Management Conference, September 10-18, Kiev, USSR.

U.S. Department of Agriculture (1973) Secretary's memorandum no. 1799, February 1.

What's ahead for fertilizers and pesticides? (1972). Farm Chem. Crop Life 135(1):22.

7

THE USES OF PESTICIDES

All analyses of the risks and benefits associated with pest control practices must ultimately rely on a knowledge of use patterns. It is ironic that the importance of these use patterns is matched by such a paucity of information about them. We will describe in this section what can, is, and ought to be known about use patterns. Most of the discussion will center on the uses of pesticides, but there will be some coverage of other control methodologies. The other four volumes in this series give more details on use patterns for particular types of pest problems.

THE MEASUREMENT OF USE PATTERNS

Gathering information about use patterns presents severe difficulties, whatever the purpose for which it is collected. Use, in contrast to production, is a diffuse activity. Whereas fewer than 100 firms engage in the production of pesticidal chemicals, their use is in the hands of literally millions of decision makers: farmers, homeowners, industrial personnel, and civil servants of the federal, state, and local governments. Use patterns are not surrounded by the kind of secrecy that is imposed on the production of pest control materials; but the very number of people involved magnifies the task of data collection. Furthermore, the Federal Insecticide, Fungicide, and Rodenticide Act, as amended (P.L. 92-516), makes no requirement or provision for the reporting of pesticide use. The only regulation on use is that it is to be in strict accordance with directions on the label. Production figures, as mentioned earlier, must be reported to the EPA.

Domestic use estimates may be made indirectly or directly, and each method has its own merits and limitations. Indirect estimation involves estimating production quantities, adding imports, and subtracting exports. The result is assumed to be the quantity used. The method does not permit discrimination of who used the material or for what purposes.

A direct determination of use patterns can be achieved by either of two basic methods: surveys of users or reporting by the users. Because of the number of users, surveys are almost certain to be based on sampling. Surveys can be performed by a government agency, private firm under contract with the government, or private firm as a service performed for another private firm. All three modes have been used and will be discussed below. The reporting method has only been used in the state of California. It is conceivable, however, that more complete reporting by users could be developed with the aid of rapid data-processing systems and, for example, return system, utilizing postcards attached to every container of pesticide.

Whatever method is used, heavy expense will attend any direct method of determining use patterns. Surveys require a battery of interviewers--who must be paid--and tolerance on the part of those interviewed (an increasingly scarce resource in all heavily regulated activities). Reporting would not require a staff of surveyors, but it would put the user to some trouble and--in the case of large users--expense. Both methods require backup by data-processing equipment. The expenses involved make it imperative that careful thought go into the data collection: For what purpose is the data wanted? Precisely what information is necessary to fulfill the purposes? What level of error can be tolerated? (The larger the tolerable error, the less costly it will be to obtain the data necessary.) How often is it necessary to collect new data? How soon after collection should the information be available for policymaking and for general public discussion? (Rapid processing of data will be more expensive, but data not made available for years after its collection may be useless, or worse than useless, if it leads to faulty policy decisions.)

A number of efforts have been made to generate information on the uses of pesticides. The following paragraphs describe the nature, strengths, and weaknesses of some such studies. In the next section, we summarize the picture of present use patterns that emerges from these studies

Surveys by the Department of Agriculture

The Economic Research Service, in cooperation with the Statistical Reporting Service, has conducted surveys of farmers' uses of pesticides for the crop years 1964, 1966, and 1971. Publication of the data from the first two reports did not occur until 4 years after the actual use--1968 and 1970, respectively. Publication of the results continues to lag years behind the actual use. These irregularities have meant, for example, that in 1973 the most recent use data available for policy decision making were from 7 years before!

Each of the surveys involved the selection of a sample of farmers and an expansion of the use patterns of these farmers into an estimated use pattern for all farmers in the nation. Data were collected so as to allow presentation of use patterns by chemical, crop, livestock protected, geographical region, and size of farm operation. Estimates were made of the total amounts of chemicals used and total number of acres treated, as well as of the percentages of farmers using chemicals and acres receiving treatment. Expenditure estimates for the pesticides were also gathered.*

The major strength of the USDA surveys is that they are produced by a public agency for public distribution and consumption. The openness of the process and its result mean that a variety of people with different perspectives on pesticides have been able to obtain and analyze use patterns. The weaknesses of these reports are (1) their publication has been slow and (2) they cover only the uses of pesticides in agriculture that are paid for directly by farmers, which is only about one-half of the total pesticide used in the United States.

The surveys have, in addition, probably been too infrequent. Development of resistance, shifts in abundance of pest species, introduction of new chemicals, and cancellation of old chemicals have all created a climate of rapid change in use patterns during the past decade. It was unfortunate that no survey was conducted between 1966 and 1971; and given past lags between data collection and

---

*Estimates of farmers' expenditures for pesticides are also gathered by the Census Bureau in the census of agriculture. Expenditures are reported by class of chemical insecticide, herbicide, etc. Our report requires knowledge of specific chemical entities, however, so we have not made any significant use of data from the Census Bureau.

publication, we cannot expect new use data before about 1978 from a 1975 or 1976 survey. While the proper spacing of surveys is debatable, surveys spaced at 2-3-year intervals would appear desirable, and national surveys at such an interval could be integrated with regional or state surveys of greater frequency.

Surveys by Private Firms for Public Agencies

The dearth of data on use patterns, especially nonfarm uses of pesticides, was probably an important element in motivating the Council on Environmental Quality (CEQ) to let a contract to the Midwest Research Institute for a study on "Production, Distribution, Use, and Environmental Impact of Selected Pesticides," published in 1974 with data for 1972. This report estimates production volumes for many pesticides and estimates the uses of the chemicals in agriculture and industry, by government agencies, and by homeowners. Persons in a wide range of occupations, including pesticide manufacturers, extension agents, government workers, and officials of a commercial pest control operators association, were surveyed; but no farmers were interviewed. Special attention was given to the manufacture, distribution, and use of 25 selected pesticides.

This report is the most comprehensive summation of national use patterns now available. Its strength lies in its thoroughness and openness to public scrutiny. Although the Committee was unable to review it critically, we point out that the necessity for the CEQ to contract for an ad hoc private use survey that includes use by federal agencies indicates a serious gap in the federal management of pest control activities. Knowledge of use patterns is sufficiently important to justify some type of federal data gathering on a regular and continuing basis by federal employees.

Surveys by Private Firms for Private Firms

As mentioned earlier, the Doane Agricultural Service of St. Louis, Missouri, surveyed farmers on an annual basis for their use patterns of pesticidal materials. Doane's surveys are confined to herbicides and insecticides used on a limited number of crops and have been made yearly from 1959 to the present. Herbicide use has been surveyed for corn, soybeans, wheat, sorghum, cotton, and pasture, and insecticides for corn, tobacco, alfalfa, sorghum, and soybeans. These reports are apparently useful to pesticide manufacturers, but--as was mentioned earlier--they are proprietary in nature, and Doane is unwilling to jeopardize its business position by releasing to the public any reports more recent than 1963 (Doane Agricultural Service, personal communication, 1973).*

The strength of the Doane reports lies in their frequency and in the rapidity with which they are produced after the actual use has occurred: considerably less than 1 year. Their weakness, of course, lies in their inaccessibility to the general public and to policymakers.

Use Patterns from Mandatory Reporting of Applications

The state of California began a mandatory reporting system for certain types of pesticide application in 1970, and its experience should be valuable should other states decide to adopt a reporting system to gather use data.

The reporting system requires the following classes of pest controllers to submit reports on their activities: licensed agricultural pest control operators and structural pest control operators, the California Division of Highways, the California Department of Water Resources, vector control agencies, state and county agricultural departments, the University of California, county road departments, irrigation districts, U.S. government agencies, reclamation districts, city and county parks, and various school districts. Private growers applying defined "injurious" materials for which a use permit is required also must report their activities. The California system thus does not register every pesticide application. For example, farmers applying "non-injurious" materials or homeowners using pesticides would not have to report these uses. Despite the omissions, the California system must record a very significant portion of the material applied. In 1973, the use of a total of 183.7 million pounds of pesticides, including petroleum oils and distillates, was recorded. The data are published by crop and by chemical.

The strength of the California system lies in its openness, thoroughness, and continuity, and in the rapidity of its publication--about 3 months from end of year to report for the year. Quarterly reports are also issued throughout the year. The only immediately apparent weakness of

*We are most appreciative that Doane has made such information as it could available to this Committee.

the system is that some uses need not be reported. The importance of the omissions is mitigated, however, by the fact that some of the compounds of environmental and healt concern are classified as "injurious" and thus must be reported by private growers.

CONTEMPORARY USES OF PESTICIDES AND OTHER PEST CONTROL TECHNIQUES*

The latest years for which any comprehensive use data are available are 1971 and 1972. The USDA estimated that in 1971 833 million pounds of pesticides were consumed in the United States. The Midwest Research Institute, in its study for the CEQ, estimated that 976 million pounds of material were consumed in 1972. The details of these esti mates by classes of pesticides are shown in Table 18.

The two estimates differ by over 140 million pounds, but it is difficult to attach any significance to the difference. The data are for different years and are not collected in a way that allows statistical estimates of reliability. We will adopt the approximation that annual use in the United States lies between 800 million and 1 billion pounds.

The use of pesticides is complicated because a number of variables affect our understanding and evaluation of use: Which chemical? Which pest? Who applies? For what purpose? Where? When? How often? The ideal presen tation of use data for a cost and benefit study would utilize purpose of use as the major organizing device. All data currently available, such as the USDA surveys and the Midwest Research Institute report mentioned above, utilize the applicator as the primary organizing theme. In the summary of contemporary use patterns that follows, we will use the applicator as the organizing theme; but, wherever possible, we will refer to the purpose of the application. (Data taken largely from Andrilenas 1974.)

Use in Agriculture by Private Farmers in 1971

American farmers purchase fungicides, insecticides, herbicides, rodenticides, defoliants and desiccants, miticides,

*The emphasis of this section will be on the use of pesticides. Data on selected uses of nonpesticidal control operations will also be presented, however.

TABLE 18 Estimates of Domestic Consumption of Pesticides

Class of Pesticide	Million Pounds Used 1971[a]	1972[b]
Fungicides	155[c]	118.9[d]
Organic	--	64.2
Inorganic	--	54.7
Herbicides	359[e]	427.7[f]
Organic	--	387.1
Inorganic	--	40.6
Insecticides	319[g]	278.8[h]
Organic	--	273.4
Inorganic	--	5.4
Fumigants and soil-conditioners	--	82.1
Penta- and tri-chlorophenols	--	68.5
TOTAL PESTICIDE	833	976.0

[a]Andrilenas (1974).
[b]Lawless et al. (1974).
[c]Excludes sulfur.
[d]Excludes sulfur, creosote, coulter, and penta- and trichlorophenols.
[e]Includes plant hormones, defoliants, and desiccants. Excludes petroleum.
[f]Includes plant growth regulators, defoliants, desiccants, and chlorates.
[g]Includes fumigants, rodenticides, and miticides. Excludes petroleum.
[h]Includes nematicides, rodenticides, miticides, and repellents.

growth regulators, and other chemicals to use in their production systems and to maintain their farmsteads. Despite uncertainties about the data, use figures suggest that agricultural applications account for only about 60 percent of all domestic applications. Table 19 shows the amounts of each type of material purchased in 1966 and in 1971 and the percentage of the total represented by each.

TABLE 19 Quantities of Pesticides Used by Farmers, 1966 and 1971[a]

Type of Pesticides	1966 Million Pounds	1966 % of Total[b]	1971 Million Pounds	1971 % of Total[b]
Fungicides				
Inorganic[c]	7.6	2	16.0	3
Organic	25.6	7	25.7	5
TOTAL[f]	33.2	9	41.7	8
Herbicides[d]				
Inorganic	4.9	1	1.8	[g]
Organic	110.4	31	226.0	46
TOTAL[f]	115.3	33	227.9	46
Insecticides[d]				
Inorganic	5.8	2	3.2	1
Synthetic organic	142.9	40	166.3	34
Other organic	0.2	[g]	0.2	[g]
TOTAL[f]	148.9	42	169.8	34
Miscellaneous				
Miticides	2.1	1	2.0	[g]
Fumigants	36.8	10	29.0	6
Defoliants and desiccants	6.1	2	17.7	4
Rodenticides	[g]	[g]	[g]	[g]
Growth regulators	3.3	1	5.6	1
Repellents	7.5	2	0.4	[g]
TOTAL[f]	55.8	16	54.7	11
TOTAL PESTICIDES[f]	353.2	100	494.1	100
Sulfur	57.1	--	112.5	--
Petroleum[e]	92.1	--	221.5	--
GRAND TOTAL[f]	502.5	--	828.1	--

[a]Andrilenas (1974).
[b]Percent of total calculated excluding sulfur and petroleum
[c]Excludes sulfur.
[d]Excludes petroleum.
[e]Used primarily in insecticidal and herbicidal preparations
[f]Numbers may not add to totals due to rounding.
[g]Less than 50,000 pounds or 0.5 percent.

The most marked change between 1966 and 1971 is the apparent increase in the relative importance of herbicides. Weed killers comprised 33 percent of the total poundage of pesticides used by farmers in 1966, but in 1971 they were 46 percent of the total. Data from the USDA use surveys cannot be interpreted as indicating trends, but production data analyzed elsewhere indicates that the increase in herbicide usage is real (Fowler and Mahan 1973). Over the past decade, the most startling and dramatic change has been the continued rise in the use of herbicides compared to other pesticides. Herbicides have largely substituted for hand labor and machine cultivation in a large number of crops.

Use of Fungicides

Sulfur is the fungicide used in the largest quantities by American farmers: 113 million pounds in 1971. Approximately three times as much sulfur is used compared to all other inorganic and organic fungicides. Approximately 38 percent of the nonsulfurous fungicides were inorganic compounds and 62 percent were synthetic organic materials. Of the organic fungicides, the dithiocarbamates, compounds containing such heavy metals as iron (ferbam), zinc (zineb), or manganese (maneb) accounted for about one-half of the pounds used; the other class of organic fungicide used in large amounts was the phthalimide group, especially captan and the guinones, such as chloranil and dichlone. Table 20 presents a summary of the fungicides used by farmers in 1971.

Most fungicides are used on fruits and vegetables. The production of citrus, for example, required 9.3 million pounds of nonsulfurous fungicide in 1971, or 24 percent of the total nonsulfurs. Apples, other deciduous fruits, vegetables, peanuts, and potatoes are other major consumers of fungicides. Relative small amounts of fungicide are applied by farmers to field crops such as cotton, corn, soybeans, wheat, and rice, although treatment of seeds with fungicide seed processors is a well-established practice. This use of fungicide does not appear in the USDA survey of farmers. The amounts involved are fairly small.

Most uses of fungicides occur in the Southeast, Northeast, Pacific states, the Corn Belt, and the Appalachian states. Only small amounts of fungicide were used on the

TABLE 20 Quantities of Selected Fungicides Used by Farmers on Crops, Livestock, and for Other Purposes, United States, 1971[a]

Class of Fungicide	Million Pounds Crops	Livestock	Other	Total	%
Inorganic[b]					
Copper sulfates	7.7	*c*	*c*	7.7	18
Other coppers	2.2	*c*	*c*	2.2	5
Mercury compounds	*c*	0	*c*	*c*	*c*
Other inorganics	6.0	0.1	0	6.0	14
TOTAL	15.9	0.1	*c*	16.0	38
Dithiocarbamates					
Maneb	3.7	0	0.1	3.9	9
Zineb	2.0	0	*c*	2.0	5
Ferbam	0.8	0	0.6	1.4	3
Others	5.4	*c*	0.4	5.8	14
TOTAL	11.9	*c*	1.1	13.0	31
Phthalimides					
Captan	6.0	0	0.5	6.5	16
Others	1.0	0	0	1.0	2
TOTAL	7.0	0	0.5	7.5	18
Dinocap, Dodine, Quinones	1.2	0	*c*	1.2	3
Phenols	0.2	*c*	*c*	0.2	*c*
Other Organics	3.4	0.4	*c*	3.9	9
TOTAL ORGANICS	23.7	0.4	1.6	25.7	62
TOTAL FUNGICIDES[b]	39.6	0.5	1.7	41.7	100
Sulfur	112.0	0.4	*c*	112.5	--
GRAND TOTAL	151.6	0.9	1.7	154.2	--

[a] Adrilenas (1974). Excludes Alaska. Figures may not add to totals due to rounding.
[b] Excludes sulfur.
[c] Less than 50,000 pounds of 0.5 percent.

southern plains, the lake states, the northern plains, the mountain states, and in the Mississippi Delta states.* The geographic distribution of use reflects the high number of acres of fruits and vegetables in the high-use areas and the climatic conditions favorable to fungal infections of crop plants.

Use of Herbicides

Three chemicals--atrazine, 2,4-D, and propachlor--constituted one-half of the herbicides used by farmers in 1971. Other chemicals used in major amounts included alachlor, trifluralin, and amiben. Virtually all (99 percent) herbicides used by farmers are synthetic organic chemicals; only 1.8 million pounds were inorganic chemicals. A summary of the major herbicides used is shown in Table 21.

Three crops (corn, 45 percent; soybeans, 16 percent; and cotton, 9 percent) account for 70 percent of all herbicides used. Wheat, sorghum, pasture and rangeland, and rice were other crops utilizing major quantities of herbicides, each requiring 3 percent or more of the total herbicides used by farmers.

Farmers show decided preferences for particular herbicides on specific crops. Atrazine and propachlor accounted for about 70 percent of the herbicides used on corn. Amiben and alachlor constituted some 40 percent of the herbicides applied for soybean production. Trifluralin and fluometuron comprised more than 33 percent of the materials used in cotton.

The heavy use of herbicides in corn, soybeans, and cotton is reflected in the geographic distribution of herbicide use. More than one-third of the herbicides are applied in the Corn Belt because of the large acreage given there to corn and soybeans. Other areas in which a high proportion of herbicides are applied are the lake states, the northern plains, and the Delta states, each area receiving around 10 percent of the total herbicides used by farmers in 1971.

---

*See Figure 3 for a map of the states belonging to each of the farm production regions. Hawaii is included in the Pacific region, even though it is not shown on the map. Alaska was excluded from the data gathered in the 1971 survey. Pesticide usage in Alaska is likely to be low and the omission is not serious.

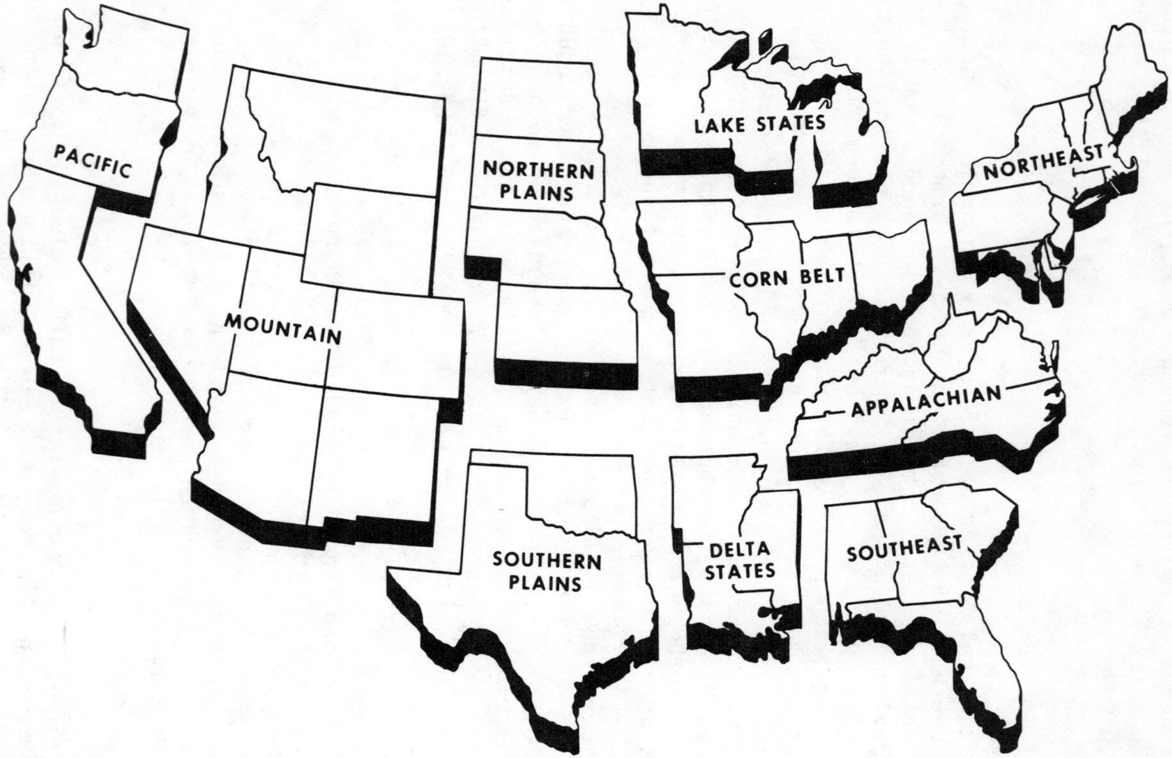

FIGURE 3 Farm production regions. Courtesy, U.S. Department of Agriculture, Neg. ERS 1399a-62(8) Economic Research Service.

TABLE 21 Quantities of Selected Herbicides Used by Farmers, United States, 1971[a]

Class of Herbicide	Million Pounds Crops[b]	Other[c]	Total	%
Inorganic herbicides	1.7	0.2	1.8	1
Arsenicals	7.8	0.1	8.0	4
Phenoxys				
2,4-D	33.3	1.4	34.6	15
2,4,5-T	1.3	0.2	1.5	1
MCPA	3.3	*e*	3.3	1
Other phenoxy	0.6	0	0.6	*e*
TOTAL	38.5	1.6	40.0	18
Phenyl urea				
Diuron	1.2	*e*	1.2	1
Linuron	1.8	0	1.8	1
Fluometuron	3.3	0	3.3	1
Other	0.3	*e*	0.3	*e*
TOTAL	6.6	*e*	6.6	3
Amides				
Propachlor	23.7	*e*	23.7	10
Propanil	6.7	0	6.7	3
Alanap	3.3	0	3.3	1
Alachlor	14.8	0	14.8	6
Other	0.8	0	0.8	*e*
TOTAL	49.3	*e*	49.3	22
Carbamates				
EPTC	4.4	0	4.4	2
Pebulate	1.1	0	1.1	*e*
Vernolate	3.7	*e*	3.7	2
Butylate	5.9	0	5.9	3
Other	3.2	0	3.2	1
TOTAL	18.3	*e*	18.3	8
Dinitro group	7.2	*e*	7.2	3
Triazines				
Atrizine	57.2	0.2	57.4	25
Propazine	3.2	0	3.2	1
Simizine	1.7	*e*	1.7	1
Other	1.4	*e*	1.5	1
TOTAL	63.6	0.3	63.8	28

TABLE 21 (continued)

Class of Herbicide	Million Pounds Crops[b]	Other[c]	Total	%
Benzoics				
Amiben	9.6	0	9.6	4
Dicamba	0.4	*e*	0.4	*e*
Other	0.1	0	0.1	*c*
TOTAL	10.1	*e*	10.1	4
Other Organics				
Trifluralin	11.4	*e*	11.4	5
Nitralin	2.7	0	2.7	1
Dalapon	1.0	*e*	1.0	*c*
Norea	1.3	0	1.3	1
Fluorodifen	1.3	0	1.3	1
Others	4.8	0.1	4.9	2
TOTAL	22.6	0.1	22.7	10
TOTAL ORGANICS[d]	224.0	2.1	226.1	99
TOTAL HERBICIDES[d]	225.7	2.2	227.9	100
Petroleum	142.8	2.8	145.5	--
GRAND TOTAL	368.4	5.1	373.5	--

[a]Taken from Andrilenas (1974). Excludes Alaska. Some materials may be used for purposes other than herbicides. Figures may not add to totals due to rounding errors.
[b]Includes crops, rangeland, pasture, and land in summer fallow.
[c]Includes fence rows, ditch banks, and other noncrop areas.
[d]Excludes petroleum.
[e]Less than 50,000 pounds or 0.5 percent.

Use of Insecticides

The 1971 crop-year survey indicates that farmers used abou 170 million pounds of insecticide in that year. The overwhelming majority (98 percent) of the material consisted of synthetic organic chemicals. Two percent of the insecticides were inorganic materials, and an even smaller fraction was composed of natural organic chemicals such as pyrethrums, rotenones, and so forth. Toxaphene, methyl parathion, carbaryl, DDT, and ethyl parathion comprised more than 60 percent of all insecticides used in 1971. Most registrations of DDT were cancelled as of December 31, 1972, so the use patterns of insecticides have undergone a profound shift since the survey was taken. The cancellation has probably increased the use of toxaphene, methyl parathion, other organophosphorus chemicals, and carbamate insecticides like carbaryl. The magnitude of the substitutuion of other chemicals for DDT is difficult to estimate because of the paucity of use data since 1971. Table 22 summarizes the uses of the most popular insecticides on crops, livestock, and for other farmstead purposes.

The bulk of insecticides used by farmers are applied to crops: 154 million pounds out of 170 million pounds, or more than 90 percent. Two chemical classes, chlorinated hydrocarbons and organophosphates, each accounted for some 60 million pounds of the insecticides used on crops. The carbamate insecticides included an additional 24 million pounds of material, approximately two-thirds of which was carbaryl. The uses of organophosphate and carbamate insecticides increased considerably between 1966 and 1971, from 40 to 71 million pounds and from 13 to 25 million pounds, respectively. Chlorinated hydrocarbon materials declined in use from 89 to 70 million pounds.

The reasons for the decline in use of the chlorinated hydrocarbons are not entirely clear, but the development of resistance in insects attacking corn (corn rootworm), cotton (bollworm and tobacco budworm), and peanuts (leaf hopper and thrips) to DDT and aldrin was undoubtedly of most importance. The amount of DDT used on cotton declined by almost one-third, from 19 million pounds in 1966 to 13 million pounds in 1971. Peanut production utilized over 2 million pounds in 1966 but less than 0.1 million pounds in 1971, a decline of nearly 95 percent. Over 14 million pounds of aldrin was applied to corn in 1966, but in 1971 about 43 percent less was used.

As has been the case for many years, cotton production utilizes the largest amounts of insecticide, 73 million pounds (47 percent) in 1971. Corn production accounts for an additional 17 percent and soybeans 4 percent; the three crops thus total more than two-thirds of the insecticides used by farmers in 1971. Other major uses of insecticides are on apples, tobacco, citrus fruits, and other fruits and vegetables.

Use patterns of insecticides like those of herbicides, show some crop-chemical specificity. For example, toxaphene, DDT, and methyl parathion found their major uses on cotton.

TABLE 22 Quantities of Selected Insecticides Used by Farmers on Crops, Livestock, and for Other Purposes, United States, 1971[a]

	Million Pounds				
Class of Insecticide[b]	Crops[c]	Livestock[d]	Other	Total	%
Inorganics	3.0	0.2	*g*	3.2	2
Botanicals and Biologicals	0.1	0.1	*g*	0.2	*g*
Chlorinated hydrocarbons					
Toxaphene	32.9	4.6	*g*	37.5	22
DDT	14.0	0.2	0.1	14.3	8
Aldrin	7.9	*g*	*g*	7.9	5
Methoxychlor	0.9	2.0	0.1	3.0	2
Chlordane	1.5	0.4	*g*	1.9	1
Endrin	1.4	0	*g*	1.4	1
Heptachlor	1.1	0	0.1	1.2	1
Endosulfan	0.9	0	*g*	0.9	1
Lindane	0.2	0.4	0.1	0.7	*g*
Dieldrin	0.3	*g*	*g*	0.3	*g*
TDE (DDD)	0.2	0	0	0.2	*g*
Strobane	0.2	0	0	0.2	*g*
Others	0.3	*g*	*g*	0.3	*g*
TOTAL	61.9	7.6	0.4	69.9	41
Organophosphates					
Methyl parathion	27.6	0	*g*	27.6	16
Parathion	9.4	0.1	*g*	9.5	6
Phorate	4.2	0	0	4.2	2
Disulfoton	4.0	0	*g*	4.1	2
Malathion	2.7	0.7	0.2	3.6	2
Diazinon	3.1	*g*	*g*	3.2	2
Azinphosmethyl	2.7	*g*	0	2.7	2
Dichlorvos	*g*	2.4	*g*	2.4	1
Ethion	2.3	0	0	2.3	1
Ciodrin	*e*	0.9	*e*	*e*	--
Bidrin	0.8	0	0	0.8	*g*
Trichlorfon	0.6	0.1	*g*	0.6	*g*
Ronnel	*g*	0.5	0	0.5	*g*
Ruelene	*e*	0.2	*e*	*e*	--
Coumaphos	*e*	0.2	*e*	*e*	--
Others	*e*	0.4	*e*	*e*	--
TOTAL	65.0	5.4	0.3	70.7	42

TABLE 22 (continued)

Class of Insecticide[b]	Million Pounds Crops[c]	Livestock[d]	Other	Total	%
Carbamates					
Carbaryl	16.6	1.2	0.1	17.8	10
Metalkamate	3.6	0	0	3.6	2
Carbofuran	2.9	0	0	2.9	2
Methomyl	1.1	0	0	1.1	1
Others	*g*	0	0	*g*	*g*
TOTAL	24.1	1.2	0.1	25.4	15
Other Synthetic Organics	0.1	0.1	*g*	0.3	*g*
TOTAL SYNTHETIC ORGANICS	151.1	14.5	0.7	166.3	98
TOTAL INSECTICIDES[f]	154.3	14.8	0.7	169.8	100
Petroleum	60.7	13.1	0.1	74.0	--
GRAND TOTAL	215.0	27.9	0.8	243.7	--

[a]Taken from Andrilenas (1974). Excludes Alaska. Figures may not add to totals due to rounding.
[b]Some materials may be used for purposes other than insecticides.
[c]Includes crops, pasture, rangeland, and summer fallow.
[d]Includes treatment of livestock and their buidlings.
[e]Separate figures are not presented for nonlivestock uses of ciodrin, ruelene, or coumaphos in Andrilenas (1974). Hence, it is also not possible to calculate "others" or "Total."
[f]Excludes petroleum.
[g]Less than 50,000 pounds or 0.5 percent.

Similarly, aldrin, metalkamate, carbofuran, and phorate were materials of choice in corn production.

The concentrated uses of insecticides on cotton and corn causes decided geographic variation in the uses of the materials. Cotton production in the humid and semi-arid South (the Southeast, the Delta states, and the southern plains) and corn production in the Corn Belt lead to about 70 percent of all insecticides used by farmers being applied in these regions. The production of cotton, fruits, and vegetables causes the Pacific states, especially California, also to be a region of comparatively high use.

The production of livestock utilizes almost 15 million pounds of insecticide, somewhat less than 10 percent of all insecticide used by farmers. The chlorinated hydrocarbons toxaphene and methoxychlor; the organophosphate dichlorvos; and the carbamate carbaryl were the materials most used on animals in 1971. Each was used in amounts greater than 1 million pounds. Other popular materials used included chlordane, lindane, DDT, parathion, malathio ronnel, ruelene, coumaphos, and ciodrin. The last four of these are used almost exclusively on livestock. Other insecticides widely used for livestock production, such as toxaphene, parathion, and carbaryl, have only a minority of their uses on animals.

The maintenance of beef cattle and dairy cattle demande the vast majority (79 percent) of all insecticides used on animals. Poultry, hogs, and sheep production accounted fo most of the remaining livestock insecticide use.

Use of Other Chemicals

Farmers use a variety of chemicals other than fungicides, herbicides, and insecticides to kill or control the growth of biological organisms. Some of these compounds are "pesticides" in the strict sense, in that they kill or destroy populations of unwanted organisms. Others, such as plant growth regulators, do not kill but rather modify the physiology of an organism so that its rate or form of growth is more congruent with man's wishes. Although pesticides and plant growth regulators may be quite distinct physiologically, they are both regulated by the Federal Insecticide, Fungicide, and Rodenticide Act (86 Stat. 978). Table 23 shows the major miscellaneous pesticides used by farmers in 1971.

<u>Fumigants</u> Farmers use fumigants to kill unwanted organisms in soil and in storage buildings holding harvested commodities. D-D mixture (a mix of various chlorinated propanes, propenes, and other compounds with three carbon atoms per molecule) is the most widely used fumigant but is closely followed by telone. Bromomethane and dibromochloropropane are other fumigants used in major quantities by farmers. Producers of Irish potatoes, other vegetables, and tobacco are the biggest consumers of fumigants (84 percent of the total).

TABLE 23 Quantities of Selected Miscellaneous Chemicals Used by Farmers on Crops and for Other Purposes, United States, 1971[a]

Class of Chemical	Million Pounds			
	Crops[b]	Other	Total	%
Fumigants				
D-D mixture	6.8	0.3	7.0	24
Telone	6.4	0.5	6.9	24
Bromomethane	0.8	5.2	5.9	20
Dibromachloropropane	3.6	*d*	3.6	12
Others	3.4	2.1	5.5	19
TOTAL	21.0	8.0	29.0	100
Defoliants and desiccants[c]				
Arsenic acid	6.1	0	6.1	34
DEF and folex	5.1	0	5.1	29
Others	6.6	0	6.6	37
TOTAL	17.7	0	17.7	100
Plant growth regulators				
Maleic hydrazide	4.2	0	4.2	76
Others	1.3	0	1.3	24
TOTAL	5.6	0	5.6	100
Miticides				
Chlorobenzilate	0.8	0	0.8	40
Dicofol	0.4	0	0.4	22
Omite	0.4	0	0.4	21
Aramite	*d*	0	*d*	1
Others	0.3	0	0.3	16
TOTAL	2.0	0	2.0	100
Repellents	*d*	0.4	0.4	100
Rodenticides	*d*	*d*	*d*	100

[a]Taken from Andrilenas (1974). Excludes Alaska.
[b]Includes crops, pasture, rangeland, and land in summer fallow.
[c]Some materials reported as herbicides may be used as defoliants or desiccants; likewise, materials reported as defoliants or desiccants may have been used as herbicides.
[d]Less than 50,000 pounds.

Defoliants and Desiccants Cotton is harvested almost entirely by machines in the United States, and cotton leaves--if mixed in with the lint--cause a loss of fiber quality by green staining. In some cotton-growing areas, farmers merely wait until frost has killed the plant and caused the leaves to dry and fall off. In other areas, the producer applies a defoliant or desiccant to accomplish the same purpose before frost comes. The most commonly used chemicals are arsenic acid, DEF, and folex. The use of defoliants and desiccants in cotton is important to the conduct of other pest control operations in the crop and is discussed in Volume III, *Cotton Pest Control*.

Plant Growth Regulators Some materials have the ability to modify such plant growth patterns as the tendency to produce "suckers" (the unwanted branches produced from roots, lower stems, or leaf axils of plants such as tobacco). Likewise, some growth regulators can influence the setting and maturation of fruits such as apples, peaches, prunes, and so forth. Maleic hydrazide, for example, is a growth regulator widely used in the control of tobacco suckers.

Miticides Mites and insects are both arthropod pests, and many insecticides have killing powers against mites as well as insects. Certain chemicals, however, are especially effective against mites alone. Farmers used 2 million pounds of these compounds in 1971, the most popular of which was chlorobenzilate (40 percent of the total poundage of miticide used). Most miticides were used on cotton and fruit crops. In many cases, the outbreaks of mites on these crops is caused by the destruction of mite predators and parasites caused by the use of insecticides against insect pests.

Use of Pesticides on Seeds, Seedbeds, Stored Crops, and Transplants

Farmers and custom operators working for farmers consumed about 10 million pounds of fungicides, herbicides, insecticides, and fumigants to treat seeds, buildings, stored crops, seedbeds, and transplants in 1971. (This figure does not include materials applied to seeds before purchase.) Fumigants and organic fungicides applied to control pests in tobacco seedbeds and on tobacco transplants accounted for about 5.2 million of these pounds.

Pesticide Use by Private Firms and Organizations Other Than Farms

A wide range of private commercial, industrial, and institutional firms use a complex variety of pesticides in the direct conduct of their business and in the maintenance of their buildings, grounds, equipment, and other facilities. Data on the use of pesticides by these establishments is even more difficult to compile than it is for agricultural firms. First, no federal agency comparable to the USDA monitors the pesticides so used. Second, the pesticides used by many nonfarm firms are at most a minor part of their expenses. Thus national and regional trade associations representing these firms are unlikely to have information on the uses made of the chemicals by their members; and even the costly and time-consuming process of surveying is not guaranteed to yield very detailed data, because often the firms themselves do not keep precise records.

The absence of good data on the use of pesticides by nonfarm firms does not mean that the use is not "essential" in that its sudden prohibition would cause unmanageable alterations in the way the firms conduct their business. The estimation of costs and benefits of pesticides and other pest control technologies in nonfarm firms is thus in need of careful attention; but the paucity of data makes the process a difficult one.

Three major classes of firms using pesticides may be distinguished: firms providing pesticides as a service or retail product, firms producing a product requiring protection from pests, and firms providing a product that is aided by pesticides.

Firms Providing Pesticides as a Service or a Retail Product

The provision of pesticides as a service or as a retail product is a specialized industry with numerous subspecialties. The uses of pesticides by these firms are performed on a contract basis to other firms and organizations in agriculture and other types of commercial activities. Services are also sold to private homeowners, nonprofit firms, and government agencies.

It is difficult to analyze "costs and benefits" for these uses of pesticides. At the simplest level, the costs and benefits of the uses from the viewpoint of the service firm can be measured simply by examining its financial statements; but from the viewpoint of the purchasing firms, matters are more complex. For example, a farm firm that purchases the services of a custom applicator presumably believes that the expense of the applied pesticides will be more than returned in higher yields, less variable yields, or lessened expenditures for some other input such as labor. The measurement of the returns actually received is a similar problem whether the farmer applies the pesticide himself or hires a custom applicator. The most complex issues of pesticides applied as a service, of course, is the question of whether the society as a whole benefits more than it loses from the activity. No general answer to this latter question can be given because of the immense heterogeneity of types of custom application made. The appropriate measurement of general costs and returns must be made on a case-by-case basis.

Pest Control Operators (PCOs) "PCO" refers to firms engaged in selling their services to homeowners, commercial firms, industrial establishments, and nonprofit institutions, including governments.

There are two major types of PCOs: those that work in and around buildings (interior PCOs) and those that work on trees and lawns (exterior PCOs). The former category contains about 11,000 firms and the latter about 15,000, with some firms falling in both (Lawless et al. 1974). Interior PCOs control a wide variety of pests, including insects, rodents, and birds. These firms performed an estimated \$600 million worth of business in 1973, of which \$30 million involved direct expenditure for pesticides (Lawless et al. 1974). The major targets of exterior PCOs are insects, various types of vertebrates, and weeds.

The National Pest Control Association--the trade association for the interior PCOs--conducted a survey of the relative importance of different organisms controlled by its member firms in 1965. The results are reproduced in Table 24. (A rating of 100 reflects major importance of the pest to every member answering; 50 indicates the pest is of minor importance to every firm or major importance to one-half the firms.) The German roach, house mouse, Norway rat, and subterranean termite are the most important pests. The major chemicals used against roaches, termites, and other insects include diazinon, chlordane, dichlorvos, dieldrin, aldrin, and heptachlor. As much as 40 percent of the business of interior PCOs is for termite control, and this activity alone utilizes an estimated 800,000 pounds, 70 percent of which is chlordane (National Pest Control Association, personal communication, 1972).

TABLE 24 Top 30 Pests Controlled by PCOs

	Ranking in 1965 NPCA Survey	Importance Rating (100 maximum)
Household insect pests		
German roach	1	96
House ants	5	76
American roach	6	66
Oriental roach	8	65
Fleas	9	64
Brown dog tick	10	61
Silverfish	11	60
Brown-banded roach	12	59
Housefly	15	53
Rodents		
House mouse	2	86
Norway rat	3	85
Birds		
Pigeon	16	50
English sparrow	24	39
Wood-destroying insects		
Subterranean termite	4	80
Carpenter ant	7	66
Powder post beetle	13	55
Outdoor pests		
Wasp and hornet	14	55
Clovermite	18	48
Bee	19	47
Cricket	23	41
Lawn ant	27	37
Sow bug	30	35
Fabric pests		
Black carpet beetle	17	50
Common carpet beetle	21	45
Webbing clothes moth	25	39
Casemaking clothes moth	26	38
Pantry pests		
Saw-toothed grain beetle	20	45
Confused flour beetle	22	45
Indian meal moth	29	36

The major chemical for the control of house mice used to be 50 percent DDT dust. Anticoagulants such as warfarin pindone, diphacinone, and fumarin form the backbone of rat control programs of the interior PCOs; other compounds that have been used include red squill, zinc phosphide, sodium fluoroacetate (1080), strychnine, thallium sulfate, and others.

Exterior PCOs encounter a different spectrum of pests in their work. Crabgrass, a variety of broad-leaf weeds, soil and tree insects, and rodents such as gophers are included. No national estimates of the major chemicals used are available, but dacthal, 2,4-D, chlordane, diazinon chlorpyrifos, and carbaryl are undoubtedly among the most prominent (Lawless et al. 1974).

Custom Applicators "Custom applicators" generally refer to firms specializing in selling their services to farm firms. Farmers spent an estimated $430 million in 1971 for such services, almost all for crop-protection purposes. Approximately two-thirds ($284 million) of this amount was for the chemicals themselves, while the remainder was for the service. Most of the material was applied from fixed-wing aircraft (62 percent), but ground rigs (35 percent) and helicopters (2 percent) were also used (USDA, personal communication of preliminary results from the 1971 crop-year survey of pesticide use by farmers, 1974).

Although some farmers own and operate their own aircraft for spray operations, most (90 percent) are operated by custom applicators. The National Agricultural Aviation Association estimates that aircraft applied about 65 percent of all pesticides used by U.S. farmers in 1971. In that year, more than 6,000 aircraft were in operation, slightly fewer than in 1970. The pilots flew about 125 million miles and treated an aggregate of approximately 180 million acres. The estimated growth rate in sales has been about 12-14 percent recently. The cost to farmers of aerial application services, excluding chemicals, was $125 million in 1971 (Fowler and Mahan 1973; National Agricultural Aviation Association, personal communication, 1974

Aerial application of pesticides has a high accident rate that constitutes a loss that must be subtracted from the benefits gained by the activity. Furthermore, drift of pesticides from aerial application is more severe than from any other type of application; the resulting environmental costs must also be subtracted from the benefits. Finally, significant health hazards to the loaders of the aircraft have been documented (see Chapter 5).

This committee could not make a thorough study of all the costs and benefits of aerial application, but it recommends that such be done. It is disturbing that there is a rapidly growing industry with a number of documented hazards not yet thoroughly evaluated for their costs to themselves and to the general society. The existence of new technologies with the potential to reduce two of the hazards--drift and danger to the loaders--makes a thorough study even more appropriate.

Janitorial Supply Houses These firms are in essence retail outlets for pesticides and disinfectants analogous to the farm chemical supply dealers; the former sell to nonfarm firms and the latter to farm firms. In both cases, the firms may sell a wide variety of materials in addition to the pesticides. There are approximately 5,500 janitorial supply houses in the United States. Some of the materials they sell in large quantities are pyrethrin, diazinon, malathion, ronnel, fly baits, and rat and mouse baits. The dollar volume of pesticide sales from these supply houses may be greater than those of the PCOs, but the poundage of materials sold is probably less, because they are sold ready-to-use, with inactive diluent already added. Some janitorial houses offer application services on a limited basis (Lawless et al. 1974).

Firms Producing a Product Requiring Direct Protection from Pests

A wide range of industries produce and market a product subject to direct attack by pest organisms. Accordingly, these companies use pesticides as an integral step in their manufacturing process to protect their final products. Insects and fungi are the most common pests, but rodents, weeds, and other classes of organisms are important to some production schemes.

The motivation for these uses of pesticides is quite similar to the uses employed by farmers: increase yield, increase quality of final product, stabilize production, or substitute for some other input such as labor. An estimation of costs and benefits from such uses would accordingly require case study examinations. This Committee did not perform any such analysis of these industrial uses of pesticides, but we present below a summary of the more important cases.

Wood Processing and Manufacturing The production of wood fiber from commerical forests is itself an industrial process utilizing pesticides. Once trees are harvested, a new complex of pest organisms, including insects and fungi, threaten the timber products. The specific nature of the pest problem depends upon whether the wood is destined for use as whole or nearly whole logs (such as utility poles or railroad crossties), lumber, or pulp and paper manufacturing.

A total of about 14 billion cubic feet of wood are removed from the American forests each year (Fowler, D. L. personal communication, 1974). A very small fraction of this total (2 percent) is treated with preservatives. The value of the treated wood is estimated at about $450 million per year. The stake society has in the success of the preservation process, however, may be much higher than would be suggested by the volume and dollar value of the treated product. More than 50 percent of the treated wood is destined for service as railroad ties and utility poles both of which play crucial roles in our transportation and communications systems. Between 20 and 25 percent of the wood is intended for use as lumber or timbers and thus will play an important role in building programs.

One pesticide--creosote, a mixture of substances distilled from coal tar--plays a preeminent role in the wood preservation industry. The Midwest Research Institute estimates that 970 million pounds of creosote were used as wood preservative in 1972. This is an amount that in pounds is apprximately equal to all other pesticides combined. Almost all creosote produced is used as a wood preservative, but creosote's use as a fuel or as a reducing agent in iron and steel production could increase in the years to come. Creosote's major competitor is pentachlorophenol, 36-38 million pounds of which were used in 1972 by wood preservers.

A variety of other chemicals are used in smaller amounts as wood preservatives: fluorochrome arsenate phenol (1.9 million pounds), chromated zinc chloride (1.8 million pounds), acid copper chromate (1.2 million pounds), and chromated copper arsenate (9.7 million pounds) (Fowler, D. L., personal communication, 1974).

The vast majority of wood treated is southern pine and Douglas fir. The location of wood-preserving plants reflects the geographic distribution of these tree species, and most plants are located in the Southeast and the Pacific Northwest. Approximately 400 companies are in the industry; most are small and privately owned, but the four top producers accounting for about 35 percent of the product are owned by large corporations (Lawless et al. 1974).

Fungal sap stains constitute a special problem in lumber manufacturing. The organisms produce a blue or brown stain that results in loss of grade (and hence value) of the lumber. The wood is dipped or sprayed with phenyl mercury compounds as a control measure. The value of the treated lumber is approximately $1 billion, and elimination of the mercury compounds would result in losses of $150 million, according to industry sources (National Forests Products Association, personal communication, 1972).

The production of paper and other wood fiber products utilizes somewhat over one-third of the harvested timber in the United States. The wood is pulped in the process under conditions favorable to the growth of slimes consisting of fungi, bacteria, or algae. Pesticides containing mercury were at one time commonly used as slimicides in paper manufacturing. For example, in the later 1950s and early 1960s, between 5 and 10 percent of all mercury produced in the United States was used in pesticides for paper and pulp mills: between 200 and 300 thousand pounds per year. Mercury use for this purpose declined steadily during the late 1960s and in 1972 only 76 pounds of mercury were so utilized (Fowler, D. L., personal communication, 1974). Such compounds as methylene-bis-thiocyanate and organo-bromine compounds have replaced the mercurial slimicides. Approximately $10 million is spent on slime control annually (Institute of Paper Chemistry, personal communication, 1972).

<u>Paint</u> The functions of paint are to protect and to improve the appearance of material to which it is applied. Paint itself, however, is subject to attack and destruction, primarily by various fungi and bacteria. Fungicides, particularly mercury salts, are therefore frequently added to paint to prevent this deterioration. Phenylmercuric acetate for latex-water-based paints, phenylmercuric oleate for oil-based paint, and phenylmercuric propionate for oil- and water-based paints are in wide usage. Two new mercurials--chloromethoxypropyl mercuric acetate and O-hydroxymercuribenzoic acid, cyclic anhydride--and copper compounds, calcium carbonate, and barium metaborate are also commonly used. Interior paints usually do not use mercury because of its toxicity. Salicylanilide, tributylin hydroxide, and tributyltin acetate are substituted because of their relatively low toxicities. Mercurials, copper compounds, and organotin compounds are used in marine

paints to prevent fouling of ships and fixed marine installations by barnacles, teredo, and other marine organisms (Lawless et al. 1974).

The importance of mercury in paints is reflected in recent statistics of mercury usage. Four million pounds of mercury were used in the United States in 1972. A substantial fraction of the metal (19 percent) was used to manufacture pesticides. Approximately 625,000 pounds, or 82 percent of the mercury used in pesticides, was used in paints (Fowler, D. L., personal communication, 1974).

Food Storage, Processing, and Manufacturing Storage of foods from harvest to harvest is important in every region of the world, especially in those regions having a significant cold or dry part of the year that separates harvests by intervals approaching 12 months. The food must be protected from pests, especially insects, fungi, and rodents, to maintain its availability and suitability for human consumption. The storage problem is compounded in highly industrialized countries by extensive operations of food processing and manufacturing. Food must not only be protected during the interval between harvests; it also must be transported, often more than once, between the farm, various manufacturers, and the ultimate consumer. Pesticides are an important element of the protective system used in the United States on stored foods and during manufacturing processes.

Most of the pesticides applied to stored food are insecticides, some of which act as fumigants. The more widely used compounds include aluminum phosphide (releases phosphine as a fumigant), methyl bromide, carbon disulfide, chloroform, carbon tetrachloride, malathion, dursban, diazinon, pyrethrins, ronnel, fenthion, and propoxur. Brewers have a special problem preventing the growth of undesirable yeasts in their equipment and use a fluorosilicate to curb such infestations. Many food processors contract their pest control work to commercial PCOs (Lawless, et al. 1974).

Cotton and Other Textiles Just as wood fibers are subjected to pest attack after timber is harvested, so cotton fibers are subject to biodegradation by pests after harvest. Cotton products exposed to the elements (tennis nets, fishing lines, sails, tobacco shade cloth, tarpaulin awnings, and tents) are particularly subjected to attack, as are products in special uses (diapers, filter cloths, etc.).

The United States consumed an estimated 400 million pounds of cotton in the above products in 1970, and approximately 4 million pounds of fungicide were used to protect them. Copper naphthenate, other copper compounds, mercurials, tributyltin oxide, etc., are the most commonly used protectants (Lawless et al. 1974).

Natural fibers such as hemp and manila are also protected with fungicides. Certain chlorinated hydrocarbon insecticides, especially dieldrin, have been used to protect fabrics and carpets from moths and other insects.

Water Water is a commodity produced by a wide range of private and public organizations for a variety of uses. Institutions and organizations producing some other product may procure and store large amounts of water. Water is subject to direct degradation in quality by unwanted organisms such as weeds, algae, fungi, and bacteria may become a breeding ground for noxious organisms such as mosquitoes and other insects. Herbicides, algicides, fungicides, and insecticides are applied to water during storage or transit to reduce the levels of all of these pests.

Sodium pentachlorophenate is used at an annual rate of about 3 million pounds to control algae and fungi in industrial processing waters and in cooling towers. Bis (tributyl tin) oxide is also useful for this purpose. Weeds can clog ditches with running water and reservoirs with standing water. A variety of herbicides are used to control these plants, but 2,4-D, dalapon, ammate, MSMA, amino triazole, and silvex are the most common, each being used in amounts over 200,00 pounds per year. A total of 3.4 million pounds of herbicide is applied on about 682,000 acres per year. The control of mosquito larvae and adults on industrial waters depends upon insecticide usage. The more commonly used materials include No. 2 fuel oil, pyrethrins, biothion, BHC, chlordane, heptachlor, dieldrin, malathion, and naled (Lawless et al. 1974).

Nurseries The production of ornamental crops in nurseries and greenhouses is really a highly specialized branch of agriculture. Because the land area occupied by this activity is not very high, however, it is better considered in the section on nonagricultural industries.

The only available estimates of pesticide use on ornamental crops concern herbicides. The USDA estimates that about 43 percent of the total ornamental acreage was treated with herbicides in 1968, an area of about 89,000 acres. Approximately $2 million was spent for this activity. No data are available on which herbicides were used (USDA 1972). Nor were data available on the use of insecticides or other pesticides on ornamentals.

Firms Producing a Product That Requires Ancillary Use of Pesticides

The characteristic common to all of the industries in the previous category was that their product required direct protection from pests. In this final grouping, we turn to firms that utilize pesticides in a variety of ways, but only as tools in their efforts to produce something; the final product itself is not particularly vulnerable to pest attack. The characteristic common to all of these latter industries is their need to control the number and kinds of organisms inhabiting a particular space in which their product resides or through which it must move. In most cases, the pest involved will be a weed of some sort; occasionally an insect, vertebrate, or other type of pest will be important.

Rights-of-Way: Transportation Vegetation management is extremely important along transportation corridors for railroads, automobiles and trucks, airplanes, and pipelines. Both vegetation management and control of vertebrates such as birds is important in and around airfields.

Weeds along railroad tracks can constitute a fire hazard shorten the life of railroad ties, reduce traction for braking, reduce drainage, and foul the ballast (a strip 12-16 feet wide made of coarse, porous cinders or gravel). Railroad companies treat an estimated 1.6 million acres with approximately 20 million pounds of chemicals each year. The most commonly used materials are salts of chlorate and borate (12.7 million pounds), 2,4-D (1.6 million pounds), MSMA (1.3 million pounds), and bromacil (1.2 million pounds) (Lawless et al. 1974).

Herbicides are commonly used as a part of highway maintenance programs and sometimes in highway construction to maintain line-of-sight conditions, prevent weeds from emerging through newly laid pavement, and to eliminate broadleaf plants in favor of grasses for aesthetic reasons. The Midwest Research Institute estimates that approximately 17 million pounds of chemicals are used on 6.1 million acres per year. The most frequently used materials are TCA, MSMA, salts of chlorate and borate, ammate, 2,4-D, and 2,4,5-T, each of which is used in quantities of over 1 million pounds per year (Lawless et al. 1974).

Birds on or adjacent to airports can cause a serious hazard to aircraft. Weeds, insects, and rodents may also create a problem in and around airport buildings. No statistics giving details of pest control operations were available, however.

Rights-of-Way: Utilities The transmission of electric power and of telephone-telegraph communications depends in part upon open corridors for wires, poles, and other devices. It is especially important that vegetation be prevented from growing into or falling upon the transmission lines. Furthermore, the corridor must be kept clear enough for easy access to the lines when they need repair. An estimated 900,000 acres are treated with 5 million pounds of herbicide each year. Approximately 70 percent of the chemicals used are 2,4-D, 2,4,5-T, or a mixture of the two (Lawless et al. 1974).

Open Space Open ground--generally covered with turf--is necessary for many of the uses to which parks, athletic fields, golf courses, decorative lawns, and cemeteries are put. Vegetation management is the key to creating and maintaining the desired qualities of these types of spaces. The USDA (1972) estimates that there are 20 million acres of turf across the nation, of which 3.8 million acres (19 percent) were treated with herbicides in 1968. The amount spent was $113 million.

Pesticide Use and Other Pest Control Activities by Governments

Federal, state, and local governments, as well as such quasi-governmental agencies as mosquito and pest control districts established for particular crops, are active in controlling pests with pesticides and with nonchemical methods. The purposes of these operations are diverse, and, accordingly, an evaluation of their costs and benefits must be made on a case-by-case examination. This section will present a basic outline of the different reasons for which programs are conducted and summarize the magnitude of the efforts.

Programs Aiding Agriculture and Forestry Two agencies of the USDA--the Animal and Plant Health Inspection Service (APHIS) and the Forest Service (FS)--are responsible for major programs of pest control at the federal level. The programs of APHIS can be regarded partly as a subsidy to the pest control operations of private farmers in the sense that they are aimed at agricultural pests, usually on private lands, and their costs are sometimes split between federal, state, and private sources. Some APHIS programs utilize pest control methods that would be difficult or impossible for private farm firms to implement. The programs of the FS are set in a somewhat more complicated economic framework than those of APHIS, because some are conducted on public lands in the national forests while others involve cooperation and cost-sharing between the FS, state, local, and private interests on nonfederal land Thus the FS operations have an element of subsidy at times while at other times they are public expenditures to protect a public resource.

APHIS activities include inspection, quarantine, and regulation designed to prevent the introduction of exotic pests into the United States and--if admitted--to prevent the spread of pests beyond areas already infested. Variou chemical, biological, cultural, and regulatory control methodologies are combined to combat a particular pest. Table 25 summarizes the chemical controls directed against the pests of major concern to APHIS. APHIS programs requiring the production of biological organisms, usually sterile-male insects or predators and parasites, are discussed in the section on production of pest control materials (see Chapter 4).

International and domestic quarantine is a major weapon for APHIS against pests. The rationale for these programs is that prevention is less expensive than cure. The agenc works with the Bureau of the Customs to monitor all incoming international traffic for plant pests. The size of this task may be indicated by a few figures for 1971:

Quarantine Inspections at Ports of Entry*	
Airplanes	305,000
Vessels	53,000
Vehicles from Mexico	38,000,000
Baggage pieces	88,000,000
Mail and packages	64,000,000

In this same year, more than 21 million pieces of cargo, carriers, and baggage from 38,000 shipments were treated or cleaned to prevent the introduction of pests. Chemical treatment, including fumigants, dusts, dips, soaks, and sprays, was given to 10,000 of these shipments. Nonchemical agents, such as steam and dry heat-sterilization, cold, and excision, were used on the remaining shipments. (APHIS also conducts programs to prevent the spread of

---

*Inspection figures are from APHIS, personal communication 1974. Interception figures are from Fowler and Mahan (1973).

TABLE 25 Selected Pest Control Programs: Pesticide, Usage, Costs, and Acreage Treated, USDA Cooperating with the States and the Republic of Mexico, FY 1972

Pest	Pesticide		Quantity (lb)[a]	Costs ($)	Acreage
Boll weevil	Malathion	(I)	689,180	415,163	558,544
	Azinphosmethyl	(I)	1,750	4,681	7,000
	Aldicarb	(I)	1,584	10,396	528
Citrus black-fly	Dimethoate (Tex.)	(I)	6,250	25,000	--
	Dimethoate (Mex.)	(I)	6,000	6,060	2,305
European chafer	Carbaryl	(I)	40	30	21
	Chlordane	(I)	1,650	660	330
Golden nema-tode	Dichloropropane-dichloro-propene	(SF)	4,895[b]	7,432	54
	Vorlex[c]	(SF)	19,360[b]	104,544	878
Grass-hopper	Malathion	(I)	555,632	351,813	868,183
Gypsy moth	Carbaryl	(I)	32,200	36,625	30,473
Imported fire ant	Mirex bait	(I)	41,494	1,957,134	11,065,153
Japanese beetle	Carbaryl	(I)	4,930	4,808	6,936
	Chlordane	(I)	198,638	46,611	8,669
White-fringed beetle	Dieldrin	(I)	720	165	144
	Chlordane	(I)	6,792	2,717	751
	Carbaryl	(I)	49,134	35,867	24,920
Witch-weed	2,4-D	(H)	376,752	140,700	376,752
	Paraquat	(H)	375	4,237	750

SOURCE: From Fowler and Mahan (1973).

CODE: (I) Insecticide, (H) Herbicide, (SF) Soil Fumigant.

[a]Active ingredient.
[b]Gallons.
[c]Trade name.

disease through interstate shipments of livestock or through the distribution of impure biologics. These are beyond the scope of this study.)

Only a few programs of the FS involve control operation on large acreages of land. Slightly more than 500,000 acres of forest in Maine were treated in 1972 with mexacarbate for control of the spruce budworm. The states of Pennsylvania, New Jersey, and New York cooperated with the FS in treating 11,647 acres with carbaryl for gypsy moth. Ethylene dibromide and lindane were used in the South and Far West for programs against bark beetles. Chlorophenoxy herbicides (2,4-D, silvex, and 2,4,5-T) were used on 172,000 acres of national forest land for weed control; most of this use was concerned with reforestation, timber stand improvement, fire hazard reduction, rights-of-way maintenance, and so forth (Fowler and Mahan 1973). Volume IV, *Forest Pest Control*, discusses these operations in greater detail.

Three agencies of the Department of the Interior--the Bureau of Land Management (BLM), the Bureau of Reclamation (BOR), and the Bureau of Sport Fisheries and Wildlife (BSFW)--conduct a variety of operations that are of importance to private agricultural firms in the western United States. The programs are, in effect, a subsidy to the farmers and ranchers utilizing the land on which they are conducted.

The BLM has responsibility for 60 percent of all federa lands, which comprises 20 percent of the nation's total. Timber, livestock forage, wildlife habitat, and public recreation facilities are among the resources managed by the Bureau. Most of BLM's pest control activities involve control of weeds on rangeland; smaller projects are design to maintain rights-of-way and to control mammals such as gophers, ground squirrels, and porcupines. In FY 1972, the project covering the largest acreage was Big Sagebrush (*Artemisia tridentata*) control on 99,000, 77,000 of which were in Oregon. The herbicides most widely used by BLM are the chlorophenoxy compounds such as 2,4-D and silvex (BLM, personal communication, 1972).

The BOR locates, constructs, operates, and maintains facilities for the storage, diversion, and development of water for the reclamation of arid and semiarid lands in the western states. Their primary pest control operation is control of weeds on irrigation facilities. There are about 60,000 miles of canals, laterals, and drains on BOR projects that could serve to irrigate up to 10 million acres. The total area actually irrigated now is slightly more than 8 million acres (about 20 percent of the total irrigated area in the western states). Most of the milage in the water conveyance and drainage system is operated by water-user organizations who buy and apply their own herbicides. The BOR, however, operates about 2,200 miles of the system, and in FY 1970 spent $162,000 for herbicides. In some areas, weed control costs on irrigation systems may be more than 25 percent of the total operating expenses. Chlorophenoxy herbicides, atrazine, dalapon, monuron, copper sulfate, acrolein, and xylene, among others, are the chemicals used for this task.

The passage of the Federal Environmental Pest Control Act (FEPCA) created an interesting problem for the BOR and other operators of irrigation systems. The BOR estimates that as much as 50 percent of herbicide use along irrigation ditches is unregistered. FEPCA prohibits the use of any pesticide except in accordance with its registration. The new law thus requires that registration of the uses be secured or the practices be altered (BOR, personal communication, 1972).

The BSFW has general responsibility for the use, perpetuation, understanding, and enjoyment of sport fish and wildlife resources. This responsibility has led to programs of production and distribution of hatchery fish, the operations of wildlife refuges, regulation of migratory bird hunting, and the management of fish and wildlife populations. One of the Bureau's programs has elicited considerable controversy in recent years. A description of it follows.

The predator control program is operated by the Division of Wildlife Services in cooperation with some 600 other organizations, including federal, state, county, and local governments; universities; and private groups. A total of $8.3 million was expended in FY 1972, $2.2 million of which was for control of coyotes and other predators in the western states. The four principal methods of killing predators formerly were baits with compound 1080 (sodium fluoroacetate) or strychnine, steel traps, and "coyote getters" (devices that expel cyanide crystals into the mouth of an animal that tugs at a bait). The Advisory Committee on Predator Control to the Council on Environmental Quality and the Department of the Interior reviewed the program in detail and recommended the discontinuance of all toxic materials (Advisory Committee on Predator Control 1972). Its recommendation was adopted by the President (Federal Register 2875 1972).

On March 9, 1972, the EPA cancelled and suspended the registrations of all products containing cyanide, strychnine, sodium monofluoracetate, and thallium sulfate. In

October 1973, the M-44 Predator Control Company of Midland, Texas, applied for registration of a sodium cyanide capsule to be used in the M-44, a spring-loaded ejector device for coyote control. The EPA did not register the M-44 but did encourage states to apply for experimental use permits for the device so that data could be gathered that may support its registration. Experimental permits have now been or soon will be issued to the Department of the Interior. Texas, Montana, California, Idaho, South Dakota, Nebraska, Kansas, and New Mexico may also apply. Control areas in which M-44s may not be used have been established so as to obtain data on the costs and benefits of the M-44 compared to other predator control methods. Such data may partially answer a major concern of the Advisory Committee on Predator Control, namely that there was little reliable data of the relative costs and benefits associated with different predator management techniques (EPA, personal communication, 1974).

Programs Controlling Disease and Nuisance Organisms Rats and mosquitoes are the pests of most public health importance in the United States today. No centralized statistics describe the magnitude of the rat problem or the control programs directed against them. Except for some federal funding of urban rat control programs, local governments or private individuals bear the responsibility for rodent control. Information on the control of mosquitoes and the results are detailed in Volume V, *Pest Control and Public Health*.

It is regrettable that so little information is available on the control of animals that do or could serve as disease vectors. The development, registration, and regulation of chemicals used in such control programs should be given special attention by the EPA. Development of data that could describe the magnitude of the control effort needed should be part of the EPA's monitoring of the vector control situation. There should be close cooperation between the EPA and other federal agencies, especially the USDHEW, in the monitoring effort.

Programs Enhancing Recreational Values Numerous federal, state, and local agencies, as well as private organizations, apply pesticides to improve the recreational quality of an area. Many of the pest control operations discussed previously, such as vegetation management in parks and mosquito control, are motivated largely by recreational values; so is the use of pesticides (fishkillers) by federal and state fish and game agency personnel.

Rotenone, antimycin, and lampricide L-30-F have all been used to kill unwanted fish so that fish considered more desirable by sportsmen could flourish without competition. The magnitude of pest control for recreational purposes, however, is not known.

The cost-benefit relationships of the recreational use of pesticides are complex, because a mixture of public and private organizations performs the operations. The dollar value of recreation in general is also difficult to measure. The amounts spent to obtain recreation of some kinds can be estimated, but such figures are often inadequate reflections of the human value of recreation. People must be licensed to catch fish, but not to watch birds. The marginal enhancement of the recreation received because pest control was conducted is even more difficult to measure than the value of recreation itself. Finally, recreational values are inherently normative judgments, and the population makes them in a heterogenous way: one man's "trash fish," for example, is likely to be another man's treasured "specimen" or yet another's valuable "resource."

Programs for Military Purposes The potential for pests, especially disease-bearing insects, to interfere with the conduct of war has been known for centuries. It has only been in the twentieth century, however, that military organizations have achieved the power to control some of the typical scourges of war, such as typhus. It is worth noting, in fact, that the insecticide DDT won its initial acclaim in the United States for its ability to control lice and mosquitoes in wartime. From the time the insecticide's properties became known in the United States in late 1942 until 1945, virtually all production of DDT was channeled into military uses.

The Department of Defense currently conducts a wide range of pest control activities in the United States and abroad. Operations include protection of buildings, equipment, clothing, stored food, and personnel.

Between 1962 and 1971, the Defense Department conducted extensive sprayings with herbicides of forest and cropland of South Vietnam. This was the first time that such pesticides had been used on a large scale in warfare. Between 1965 and 1971, the period of heaviest spraying, the United States treated at least once an estimated 3.58 million acres of South Vietnam, 8.6 percent of the country, with 2,4-D, 2,4,5-T, picloram, and cacodylic acid. The controversy over the defoliation and crop destruction programs was extensive in political and ethical terms, and questions were raised as to the military value of the herbicide use (NAS 1974).

The costs and benefits of the military use of pesticid[e]s are perhaps the most difficult of all to evaluate. Natio[ns] guard their warmaking prerogatives closely, and admonitio[ns] about the environmental and economic soundness of wartime activities are seldom heeded; the costs of failing to achieve victory appear overwhelmingly large, and nations will spend extraordinary sums to avoid them. Furthermore the environmental impact of military pest control operations is generally one of the smaller efforts of combat; the enormous diversion of resources and people into military activity dwarfs any undesirable toxicological effect of pesticides. The massive tactical use of herbicides, in fact, is probably the only example of the military use of pesticides that raises significant questions about the environmental impact of a pest control operation.

This committee did not attempt to examine military pesticide use in any detailed manner. It notes, however, that the use of herbicides in Vietnam created a revulsion among some Americans that has probably spilled over into emotional opposition to the use of those compounds in the United States itself. It would be regrettable, however, if such emotional opposition were to result in the disappearance of a useful chemical that in all other ways was judged to return benefits commensurate with its risks. 2,4,5-T, for example, is a controversial compound in part because of questions raised about impurities present as a result of its manufacture, but in part also because of its military history. We regard the question of impurities as a necessary part of a difficult cost-benefit calculation.

Pesticide Use by Private Home Residents

Home residents use pesticides on the outside to beautify their lawns and gardens, protect valuable trees, and prevent structural damage to their buildings. Inside, they use pesticides to eliminate noxious organisms, usually insects. Some pesticides used in and around homes are applied by pest control operators. The residents themselves, however, purchase and apply many materials. Such application is of considerable environmental importance, because for the average citizen it is the point during which he encounters the highest concentration of pesticide The characteristics of pesticide use patterns in the home and on the garden play a significant role in incidental human exposures and environmental pesticide pollution in a urban setting.

In a study of the role of social class and DDT pollution in Dade County, Florida, significant correlations of DDT and DDE levels with social class were demonstrated, higher levels being found in the less affluent (Davies et al. 1972). Overcrowding, inadequate fly screening, garbage accumulation--all contribute to a domestic pesticide problem that in the past had been managed by the application of DDT in the home environment. Domestic pest control use practices thus played a major role in incidental human DDT pollution in Florida.

Horticultural pest control practices have also resulted in significant environmental contamination. In an ongoing air-monitoring survey of pesticides in the air of south Florida, several pesticides most commonly used in the garden have been detected within the last 2 years. In air samplings from a residential area of Dade County, dichlofenthion (VC-13) and dursban, two pesticides primarily used for horticultural control, were identified in all six air samples collected on a monthly basis in 1974; and, in addition, diazinon was identified in four of six collections. These findings were in striking contrast to those observed in 1971 when air samples were taken in similar localities within the greater Miami area, and at that time DDT and its metabolites were regularly identified in every sample. In contrast, this pesticide and its metabolites have not been found in 1973 and 1974 in surveys taken in the same area (Davies, personal communication, 1974).

Despite its importance, the residential use of pesticides is a type of use virtually undescribed by any quantitative data. The size of the home and garden pesticide market is significant. In 1970, the total retail value of all pesticides sold in the United States was estimated at $1,545 million, of which $298 million (19 percent) was in the residential category (Farm Chemicals and Croplife 1972). The percentage of pounds of pesticide going into the home and garden market is probably considerably less than 19 percent, because the materials are sold at a higher rate per unit small packages. Sale does not necessarily mean use for the homeowner, because home packages are not fully used.

Only one thorough survey has been performed on the home uses of pesticides: It sampled Lansing, Michigan; Dallas, Texas; and Philadelphia, Pennsylvania, and was performed for the EPA (Rumker et al. 1972). Slightly more than 750,000 pounds were used in the suburban areas of these three cities in 1971. Homeowners used the bulk of the material--609,000 pounds. The portion used by homeowners consisted of 122,000 pounds of herbicides, 429,000 pounds of insecticides, and 69,000 pounds of fungicides. Improvement of lawns and gardens was the major objective of the use by residents. Lawn insects elicited the most treatment in Dallas, but lawn weeds dominated in Philadelphia and Lansing. The most commonly used herbicides were such chlorophenoxy materials as 2,4-D and chlordane. Dicofol, dimethoate, and malathion were the insecticides used in largest quantities. The survey did not report the name of any specific fungicides used in large amounts.

Three findings of the survey are of particular interest First, the timing of application of most residential pesticides is very condensed: The last weekend of April, the four weekends of May, and the first weekend of June are the most probable dates of application of all materials. Second, the rates of application on lawns are apt to be rather high compared to many agricultural uses, between 5 and 10 pounds total pesticide deposited per acre. Third, the homeowner faces a complex variety of pesticides when he goes to his retail outlet. Different brands or chemicals and combinations of two or more chemicals create a difficult choice, even for the expert. Furthermore, many labels contain only trade or chemical names, whereas recommendations from government agencies are usually in terms of common names. As a result, the probability that the homeowner will make an informed decision is rather low. By comparison, the attractiveness of the packaging and the exhortations of the clerk will be relatively important.

The Geographical Extent of Pesticide Use

The spatial pattern of pesticide use will in part determine the magnitude of the hazards to humans and other nontarget organisms. The greater the overlap between the area to which pesticides are applied and the area inhabited by people, the higher the "health" hazard; similarly, the greater the overlap between the area treated and the area inhabited by any other valued nontarget species, the greater the "environmental" hazard. The ability of pesticides to drift or move in some other way from their original sites of application complicates the assessment of pollution dangers. Nevertheless, knowledge of where chemicals are first applied is essential to an evaluation of the potential for toxicological hazard.

It would be most desirable to have indices indicating overlap between areas treated with pesticides and areas inhabited by species valued by man. No such measurements currently exist. It would be appropriate for the EPA to give consideration to the development of such information as a part of their overall responsibilities for monitoring pesticide residues in the environment.

The USDA has collected information on the acreage treated with chemicals as part of its surveys on the uses of pesticides. Such information is quite valuable, because agriculture is the single industry using pesticides in largest amounts over the widest area. The following paragraphs give a brief summary of the results of the 1971 crop-year survey (USDA, personal communication, 1974); selected information on areas treated by other industries is also included.

Herbicides

In the aggregate, 45 percent of the farmers used herbicides in 1971; and 17 percent of the crop acreage, including pasture (159 million acres), was treated. There is a very large variation in the proportion of farmers treating their land with herbicides and the proportion of land they treat, which depends upon crop grown, location of farm, and size of farm operation. The proportion of acreage treated varied from about 1 percent for hay crops to about 95 percent for rice. In cotton, corn, rice, and peanuts, herbicides were applied to more than 75 percent of the acres grown; more than one-half of the acreage devoted to soybeans, sugar beets, and Irish potatoes was treated.

Farmers with large operations tend to use more herbicides than those with small farms. Fifteen percent of farmers reporting sales of less than $2,500 per year used herbicides, but close to 80 percent of farmers with sales of over $40,000 reported using the materials. Table 26 summarizes the number of acres of each crop grown, the number of acres treated with herbicides, the percent of acres treated, and the quantities of herbicides used for selected crops.

Insecticides

A lower proportion of U.S. croplands was treated with insecticides than with herbicides: 57 million acres, or 15 percent of the total. Cotton, peanuts, and fruits were the heaviest *relative* users, with more than 60 percent of the acreage treated in each case. Corn and soybeans are intermediate, with 28 percent and 39 percent, respectively;

TABLE 26 Extent of Use of Herbicides in Selected[a] Crops (1971) and in Selected Nonagricultural Industries (1972)

	Acres Grown[b] (million)	Acres Treated[b] (million)	Percent Treated[b]	Quantity Used[c] (million pounds)
Crops				
Corn	74.1	58.5	79	101.1
Soybeans	43.5	29.6	68	36.5
Wheat	53.8	22.1	41	11.6
Other grain	37.9	11.8	31	5.4
Cotton	12.4	10.1	82	19.6
Sorghum	20.8	9.5	46	11.5
Pasture	563.5	5.6	1	8.3
Other field crops	6.2	2.5	40	6.1
Rice	1.8	1.7	95	8.0
Peanuts	1.5	1.4	92	4.4
Other vegetables	3.3	1.3	40	3.4
Sugarbeets	1.4	1.1	75	3.0
TOTAL, ALL CROPS (Except pasture, including summer fallow)	368.9	153.0	41	217.4
TOTAL, ALL CROPS (Including pasture and summer fallow)	932.2	158.7	17	225.7
Noncrops[d]				
Water management		0.7		3.4
Nurseries	0.207	0.089	43	
Railroad rights-of-way		1.6		20.0
Highway rights-of-way		6.1		17.0
Utility rights-of-way		0.9		5.0
Turf	20.0	3.8	19	
APHIS/witchweed		0.4		0.4
National forests	183.0	0.2	0.1	
BLM/big sagebrush		0.1		

[a]Crops and crop groups in which 1 million acres or more were treated are included.
[b]U.S. Department of Agriculture, personal communication, 1974.
[c]Andrilenas (1974). Excludes petroleum.
[d]All noncrop uses were discussed earlier in this chapter.

legumes and small grains are not extensively treated. In terms of volume, cotton, with 73.4 million pounds, and corn, with 25.5 million, account for nearly two-thirds of the insecticides applied to U.S. croplands.

No nonagricultural uses of insecticides extend over large land areas except for forestry, mosquito control (see Volume V, *Pest Control and Public Health*), and perhaps turf insects. In some government programs, large acreages are treated with insecticides (Table 27), but these programs should be considered agricultural. The operation

TABLE 27 Extent of Use of Insecticides in Selected Crops, 1971

Crop	Acres Grown[a] (million)	Acres Treated[a] (million)	Percent Treated[a]	Quantity Used[b] (million pounds)
Corn	74.1	20.5	28	25.5
Sorghum	20.8	8.1	39	*c*
Cotton	12.4	7.5	61	73.4
Wheat	53.8	3.8	7	1.7
Soybeans	43.5	3.5	8	5.6
Alfalfa	27.5	2.2	8	2.3
Other vegetables	3.3	1.9	56	8.3
Peanuts	1.5	1.3	87	6.0
Other fruits and nuts	1.7	1.2	71	6.3
Other grain	37.9	1.1	3	7.5
Irish potatoes	1.4	1.1	77	2.8
Citrus	1.2	1.0	88	3.0
ALL CROPS (Except pasture, including summer fallow)	368.9	56.9	15	153.9
ALL CROPS (Including pasture and summer fallow)	932.2	56.9	6	154.3

[a] U.S. Department of Agriculture, personal communication, 1974. Crops and crop groups in which 1 million acres or more were treated are included.
[b] Andrilenas (1974). Excludes petroleum.
[c] Included with "other grain."

TABLE 28 Extent of Use of Fungicides on Selected Crops, 1971.

Crop	Area Grown[a] (million)	Acres Treated[a] (million)	Percent Treated[a]	Quantity Used (million pounds)[b] Excluding Sulfur	Including Sulfur
Peanuts	1.5	1.3	85	4.4	30.4
Apples	0.5	0.4	67	7.2	8.3
Citrus	1.2	0.7	58	9.3	33.8
Other deciduous fruits	0.7	0.4	54	3.8	12.0
Irish potatoes	1.4	0.7	49	4.1	4.1
Other fruits and nuts	1.7	0.8	46	3.1	34.3
Other vegetables	3.3	0.6	18	5.7	10.9
Sugarbeets	1.4	0.2	13	[c]	[c]
ALL CROPS (Except pasture, including summer fallow)	368.9	7.6	2	29.6	151.6
ALL CROPS (Including pasture and summer fallow)	932.2	7.6	0.8	39.6	151.6

[a] U.S. Department of Agriculture, personal communication, 1974. Crops and crop-groups in which 10 percent or more of the acreage was treated were selected.
[b] Andrilenas (1974).
[c] Aggregated with other field crops.

largest in aerial extent involves the use of mirex against the imported fire ant (*Solenopsis sacvissima*) on 11 million acres in the southeastern United States.

Fungicides and Nematocides

Only a small proportion of the aggregate farmland in the United States is treated with fungicides and nematocides. An estimated 6 percent of the farmers used fungicides on 2 percent of the crop acres excluding pasture and rangeland (7.6 million acres). Fungicides were applied most extensively to peanuts, fruits, and vegetables; and nematocides had their heaviest use on tobacco, peanuts, sugarbeets, Irish potatoes, other vegetables, and fruits. Fungicides were used most extensively in the humid areas of the Southeast, Northeast, and Delta states, and in the Pacific region (see Figure 3). This geographic distribution partly reflects the propensity for fungal infections to develop under warm, humid conditions and it partly reflects the distribution of fruit, vegetable, and peanut acreage. There was considerable variation in the proportion of acres of a given crop treated from region to region. For example, 85 percent of the apples were treated with fungicides in the Northeast, but only 19 percent were in the Pacific states. Similarly, 100 percent of the Irish potato acreage in the Southwest were treated, but only 47 percent were in the northern plains. Some crops, such as corn, wheat, and other grains, are almost never treated with chemical fungicides, while 85 percent of the peanut acreage is treated. Except for Irish potatoes, there is little or no tendency for fungicide use to vary with size of farm. In the case of the potatoes, there is some evidence that the smaller farmers tend to treat their crops less extensively than do the larger ones. Table 28 summarizes the extent of fungicide use on selected crops. No evidence indicates uses of fungicides or nematocides on substantial areas of land other than in agriculture.

REFERENCES

Advisory Committee on Predator Control (1972) Predator Control--1971. A report to the Council on Environmental Quality and the Department of the Interior.

Andrilenas, P. A. (1974) Farmers' Use of Pesticides in 1971: Quantities. Economic Research Service, U.S. Department of Agriculture, Agricultural Economics Report No. 252.

Davies, J. W., W. F. Edmundson, Raffonelli, J. C. Cassady, and C. Morgade (1972) The role of social class in human pesticide pollution. Am. J. Epidemiol. 90(5):334-341.

Farm Chemicals and Croplife (1972) Willoughby, Ohio: Meister Publishing Company. 22 pp.

Federal Register 2875 (1972) Executive Order No. 11643. Animal damage control on federal lands. (Issued Feb. 8)

Fowler, D. L., and J. N. Mahan (1973) Pesticide Review 1972. USDA, Agricultural Stabilization and Conservation Service.

Lawless, E. W., A. F. Meiners, K. A. Lawrence, G. L. Kelso, and R. von Rumker (1974) Production, Distribution, Use and Environmental Impact Potential of Selected Pesticides. Report of the Midwest Research Institute to the Council on Environmental Quality. 404 pp.

National Academy of Sciences (1974) The Effects of Herbicides in South Vietnam. Part A: Summary and Conclusions. Washington, D.C.: National Academy of Sciences.

Rumker, R. von, R. M. Matter, D. P. Clement, and F. K. Erickson (1972) The Use of Pesticides in Suburban Homes and Gardens and Their Impact on the Aquatic Environment. Submitted to the Office of Water Programs, Environmental Protection Agency.

United States Department of Agriculture (1972) Extent and Cost of Weed Control with Herbicides and an Evaluation of Important Weeds, 1968. ARS-H-1. pp. 149-164.

8

PEST CONTROL AND THE LAW

PURPOSES OF REGULATION

The laws and policies that directly or indirectly influence pest control activities have developed over many decades. They pursue diverse objectives, and their administration has been spread over numerous agencies at both federal and state levels. The result is a regulatory network that is complex, both in purpose and operation. Moreover, there are instances in which unintended, unanticipated, and undesirable impacts of the regulations have arisen.

The laws relating to pest control have four major purposes:

1. protection of pesticide innovators and manufacturers;
2. protection of pesticide consumers;
3. protection of humans, and other organisms in which man is interested ("the environment") from pesticide contamination; and
4. protection of individuals from various types of uncontrolled pests.

Individual laws frequently have provisions that relate to more than one of the above purposes. FEPCA, for example, deals with the first three types of protection. Additionally, the administration of a single objective may be shared among several governmental agencies. For example, responsibilities for protection of humans and other organisms from pesticide contamination are spread among the EPA and the departments of Agriculture; Health, Education, and Welfare; Interior; Labor; and others. Finally, a given individual may be subject to multiple regulation. A grower of vegetables, for example, is subject to regulation regarding the type and quantity of pesticide he can apply and to the levels of pesticide residues and of insect parts that can be tolerated in his marketed produce.

The following paragraphs summarize the types of regulation now imposed by the federal government in the pest control area. We will trace the interactions between the different laws and examine the major costs of the regulatory network in so far as data are available. The account focuses on the individual regulated, in an attempt to show how the rules affect the conduct of his activities.

INNOVATORS AND MANUFACTURERS OF PEST CONTROL TECHNOLOGIES

Frequently, the innovators of knowledge or technology are the same persons or groups who are engaged in manufacture. As described elsewhere in this volume, they can operate in either the public or the private sector; in general, the relationship between innovation and manufacture is closer in the private than in the public sector.

Innovators belong to three major classes. The first includes those who design or develop specific compounds or devices that kill or repel unwanted organisms. These are generally suited for patenting, production, and sale in the private sector so as to yield a profit to the innovator and others. There may nevertheless be some involvement of public sector scientists. The second class includes the innovators who develop items resistant to pest attack; plant breeders who develop cultivars resistant to insect or fungal attack represent the predominant example, but the category would also include scientists who develop synthetic fabrics resistant to microbial degradation and a few others. Public sector scientists, especially those in the SAES, are heavily involved in this class of activities.

The third category includes innovators of alternative methods of accomplishing objectives where pest attack is a present problem. Agronomists, for instance, have developed alternative cultural practices that prevent pest attack; such activities, in which public sector scientists have predominated, are good examples of this kind of innovation.

Patent Protection

Whichever type of innovator or manufacturer is involved, a variety of laws affect the activity being conducted.

The fruits of innovative activity, especially private, can be protected under the patent laws where required conditions (e.g., novelty) are met. Publicly financed innovation can also enjoy such protection; but, because public agencies generally lack the authority to manufacture items for sale, discoveries in the public sector are frequently not patented. If they are, they often encounter difficulties in arranging for a private firm to manufacture the invention. There is now a special patent type law to protect varieties of plants developed: the Plant Variety Protection Law of 1970 (Public Law 91-577). This law may be useful in encouraging the development of cultivars resistant to pest attack.

The rate of innovation in the development of new pesticides is influenced, among other factors, by the right of the developer-manufacturer to appropriate the knowledge that produced the product. The federal patent laws appear to give adequate protection to innovators. We endorse the general conclusions of others who have studied the problem that additional protection beyond that conferred by the patent laws is undesirable. Although the committee has not undertaken a detailed study of the complex nature of patent law, one aspect deserves special mention. Federal patent laws confer a 17-year monopoly on those who advance the state of the art. The laws are properly biased toward those who score a major breakthrough, as opposed to those who make marginal improvements. For example, patent applications on improvements run the risk of being rejected under the equivalency doctrine, but the laws appear to have been interpreted more favorably toward those who have previously made a major breakthrough.

FEPCA is intended to limit protection of knowledge to that protection afforded by the patent laws. This position, however, was taken only after a series of compromises between proposals of the Senate Commerce Committee and the Senate Agricultural Committee. The latter eventually abandoned its position that a manufacturer had no duty to disclose unprotected, as well as patent-protected, data. FEPCA thus reflects a balance between the total disclosure theories of the Senate Commerce Committee and the nondisclosure theory of the Senate Agricultural Committee.

Manufacturers must disclose their registration statements to interested third parties and to competing manufacturers who wish to rely on it for the purpose of registering a product. This duty is limited by the manufacturer's right (1) to designate portions of the data as trade secrets and (2) to demand reasonable compensation from second registrants who wish to rely on his registration data.

Common law protects trade secrets on theories of property or the protection of a confidential relationship between the producer and his employees. Some test data would potentially fall within the protection of common law, because it is knowledge not generally available to outsiders. This confers a competitive advantage on the claimant, because the information is not easily duplicated by others without considerable expenditure. Section 10 of FEPCA gives the manufacturer the right to designate portions of his registration data as trade secrets. These portions are submitted separately from other material required under the act and may not be disclosed by the administrator "except . . . when necessary to carry out the provisions of this Act. . . ." In such instances the trade secrets may be disclosed only to other government agencies or at a public hearing. It is well settled that a manufacturer has no right to the exclusive possession of trade secrets against a valid governmental interest in disclosure.

The Senate Commerce Committee was concerned that nondisclosure would create protections beyond those of both the patent laws and the common law of trade secrets and would result in the waste of research dollars on duplicative testing. Section 3(c)(1)(6) provides that a manufacturer must offer to disclose his test data to all registrants who offer reasonable compensation. Those portions of the data designated as trade secrets do not, however, have to be disclosed. If the parties cannot agree on reasonable compensation, the administrator may fix the amount. All disputes over what is a trade secret and compensation are subject to judicial review.

The doctrine of "nonessentiality" in FEPCA also probably has a positive impact on the rate of innovation, but perhaps at the cost of giving inadequate consideration to alternative pest control techniques. The dilemma is that steps to avoid these latter costs may in turn inhibit innovation. The trade-off between the rate of innovation and the use of socially most-efficient pest control techniques is one that should be given careful attention.

As discussed more fully below, the EPA is charged with the responsibility to register pesticides. In this process, the EPA does not evaluate the efficacy of pesticides other than to ask if the material will do what is manufacturer claims. Thus, for example, the EPA would not refuse to register a pesticide on the grounds that increasing amounts of it will be required to maintain the same level of pest elimination because of genetic resistance to the pesticide or because of the disruption of natural controls.

To some extent, market forces may operate to encourage users to shift to pest control practices that do not involve chemicals (or that require less use of chemicals) and thus do not encounter genetic resistance. (After all, such a shift would be rational because pest control costs would be lowered.) However, such nonchemical or less-chemical alternatives may be perceived by individual users as entailing a risk sufficiently greater than continued reliance on chemical controls as to discourage a shift away from use of the latter.

It could be argued that the law should concern itself more directly with the merits of alternative pest control techniques. In some ways concern over environmental contamination has deflected attention away from equally important subjects that figured prominently in the deliberations of this committee: pesticide efficacy in the face of genetic resistance and disruption of natural control. Some consideration of the alternatives is, however, required under the existing risk-benefit approach *whenever a risk has been demonstrated*; the existence of a safer method of control will change the risk-benefit analysis (Shell Chemical 1974).* Some have argued that confining consideration of alternatives to instances where health and environmental hazards are at issue is too narrow and that alternatives should be considered in a wider range of situations.

If alternatives must be considered in many if not most registrations, the burden would be placed on the registrant to prove his product is better, both in terms of efficiency and freedom from health and environmental hazards, than existing registrations. This shift in the burden of proof by making essentiality a requirement for registration would undoubtedly increase the costs of registration. It would arguably also discourage innovation at a time when such innovation in the industry is needed. Nor does the agency require labels on pesticides to compare control by the pesticide with control by other means. FEPCA carried forward the principle that a pesticide is entitled to registration and use if preestablished standards are met [Federal Environmental Pesticide Control Act of 1972 (1973)] by the provision that lack of essentiality shall not be made a criterion for denying

*The EPA's decision was affirmed in 1975 by the District of Columbia Circuit Court (EDF v. EPA 510 F.2d 1292).

registration of a pesticide. The Senate Commerce Committee, more environmentally oriented than the Senate Agricultural Committee, had sought to make an explicit EPA duty to study nonchemical methods of pest control such as integrated control (Hearings 1971). The Senate Commerce Committee also unsuccessfully proposed an amendment to make lack of essentiality a criterion for denying a registration. The meaning of essentiality is not clear:

> The "doctrine of essentiality" has different meanings to different people. To some, it would forbid the registration of any pesticide if there exists an alternative which is just as safe and effective as that pesticide--since under those circumstances the pesticide is not "essential" for the well-being of society. Under another reading, the doctrine merely mandates consideration of the availability of less hazardous substitutes as one of the interests to be balanced in determining whether a pesticide should be registered [Senate Report No. 92-970 (1972)].

The EPA takes the position that the term means that a pesticide will be registered only on the basis of the information that the manufacturer submits.

Regulation of Activities

Although the products of innovation are protected, the innovator is not completely free to produce and sell his product in the marketplace. With the Federal Insecticide Act of 1910, manufacturers of devices and compounds to kill pests became subject to labeling requirements designed to ensure the purchaser some protection that the product is both effective and safe. Since 1938, the purchasers of food and feed have received protection from pesticide contamination through the Federal Food, Drug, and Cosmetic Act (FFDCA). The labeling laws require the registration of pest control devices and compounds, and FFDCA requires the imposition of tolerances on pesticide residues in or on agricultural commodities traded in the marketplace.

Experimental Use Permits, Temporary Tolerances, and Registration

The first impact of regulation upon innovators and manufacturers occurs when they wish to test their inventions on a large scale. An experimental use permit and temporary tolerance or exemption from tolerance must be obtained from the EPA.

The major control over the activities of innovators and manufacturers is exercised by the EPA's registration process. The major purpose of this process is to protect the pesticide consumer and the general public and the environment from fraudulent or unsafe compounds or from pesticide contamination. In the past 2 years, the protection of workers contacting pesticide-treated foliage has been added to the concerns motivating regulation; this will be discussed under regulations affecting the pest control decision maker. Because of the fundamental role of the EPA in the registration process, a description of the organizations of the agency will be given first, followed by a discussion of the major laws affecting registration, especially the Federal Insecticide, Fungicide, and Rodenticide Act (FIFRA), as supplemented by FEPCA.

<u>The Environmental Protection Agency (EPA)</u> The EPA was created on December 2, 1970, by Reorganization Plan No. 3 of 1970. It inherited a variety of research and regulatory programs that were previously distributed among several federal departments; the most important legacy for our purposes is the EPA's administration of the regulatory activities affecting pesticides and other "economic poisons." The EPA's authority to regulate pesticides resides largely in FIFRA, FEPCA, and the FFDCA.

The responsibilities mandated by FIFRA are split between the national EPA headquarters in Washington, D.C., and 10 regional offices. Broad policies for pesticide regulation are usually formulated in Washington, while enforcement is regional. The FFDCA allows the EPA to set tolerance for residues of pesticides in or on food and feed, and those tolerances are enforced by the FDA and the USDA.

The Office of Pesticide Programs, responsible for most of EPA's pest control responsibilities, is one of four offices responsible to the Assistant Administrator for Water and Hazardous Materials Control. There are four divisions of the Office, each with a special set of responsibilities in the area of pest control regulation:

- Criteria and Evaluation Division. Responsible for formulating the standards and criteria by which pesticides are registered, and by which tolerances of pesticide residues in or on food are set. The efficacy of pesticides and their effects on human and environmental health are considered in setting standards. This division also reviews currently registered materials and their tolerances, identifies which area needs research and monitoring efforts, and provides technical expertise to other parts of the EPA and of the government in the area of pesticides.

- Registration Division. Conducts the actual registration of pesticides and sets the tolerances for residues that can remain in or on food or animal foodstuffs. This division also specifies the sampling procedures to ensure accurate label description of the composition and potency of commercially available pesticides. The Office of Enforcement, under the Assistant Administrator for Enforcement and General Counsel works closely with the Registration Division in enforcing the requirements of FIFRA.
- Technical Services Division. There are two major types of functions conducted by this Division. First, it stores technical information on pesticides and makes it quickly available to other parts of the EPA, the federal government, and the states when asked. The types of information stored include data on registrations and tolerances of different pesticides and reference material of a highly specialized nature. Because this Division utilizes computer facilities for storing data, it also makes these computation facilities available for analyses and studies conducted by other divisions of the Office of Pesticide Programs. Second, the Technical Services Division conducts a monitoring program of pesticide residues in air, water, soil, crops, livestock, wildlife, and humans. More will be said of this monitoring function in a later section.
- Operations Division. This Division's major task is to enhance the conduct of the EPA's pesticide programs in other areas of government. To this end, the Division helps the states develop technical expertise and legislation in the pest control area, participates in federal interagency activities in the pesticide field, and conducts investigations of accidents and unplanned incidents involving pesticides.

Figures supplied to us by the Office of Pesticide Programs indicate that the approximate cost of the pesticide regulatory program in FY 1974 (July 1, 1973, through June 30, 1974) was $21.6 million and 845 people were employed in these tasks. Details of these expenditures for FY 1972 to FY 1974 are shown in Table 29.

TABLE 29 Expenditures by the Environmental Protection Agency for the Regulation of Pesticides, Fiscal Years 1972-1974[a]

	1972		1973		1974	
	Positions	Thousands ($)	Positions	Thousands ($)	Positions	Thousands ($)
Registration and tolerance	240	4,744	372	7,118	380	6,876
Monitoring and surveillance[a]	178	6,910	91	5,047	58	4,052
Technical information and assistance[b]	41	924	100	2,470	201	7,066
Manpower planning and training	9	200	9	200	--	--
Enforcement[b]	80	1,000	80	1,000	169	2,800
Program management and support[c]	--	--	--	--	37	796
TOTAL	548	13,778	652	15,835	845	21,590

[a]Taken with modification from Environmental Protection Agency, 1974. "Budget Planning Summaries for EPA, Pesticide Programs Activities, FY 72-74. (January, 1974, Estimates)." Unpublished document.
[b]A portion of these funds are used to monitor the quality of pesticides on the market. Other expenditures are for monitoring of residue levels in soil, air, water, food, feed, wildlife, livestock, and humans; and other activities.
[c]Expenditures for these activities occur both in Washington and in the regions.
[d]Program management and support funds are included in other line items in FY 1972 and FY 1973.

Registration under FIFRA FIFRA was passed in 1947 as a piece of limited consumer protection legislation. The basic assumptions were that the primary uses of pesticides were beneficial and that users could and would react rationally to information about the pesticides' impact. Thus pesticide regulation was confined to labeling requirements designed to ensure that users had accurate information about the efficacy of the pesticide on target species, the impact on nontarget species, and health hazards to humans. In particular, FIFRA required (1) the registration of economic poisons or chemical pesticides prior to their sale or movement in interstate or foreign commerce; (2) the prominent display of poison warnings on the labels of highly toxic pesticides; (3) the coloring or discoloring of dangerous white-powdered insecticides to prevent their being mistaken for edible foods; (4) the inclusion of warning statements on the label to prevent injury to people, animals, and plants; and (5) the inclusion of instructions for use to provide adequate protection for the public. Registration was relatively nondiscretionary, because the standards to be met were fairly clear. Not until 1964 was a manufacturer prohibited from registering a challenged pesticide "under protest."

This consumer protection legislation ignored such environmental problems as the remote and cumulative impacts of pesticide application upon nontarget species and the risks posed to future generations by the application of mutagenic and teratogenic chemicals. The controversy over DDT focused attention on the environmental protection defects in pesticide regulation (Rodgers 1970) and spurred the Federal Environmental Pest Control Act of 1972 (FEPCA). FEPCA extended federal jurisdiction to intrastate commerce and expanded control by labeling to require discretionary review of a proposed registration to determine whether a pesticide poses unacceptable environmental risks. (For a survey of FEPCA, see Butler 1974.) The critical section provides that:

> The Administrator shall register a pesticide if he determines that, when considered with any restrictions imposed under subsection (d)--
>
> (A) its composition is such as to warrant the proposed claims for it;
>
> (B) its labeling and other material required to be submitted comply with the requirements of this Act;
>
> (C) it will perform its intended function without unreasonable adverse effects on the environment; and

> (D) when used in accordance with widespread and commonly recognized practice it will not generally cause unreasonable adverse effects on the environment.
>
> The Administrator shall not make any lack of essentiality a criterion for denying registration of any pesticide. Where two pesticides meet the requirements of this paragraph, one should not be registered in preference to the other. . . ."

The feature of the new act that most directly affects innovators and manufacturers is the requirement that the EPA undertake a risk-benefit analysis before pesticides are registered. The legislation affirms federal judicial developments that occurred simultaneously with the consideration and passage of the legislation. In 1971 the District of Columbia Circuit Court of Appeals* required the Administrator to make a risk-benefit analysis to determine if the DDT registrations should be cancelled, and most registrations were subsequently cancelled. A risk-benefit analysis is a social judgment as to whether the potential costs of an activity are greater than its benefits, which are generally known and quantified. A risk-benefit analysis requires an agency (1) to make the inferences about future impact that are justified by existing scientific knowledge, (2) to determine possible harmful risks and estimate their magnitude, and (3) to compare these with the benefits to society that may result from continued usage.

In an EPA risk-benefit analysis, greater weight is given to the risk that human health will be impaired than to risks that ecosystem stability will be disrupted where the impact on human health or on the future human use of the resource is not known. The weight assigned by the EPA to risks to human health compared with that assigned to risks to ecosystem stability is consistent with the Western rational tradition. A "problem" exists only

---

*EDF v. Ruckelshaus, 439 F.2d 584 (D.C. Cir. 1971) held that when the administrator had concluded that a substantial question existed as to whether a pesticide poses an "imminent hazard," he had a duty to initiate cancellation proceedings and articulate the factors relevant to determining the magnitude of the anticipated harm and the likelihood of its occurrence. See also EDF v. EPA, 465 F.2d 528 (D.C. Cir. 1972) (aldrin and dieldrin).

where the health of human beings is threatened or a future human use of a resource might be foreclosed. It is beyond the scope of this report to challenge this view directly or to suggest radical philosophical departures from it. We merely point out, as the EPA partially recognizes, that the costs of technological advancement have historically been defined very narrowly and that recent thinking about the value of relatively undisturbed ecosystems *qua* ecosystems is undergoing a fundamental reevaluation. Therefore, we recommend that the EPA view the variables relevant to risk-benefit analysis as dynamic and that the concept of cost be subject to expansion as societal attitudes toward our natural resources change.

In the end, a reasoned decision that explicitly confronts the various trade-offs must be made. The Administrator is not--under the substantial evidence rule that governs review of administrative decisions made on a record--bound by the limits of scientific inference, but is entitled, so long as there is evidence of significant risk, to conclude that society should not bear it. A federal circuit court of appeals has held that the Administrator need not make a risk-benefit analysis to decide if a pesticide should be cancelled if the evidence establishes that human health will be impaired only when a properly labeled pesticide is not used according to label instructions (Stearns Electric Paste Co. v. EPA 461 Fed. 293, 7th Cir. 1972).

FIFRA's strict registration requirements are potentially inconsistent with research and innovation, because a pesticide must often be used before its effectiveness and impact are known. Section 5(a) attempts to resolve this conflict by authorizing the issuance of experimental use permits. The permit gives the right to use a pesticide, subject to conditions, in order to accumulate the information necessary for registration. The administration of this section will be controversial, as environmentalists are concerned that the experimental use permit may be used as a form of in lieu registration. If the EPA responds to this criticism by requiring information that approaches the requirements for a registration, public and private researchers will urge that the standards are too restrictive. The regulations are general and do not resolve these conflicts. Permits are classified as either general or specific. Residues on food or feed are regulated, but apart from this, the type of information that must be submitted before a permit can be issued is not specified. Only a description of the nature and geographical area of the proposed experimental program is required.

Regulation of New Plant Varieties

The FDA has proposed new regulations governing the production of plant cultivars that differ from traditional varieties in levels of toxic substances, with the intent of protecting consumers from encountering increased concentrations of substances that might pose a health threat. In an era in which there is considerable potential for the genetic manipulation of traditional agricultural crops, such regulations are not unreasonable. Certainly consumers deserve protection from hazardous substances occurring naturally in the plant. Before any such rules are promulgated, however, the potential for unanticipated regulatory consequences should be examined for their impact upon the development of plant varieties resistant to pests, or with desirable nutritional attributes.

Plant variants resistant to insect attack, for example, often owe their protected status to higher concentrations of some natural chemical than protection through repelling or killing insect herbivores. We should know if the same chemical poses any threat to human consumers of the plant, but rules constructed for testing should not make the search for and development of such variants prohibitively expensive. Otherwise we may eliminate potential improvements in the food value of crops, as well as one of the most powerful sources of alternative pest control strategies known: the development of resistant varieties. Regulations for testing such varieties should embody the principle that costs and benefits to the consumer associated with the use of pesticides should be weighed against those associated with eating plants having an altered natural chemistry.

We are disturbed that little attention has been given to the potential impact of the FDA's proposed rules on the development of new pest control tactics and recommend that the USDA and the EPA work with the FDA to explore the problem in more detail before final rulings are made.

Protection of Fair Trade Practices

The Federal Trade Commission (FTC) is empowered by the Federal Trade Commission Act to identify and prohibit unfair methods of competition and unfair or deceptive acts or practices. The Commission issues rules applying to a whole industry, or orders against particular companies within the industry. The Commission has proposed general rules regarding trade in pesticides on three occasions, but as yet no final rule has been promulgated (Federal Register 1968, 1970). The Commission issued proposed complaints against three companies for engaging in deceptive advertising of pesticides on October 25, 1972. Two of the companies--Union Carbide and Hercules--have signed consent agreements and agreed to change their practices. The third--FMC Corporation--is contesting the complaint in an administrative hearing.*

The assumption behind the activities of the Commission is that government regulation of companies is necessary to maintain competition within the industry and thus foster innovation and the lowest possible prices, both of which are beneficial to pesticide consumers and the general public. The antitrust laws, also applicable to pesticide companies, have a similar thrust. The assumptions behind the patent law are of course precisely the opposite: Companies need monopoly control over their innovations, at least for a period of time. Such monopoly control is assumed (in the long run) to have beneficial results for pesticide users and the general public in that useful new technologies are fostered that offer an economic gain to society as a whole. These opposing forces are reconciled only if a balance is struck between the need to foster competition and the need to protect innovations so as to foster them.

Our Committee was unable to examine any of these assumptions in a rigorous manner. In particular, there are a number of important questions to which we simply have no answer:

1. Are the economic interests of pesticide users and the general public best served by having a large number of products from which to choose, each of which is vigorously promoted by its manufacturer?

2. Antitrust laws designed to encourage competition between companies keep companies from engaging in cooperative research efforts. The high costs of registering a

---

*We made no effort to determine the costs of FTC administering regulations, but they are probably small: No more than 10 persons have been assigned to pesticides at any given time (Federal Trade Commission, personal communication, 1974).

new pesticide might suggest that pooled research efforts could return a public benefit in the form of new pesticidal materials. What validity is there to this argument?

3. Given the tendency of some pesticides---especially insecticides--to decay in efficacy due to emergence of resistant pests, do the patent laws merely provide a mechanism for a company to maintain a monopoly position over a chemical until the material is of little use? Even if there is no evidence for a "yes" answer to the above question, what would the case be if patent protection were extended so as to foster more innovation in the pesticide field?

4. Have we in fact achieved a "balance" between our efforts to foster competition and to encourage innovation through the patent laws? If not, is the reason in the structure of the laws and policies or in the economic structure of the corporations with competence in the field? Or does the answer lie in the scientific properties of pests and pesticides?

5. Finally, does the present incentive structure offer adequate encouragement to innovators outside the traditional center of the pest control industry?

The above questions do not begin to exhaust the points in need of clarification; they are only exemplary. In view of the need to create options in the field of pest control practices, we reiterate our earlier recommendation that the EPA and USDA give priority attention to studies of innovation in pest control technologies. We are deeply concerned that the economic, legal, and social atmosphere currently in the United States is not conducive to beneficial innovative activities. We require a better knowledge of the efforts of societal constraints and guidance on innovation. Government agencies should take the lead in funding such studies by members of their own staffs and by competent researchers outside the government.

Protection of Water Quality

The EPA has proposed effluent standards for toxic pollutants as required by section 307(a) of the Federal Water Pollution Control Act Amendments of 1972 (FWPCA) (Federal Register 1973). Section 307(a)(1) provides that:

> The (EPA) Administrator shall . . . publish . . . a list which includes any toxic pollutant or combination of such pollutants for which an effluent

> standard (which may include a prohibition of the discharge of such pollutants . . .) will be established under this section. The Administrator in publishing such a list shall take into account the toxicity of the pollutant, its persistance, degradability, the usual or potential presence of the affected organisms in any waters, the importance of the affected organisms and the nature and extent of the effect of the toxic pollutant on such organisms.

This section further requires the EPA Administrator to propose and establish effluent standards. Proposed regulations have been published for nine toxic pollutant classes, including four categories of pesticides. The pesticide groups include aldrin-dieldrin, DDE-DDD-DDT, endrin, and toxaphene. The regulations apply to pesticide manufacturing and formulating facilities that are considered point sources discharging into navigable waters; they designate pollutant weights and concentrations permitted for discharge from above facilities into streams, lakes, estuaries, and coastal waters of various flow rates. These effluent limitations are to be eventually incorporated into permits issued under section 402 of the Act, the "National Pollutant Discharge Elimination System."

The EPA has also given "advanced notice of proposed rule-making" pursuant to section 311(b) of FWPCA, according to which a number of chemical pesticides will be regulated (Federal Register 1974). This section provides that:

> The (EPA) Administration shall develop, promulgate and revise . . . regulations designating as hazardous substances, other than oil . . . , such elements and compounds which, when discharged in any quantity into or upon the navigable waters of the United States or adjoining shoreline or the waters of the contiguous zone, present an imminent and substantial danger to the public health or welfare, including, but not limited to, fish, shellfish, wildlife, shorelines, and beaches.

A determination of whether the hazardous substance can actually be removed must accompany each regulation.

Substances have been selected for regulation on the basis of their lethality, at various concentrations and exposure times, to test populations of animals and aquatic flora. Discharges of these substances fall into two categories: (1) transportation and storage incidents in which there is normally no release of the substance except in spill situations; and (2) industrial production facilities in which a named hazardous substance is released in quantities exceeding those of normally permitted effluent (as in effluent limitations and guidelines established under other sections of FWPCA; particularly section 307). Rules will now be developed prohibiting discharges of designated quantities of hazardous chemicals and requiring removal of these chemicals when they are released into navigable waters.

Many pesticides reach bodies of water by runoff and seepage. These discharges are classified as nonpoint sources. For reasons of cost and the lack of adequate technology, nonpoint sources are not now regulated by a permit program. The Federal Water Pollution Control Act Amendments of 1972, however, require the states to develop programs to reduce nonpoint sources of pollution, and this may result in an increased control of pesticide use.

Transportation of Pesticides

The laws governing the transport of pesticides from the manufacturer to the ultimate user were discussed briefly in the section on Transportation, Storage, and Accidents. Certain laws govern the types of packages that must be used for pesticides, the methods by which they must be transported, and the information to be reported in the event of an accident. The intent of these laws is to protect transportation workers and the general public from the hazard of accidentally spilled toxic substances and to protect the environment from pesticide contamination.

Aerial application of pesticides is treated--for legal purposes--as a special type of transportation operation. The Federal Aviation Administration (FAA) of the Department of Transportation has the power to regulate aircraft and pilots that apply pesticides by certifying the operators. An applicant must pass tests on his knowledge of agricultural materials and their application by aircraft. The FAA further prohibits the application of any pesticide in a manner contrary to the directions for use on the label.

Certification of applicators and enforcement of their day-to-day operations is conducted in a decentralized manner by the 11 regional offices of the FAA. General Aviation Inspectors of the FAA--the law-enforcement officials of the agency--are busy with a variety of different enforcement problems and do not account for their time by specific task, such as enforcement of agricultural aircraft regulations. Only a very small portion of FAA enforcement activities involves aerial applicators, however, and accordingly the administrative costs of the regulatory activity are probably rather small (FAA, personal communication, 1974).

Annual reports on accidents involving aerial application of pesticides are made by the National Transportation Safety Board (NTSB), an independent agency housed within the Department of Transportation. The Board spends between $4 and $5 million per year for accident investigation. Analysis of accidents involving aerial applicators is only a small portion of the Board's work, and the costs of analyzing such accidents are therefore rather small (NTSB, personal communication, 1973).

REGULATION OF THE PEST CONTROLLER

The pest control decision maker is faced with a variety of legal prescriptions and proscriptions as well as the network of economic and social forces that were discussed earlier. The laws governing the pest controller have two kinds of impact: First, they control his use of pesticides; second, they may require, in some cases, that he control pests whether he wants to or not. Both types of laws are motivated by a desire to protect workers, consumers, the general public, and the environment from the damages inflicted either by pesticides or by pest organisms; but the desire to protect from pesticides may not always be compatible with the desire to protect from pests.

Direct control of pesticide use is mandated by two provisions of FEPCA. First, all pesticides are classified for "general" or "restricted" use at registration. Materials for general use may be used by any person, but restricted materials may be used only by or under the direction of certified applicators. Second, all uses must be in strict accord with the directions for use on the registered label. Penalties are provided for individuals who use a pesticide in a manner not consistent with the directions.

Regulation of Residues on Food

Indirect control of pesticide use is achieved by regulating the minute amounts of pesticides remaining in or on food products due to chemical treatment in the production, processing, or storage of foodstuffs. The FDA of the USDHEW has major responsibility for surveying pesticide residues in food as it passes from farm to consumer and for preventing food having more than the tolerance limit of residues from reaching the consumer. The tolerance limits are authorized by the Federal Food, Drug, and Cosmetic Act (FFDCA). The FDA was responsible for establishing these tolerances from 1938 until 1970, when the task was given to the EPA. The FDA's role is now strictly one of enforcement.

The Bureau of Foods of the FDA currently conducts three programs. The Total Diet Studies Program is designed to monitor a "market basket" of food that represents the recommended diet for 2 weeks of a 15-20-year-old male. Retail samples of approximately 117 items are purchased once a year in each of 30 different geographic regions in the country. Items are analyzed for pesticides and various other contaminants: polychlorbiphenyls, arsenic, lead, selenium, cadmium, mercury, and zinc.

The Pesticide Program is the second major effort of the Bureau and involves selecting a minimum number of samples of various foods, analyzing them for residues, calculating the variation in the sample, and then inferring the level of contamination in the entire food supply. This program allows the Bureau to monitor the general level of contamination and at the same time allows the Field Offices of FDA flexibility to act on geographically isolated problems that are detected through field observations, suspicions, or the statistically inferred level of contamination. District Offices of the Bureau are directed to give priority attention to commodities produced in significant amounts in that district.

The final program--the Retail Sampling Survey--enables the FDA to do exploratory surveys for a variety of problems, one of which is the occurrence of illegal pesticide residues. (See Tables 31-33 in Chapter 9 for a list of chemicals monitored by the FDA.)

The FDA estimates that it spent a total of $3.4 million on pesticide efforts in FY 1973. The vast majority of these funds (more than $3.0 million) were for enforcement activities, and only a small amount (less than $0.2 million) was spent on research efforts. A total of 186 positions were allotted to pesticide work in that year.

Before the EPA assumed responsibility for setting tolerances, the FDA sponsored research work on the pathological effects of pesticides. The last of these studies was completed in FY 1973, and the FDA's sole research efforts are now designed to improve their analytical abilities for pesticide residues. They are particularly interested in developing analytical procedures that are practical for detecting many different chemicals in one analysis and yet precise enough for legal action if a violation is detected.

Monitoring pesticide residues in meat, poultry, and broken egg products is the responsibility of the USDA. Analyses of meat and poultry are conducted by the Animal and Plant Health Inspection Service, while those on egg products are performed by the Agricultural Marketing Service. The former Service estimates meat and poultry monitoring to cost about $140,000 per year, and the latter estimates egg products monitoring to be about $22,000 annually. The chemicals monitored in these two programs are listed in Tables 31-33 in Chapter 9 (USDA, personal communication, 1974).

Occupational Health Regulations

A second indirect way of controlling the uses of pesticides arises from the requirements of the Occupational Safety and Health Act (OSHA) that the work environment be free of undue hazards. Responsibilities are located through OSHA in the DOL and through FEPCA in the EPA. The jurisdictional and data-gathering problems in this area are serious, and are discussed in Chapter 1.

Regulation of Pest Parts in Food

The protection of the general public against pests, pest parts, and products is complex and in a state of rapid flux at the present time. This Committee was able only to highlight the subject by noting that there are two fundamentally different problems involved that are not always distinct in practice. The first relates to the protection of consumers from agricultural produce that is grossly contaminated with pest organisms; in general, such contaminating organisms are dead. The second is posed by the person who allows pests to propagate unreasonably on his own land and thus creates a pest problem for his neighbors; this problem is of considerable importance to the implementation of programs for integrated pest management.

Protection of consumers from food contaminated by pests is the responsibility of the FDA. Among the contaminates are insects and insect parts in food or feed, microorganisms, and such rodent by-products as hair or fecal material. An especially serious problem of contamination is receiving increasing attention, that of contamination of food and feed with a highly toxic fungal by-product, aflatoxin. The prohibition against selling foods that are so contaminated as to be unwholesome is undoubtedly an important protection for all people who buy food and feed products.

It should be noted, however, that in some cases the desire to protect consumers from contamination by insect parts clashes directly with the desire to protect them from pesticide residues. For example, in certain vegetable crops, especially those largely sold to frozen-food processors or canners, FDA regulations regarding the inclusion of insect parts are the primary determinant of pest control practices. The growers of brussels sprouts in the coastal areas of northern California, for example, contend that no chemical pesticides are required to maintain crop yields at their present level. They regularly contract for chemical pest control services, however, because the packers inspect the crop before delivery to ensure that it will meet FDA standards. We are unaware that the FDA insect-parts standards on this or any other food crop, however, are made on the basis of continuous studies of actual consumer preferences. (Indeed, it is difficult to see how such data could be generated: The current standards are high enough so that almost no complaints are apparently produced.)

Regulation of Seed Purity

Commercial seed producers are under some restrictions as to the purity of their products; that is, they must not sell seed that contains too high a proportion of weed seeds. These requirements for purity act as an incentive for farmers to use herbicides so as to assure the presence of few or no weed seeds. Seed purity laws also provide an example in which the desire to protect consumers against contaminated produce is important in protecting the general public against the propagation of pests, because contaminated seed will clearly provide a much more severe weed problem than will pure seed. It is unclear whether we are using the pure seed regulations to their maximum extent as a prophylactic weed control measure.

Collective Action

Laws requiring landowners to follow specified practices so as to keep their land from becoming a source of pests for the entire neighborhood are currently in a state of rapid change. The oldest laws of this type are, of course the quarantine laws that attempt to isolate pests in limited geographic areas by controlling the flow of produce and other items that act as pest carriers between geographic regions. More recently, serious efforts have been made to form institutions that can require farmers to follow certain practices, so as to provide a net benefit to all farmers in the region in the form of reduced pest control expenditures.

New pest management strategies may have to be followed uniformly over a wide area in order to be effective; and, therefore, they may require the imposition of some form of government action. The law has been generally tolerant of "free riders" who benefit from activities taken by others in their own self-interest. Thus a property owner who undertakes an improvement or activity that serves a public purpose has no right to recover the costs of the improvement or activity--even though the activity benefits other owners, and even though it could have been undertaken by a public agency and the costs recouped through special assessments or *ad valorem* taxes levied against the beneficiaries (Dawson 1974). In short, one property owner cannot conceive a scheme to benefit others and impose it on the benefited class without the help of legislatively authorized collective action. Collective action for the protection of a crop would probably take the form of a pest management program formulated by a state agency with grower input and subject to a grower referendum in which the majority has the power to bind the minority. Special districts, formed by grower initiative, have long been important institutional mechanisms in pest control; but their use has been primarily confined to financing or enforcing chemical, cultural, and other abatement activities. These districts have engendered little opposition, as the benefits relative to taxes or assessments have been evenly distributed and well perceived by those subject to the taxing power. More sophisticated problems will be presented by the imposition of new pest management strategies. Although in theory the benefits will accrue to all those growers included within a district, the risks are greater and thus grower resistance may be expected. Nonetheless there is firm legal precedent for the imposition of new pest management strategies. In this way, pest management problems seem analagous to oil and gas conservation programs that have long been held constitutional on the theory that oil and gas is a common pool resource and that collective restriction of the right to capture is necessary to secure each owner's right to a minimum level of enjoyment of his common property rights. Similarly, it can be argued that all cotton farmers in a defined area are part of a common pool, in that the failure of one farmer to participate in the program jeopardizes a commonly held resource--namely, freedom from infestation. This rationale, incidentally, supports state-imposed programs as well as those subject to a grower referendum.

Government agencies that act to control pests on either public or private land face a unique set of regulations. Not only must they strictly follow the directions for use of pesticides as given on the label--a provision of FEPCA--but they must also follow the procedures outlined in the National Environmental Policy Act (NEPA). Specifically, if their project is of sufficient scale, the agency must file an environmental impact statement analyzing the costs and benefits of their intended action compared with alternative methods of achieving the same goal. Thus, for example, the Forest Service and the Animal and Plant Health Inspection Service must file statements for major programs, such as spray operations against the spruce budworm or witchweed. It is not clear whether the requirements of NEPA have actually resulted in any alterations of government pest control operation, but the requirement for serious evaluation of alternative strategies is surely to be welcomed. The courts have repeatedly held that NEPA does not alter the missions of the federal agencies to which it applies, but agencies are now required to balance the benefits of an activity against its environmental costs and make a reasoned choice. Under the decided cases it is unclear whether an activity with substantial environmental costs can proceed on the grounds that the benefits outweigh the costs; but some courts are imposing a duty to mitigate the environmental impact when the statement discloses that the costs are substantial.

REGULATION OF APPLICATORS AND CONSULTANTS

Pesticides must be registered for "general" use or for "restricted" use under FEPCA. Any person may apply a pesticide registered for "general" use so long as the application is in accord with the directions for use on the label. Pesticides registered for restricted use can be applied only by or under the direct supervision of a "certified" applicator. States operate certification programs to ensure that persons handling restricted use materials have a certain minimum level of competence. States may provide separate guidelines for certification of private applicators, such as farmers and commercial applicators who apply pesticides on a for-hire basis. All state programs for certification must comply with minimum standards set by the EPA. Should a state program fall below the minimum standards, the EPA may withdraw its approval of the program.

The intent of "general" and "restricted" registration, and of applicator certification, was to protect applicators and other workmen in the area of application, the general public, and the environment. The provisions of FEPCA relating to classification of pesticides and certification of applicators will not take full effect until the fall of 1976, but we can anticipate that the new rules will raise the level of caution under which specific, dangerous chemicals are used. A more difficult problem is the legal framework under which consultants in pest control matters should work. Although pest control consultants may well be certified commercial applicators under the provisions of FEPCA, these individuals and society face a variety of problems that are not covered by FEPCA. Among the more important policy questions still to be answered are (1) how are such individuals to be encouraged as a component of systems of integrated pest management and (2) under what standards of liability should they operate. These problems are considered in more detail in Volume III, *Cotton Pest Control*.

LIABILITY CONSIDERATIONS

If alternative pest management strategies are deemed socially desirable because they will decrease reliance on chemical pesticides, the role of the legal system in encouraging a pest consultant and management industry is important. Rules governing the liability of pest consultants and managers where damage occurs as a result of the application of a recommended strategy may play a crucial role in determining the growth of the industry. They may also affect its quality, since liability rules may influence the level of innovation within the industry. If pest management consultants are subjected to stringent liability rules, the industry may become conservative; and departures from chemical control strategies will be sparingly recommended. On the other hand, less-stringent liability rules should encourage more innovation.

The distinction between a pest manager and a consultant should be emphasized. By contract the pest consultant merely supplies information to be used by the farmer as the latter chooses. No reliance is induced, and thus there is a limited basis for liability. A pest manager, on the other hand, induces reliance on the purchase of services; there is thus a stronger basis for liability if crop damage is caused by the activities of the manager.

Liability for injuries caused by an activity is imposed on three grounds: (1) the intentional infliction of harm, (2) the negligent infliction of harm, and (3) strict liability. The first is of no interest in this volume; no one would disagree that a pest manager who relied on advice known to be false should be liable for the resulting crop damage. The critical choice is between the second and third standards.

Our conclusion is that the appropriate standard is negligence rather than strict liability. Negligence is commonly defined as "conduct which falls below the standard established by law for the protection of others against unreasonable risk of harm." Generalizations about negligence are difficult, because the responsibility for judging whether an activity is negligent is shared by jury, trial judge, and appellate court; and there is a strong tendency to make particularized judgments where the equities of the individual plaintiff are given great weight and precedential consistency relatively little. Negligence law rests on the assumption that the appropriate basis for liability is fault. A plaintiff who has been harmed must establish cause in fact as well as fault. Negligence is sometimes seen as a theory of limited liability, although it has been argued that a manufacturer has the same incentive to invest in accident reduction innovations under negligence as under strict liability rules (Posner 1972).

Negligence rests on a finding that harm was caused by subjecting a person or property to an unreasonable risk of harm. Generally an actor owes others a duty not to subject them to unreasonable risks. The unreasonableness of the risk is defined by objective rather than subjective standards, and the crucial inquiry centers on the knowledge that the actor had or should have had at the time an activity was undertaken and on the comparative costs of each parties' ability to avoid the injury. A person offering skills is held, at a minimum, to the skill possessed by the general class of persons engaged in that line of activity; thus it is often said that the injury must be foreseeable to a reasonable man. Besides the likelihood of harm, the unreasonableness of the risk depends on the seriousness of the injury and the value of the interest to be sacrificed if the injury is to be avoided. The issue with respect to the latter factor is the comparative burden that would be imposed on each party if responsibility for avoiding the risk were imposed on him or her. In theory, a court is not allowed to absolve an actor because of the value of an enterprise to society as a whole; it can do so only if the cost of precaution is much higher than the benefits.

Strict liability is a theory of liability expansion. Liability is imposed if there is cause in fact; fault need not be established, but limiting conditions do exist. The classic theory imposed strict liability on the costs of an activity on a nonnegligent actor because the activity was more dangerous than ordinary activities. The classification of activity as abnormally dangerous is based on a number of factors, such as whether the activity is a matter of common usage and the value of the activity to the community. Pesticides have been so classified, but one can question whether this is justified (Harper and James 1956). The rationale for the theory has never been made clear and the doctrine has been supported by three different and somewhat inconsistent policies.

The first is that certain activities pose such a high risk of injury, even when they are carried out with due care, e.g., without negligence, that they should only be undertaken if the actor is willing to pay for injuries to others suffered as a result of the activity. The theory seems a rough probabilistic judgment that there will be a certain number of injuries no matter how much care is taken. Consistent with this theory, a plaintiff must establish that the activity caused the injury. More importantly, he or she must show that the injury falls within the scope of injuries that could reasonably be anticipated to occur as a result of the activity. A second reason is that a higher standard of liability is needed to induce the necessary modifications of an activity to make it safer; but, as mentioned above, doubt has been cast on this rationale. A third reason is a vague and somewhat inconsistent notion that losses ought to be distributed over as wide an economic base as possible, and that all purchasers of a product ought to be forced to join an insurance pool. The premiums are the increased prices a manufacturer charges to consumers.

The choice of a liability rule depends on the degree to which an industry should be subsidized. Courts would infrequently think of themselves as making such a decision, but it is clear that the standard of liability in a case that will have precedential value is to grant or withhold a partial subsidy. Sometimes there is implicit judicial recognition of the impact. For example, in the 1940s California adopted a liberal rule of consequential damages for property owners who lost profits because highway access to their tracts was made more difficult as the result of the construction of a freeway. Subsequently, the state highway department convinced the state supreme court to adopt a rule that shifted more losses to the property owners so that freeway construction would not be prohibitively expensive.

None of the traditional reasons for adopting strict liability seem to apply to the pest consultant and management situation. The production and sale of information in this context is different from the production and sale of goods. Producer and seller stand on a relatively equal footing in terms of knowledge, and thus the buyer can adequately appraise the risk of following the advice given. Farmer-buyers do not need the level of protection afforded other consumers, because the former can be reasonably expected to expend the resources to inform themselves about the product they are buying. Loss-spreading among as broad a base as possible would not be accomplished by the adoption of a standard of strict liability, because the pest management industry is small and not heavily capitalized. The existence of insurance would not solve the problems of a small resource base. If anything, this suggests that farmers are a better risk-bearing group.

REFERENCES

Butler, W. (1974) Federal pesticide law. *In* Federal Environmental Law, E. Douglin and T. Gilbert, eds., 1232.

Dawson (1974) The Self-Serving Intermeddler, 87 Harv. L. Rev. 1409, 1421. Green Tree Estates, Inc. v. Fursterburg, 21 Wis. 2d 193 124 N.W. 2d (1963).

Federal Environmental Pesticide Control Act of 1972, A Compromise Approach, 3 Ecology L.Q. 277, 292 (1973).

Federal Register, January 24, 1968, Vol. 33, No. 16.

Federal Register, August 11, 1970, Vol. 35, No. 155.

Federal Register, December 27, 1973, Vol. 38, No. 247, p. 35388.

Federal Register, August 22, 1974, Vol. 39, No. 164, p. 30465.

Harper and James (1956) Vol. II, The Law of Torts, Paragraph 14.16 and 1968 Supplement.

Hearings on S.232, S.272, S.660, and S.745 (1971) Before the Subcommittee on Agricultural Research, and General Legislation of the Senate Committee on Agriculture and Forestry, 92d Congress, 1st Session 67.

Posner, R. A. (1973) Economic Analysis of Law 90. Boston: Little, Brown & Co.

Rodgers (1970) The Persistent Problems of Persistent Pesticides: A Lesson in Environmental Law, 70 Colum. L. Rev. 567, 578-80.

Senate Report No. 92-970 (1972) 92d Congress, 2d Session 4.

Shell Chemical (1974) EPA Docket No. 145, et al., October 1, 6 Environmental Reporter Cases 2047 (Aldrin-dieldrin decision).

9

MONITORING AND RESEARCH ON EFFECTS OF PESTICIDES

THE NECESSITY OF MONITORING

The use of pesticides requires the injection of the materials into the environment. Even with the best methods of application, only a portion--probably quite a small portion in most cases--reaches the target species and works its intended effect. The remainder never accomplishes any useful work. The application of "excess" material is not necessarily "overapplication." Rather, the applicator has insufficient time and equipment to search out and kill individual pest organisms, so he instead applies pesticide so that it reaches a critical concentration over a broad area, even though the pest may not be present throughout. For example, a few grams of the right insecticide would kill 100,000 corn rootworms if all the insects were located in one square meter of soil. Rootworms are not so concentrated, however, and the farmer instead chooses to apply several times as much material over many hectares of land. The unused pesticide may remain in the environment, where it may produce undesired effects. It can move from place to place and be transformed into different chemical entities, although in many cases we can neither predict the actual pattern nor say how the material will affect nontarget organisms once movement or transformation has occurred.

The uncertainty of the ultimate effect on nontarget species of materials applied as pesticides is the major motivation for both monitoring and research on the effects of pesticide residues. All programs of monitoring and many programs of research are publicly financed and are consequently a cost of pest control that is born by the taxpayer, rather than the applicator, buyer, merchandiser, or manufacturer of the material. It is worth noting at this point that the existence of public monitoring and research programs is an indication that the formal registration procedures for pesticides indeed do not answer all questions about pesticidal toxicity. From this viewpoint, it is evident that the registration protocols are incomplete. Formal registration or label approval is the first step in registration, and it is privately financed; monitoring and research constitute a continued examination of the pesticide's registration, and they are publicly financed.

THE BENEFITS OF MONITORING

The costs of monitoring programs are born diffusely by the general tax-paying population, and the benefits are enjoyed by the same group. The form of the benefits are qualitative, and it would be difficult or impossible to quantify them. In part they take the form of "anxiety-reducers": To the extent that pesticide use places residues with uncertain effects in unknown locations in uncertain form, monitoring can tell one where the residues are and what their chemical form is. Such knowledge helps set boundaries on the types of problems that might be caused by the residues, and the derived certainty of this knowledge has a value.

A second type of benefit is the potential for a monitoring program to serve as an early-warning system for the buildup of residues in specific areas that could lead to damage among nontarget organisms. Successful operation of monitoring programs in the latter capacity depends upon (1) concomitant knowledge of what constitutes a "dangerous level, (2) the source of the input, and (3) adequate socia legal institutions that can control or curtail the source of the pesticide and thus lead to a reduction of the level of residues. The dollar value of such predictive monitoring could be quite large. For instance, if monitoring predicted that residues of a particular pesticide were increasing in a commercial fishery and it was evident that the residues would soon exceed the tolerance set for that pesticide, then corrective action in the use of the chemical could lead to saving the fishery as an economic resource.

The fact that monitoring can yield results that have an actual or potential real value in and of themselves should not obscure a more fundamental notion: The cost of monitoring must be subtracted from the overall benefits derive from the use of pesticides in any calculation of costs and benefits. In other words, the benefits of monitoring mentioned above are ficticious, in the sense that if pesticid were not used at all or if the tests needed to secure label approval were totally satisfactory to all concerned, there would be no need for monitoring.

THE RATIONALE OF MONITORING

Monitoring--the systematic gathering across time, space, and media of the concentrations of specific chemical species--is a very new type of scientific activity that is rapidly evolving as experience is gained. Its costs are sufficient to demand careful thought for construction, maintenance, and operation of programs.

The interpretation of data generated should tend to modification of the programs so that information gathered is useful. In this section we turn to a discussion of the principles on which monitoring systems should be constructed and to the relationships monitoring programs can have to other types of government programs. In a following section, we will briefly survey the current programs of monitoring and present estimates of their costs. It was beyond the scope of this study to thoroughly evaluate the adequacy of the current monitoring efforts, but we will conclude with some very general remarks that we hope will be useful in the on-going and continuing efforts to improve the usefulness of the activities.

PRINCIPLES FOR CONSTRUCTION OF MONITORING NETWORKS

Because monitoring programs are a relatively new type of scientific program, it is not yet possible to ennuciate detailed and broadly accepted sets of principles that should be applied in their construction. Nevertheless, it is possible to discuss some general rules relating to these questions: (1) How are sites for monitoring stations to be selected? (2) What materials or organisms should be analyzed? (3) How are materials to be collected for analysis? (4) How are laboratory analyses to be checked for quality control? and (5) What chemicals should be monitored on a routine basis?

A recent review paper on pesticides and wildlife (Moriarty 1973) emphasizes some points of particular importance that must be considered in answering questions 1, 3, and 4 above. Detection procedures, at least for the chlorinated hydrocarbon insecticides are now fairly sensitive and reliable. Nevertheless, analyses performed by different individuals or by different laboratories may differ by a factor of two. Moreover, there are serious problems of standardization among laboratories. The distribution of residues in organisms from a given population is highly skewed: A large number of individuals may have no residues, and a very few individuals may have very high concentrations.

The implications of these two generalizations for monitoring systems are straightforward. First, trustworthy analytical results require dedicated, competent people in the laboratory and a formal system of cross-checking results to maintain high standards of accuracy. Second, because the distribution of residues in a monitored populati(of organisms is not likely to be normal, great care must be exercised in the selection of individuals from the population for analysis. If the selection technique is biased either in favor of individuals with high residues or those with low residues, a false measurement will be obtained of the true levels of residues in the population. Furthermore, if individuals from the population are composited before analysis, the actual distribution of residues will be obscured. Comparison of average residue levels between different populations--in time or space--may be unwarranted if analyses are performed on composited individuals drawn from a population in which the actual distribution of residues is not normal. The point about the distribution of residues in the population is of considerable importance, because several established monitoring programs rely on compositing individuals before analysi

The selection of materials and organisms to be monitore presents some difficult theoretical and practical problems. It seems clear that the major inorganic portions of the biosphere should be monitored: soil, water, and air. Selection of organisms, however, is more complex: We would want organisms from different trophic levels--from terrestrial, freshwater, estuarine, and saltwater ecosysten as well as those of special aesthetic or economic importance to man. The desire to monitor these different types of organisms must, however, be tempered by their availability. For example, the bald eagle might be desirable t monitor because of its high trophic level or because of its aesthetic significance as a political symbol in the United States. The bald eagle, however, is an endangered species; and it would seem unwise to kill rare birds merel to monitor them for pesticides.

The selection of chemicals to monitor also presents difficult theoretical and practical problems. We would of course like to restrict our monitoring to those chemicals that pose a threat to man and his environment. Knowledge of how molecular structure relates to environmental hazard, however, is still very primitive; and it is not yet possible to identify candidate compounds for monitoring solely on the basis of their molecular structure. The following properties of compounds should be considered:

Property	Indication for Monitoring
Toxicity	High acute toxicity; known chronic toxicological effects such as mutagenicity, carcinogenicity, or teratogenicity; broad spectrum; many sublethal effects such as induction of sterility, effects on behavior, reduced resistance to disease, subliminal neurological impairment.
Persistence	Long lasting (greater than 1 year)--either pesticide or its breakdown products and metabolites.
Mobility	High water or lipid solubility; high vapor pressure; high but reversible adherence to soil particles.
Use	High amounts; near highly valued ecosystems; during breeding times of wildlife; etc.
Purity	Low: contaminated by other toxicants (e.g., 2,4,5-T contaminated with dioxin) or a mixture of substances (e.g., toxaphene).

The difficulty or relative expense of analyzing material for a given chemical will inevitably exert an influence on the monitoring efforts for that chemical. Monitoring agencies, however, should be firm in arguing that expense alone should not be the sole criterion for the effort exerted. As will be discussed below, the money currently expended on monitoring is so low that an increase in monitoring expenditures is not likely to make the total monitoring expenditures very large relative to total government expenditures.

RELATIONSHIP OF MONITORING PROGRAMS TO OTHER TYPES OF GOVERNMENT PROGRAMS

In the absence of information or theory about the potential problems that pesticide residues might cause, a random sampling of organisms, soil, water, and air would be as useful as any other scheme of monitoring. Fortunately we do not operate in such a vacuum: The established monitoring programs have a relationship to knowledge already gathered about pesticide residues and their effects and also to other government programs that may have been started for reasons quite independent of any concern about pesticides. Nevertheless, a listing of how monitoring programs might relate to other government work is useful, because it will point out pathways by which current programs could be improved. At the most general level, monitoring could be related to research efforts, efforts to track the source of entrance of pesticides into the environment, and regulatory schemes.

Monitoring networks could be an integral part of research efforts on the effects of pesticides. There are four major questions about the effects of pesticides that research attempts to answer:

- What is the efficacy of the chemical for its intended use? This question is answered rather carefully during the formal registration process. No monitoring programs are currently linked to research of this type.
- What are the levels of pesticides and their metabolites in the environment now, and what are their trends in time? Most monitoring systems are designed at least in part to answer this question.
- What are the effects of pesticide residues and their metabolites on nontarget organisms? The determination of effects is a difficult task at best, and monitoring can contribute only part of the evidence necessary to demonstrate cause and effect relationships. Controlled laboratory experiments and epidemiological surveys must contribute other information. Some monitoring programs are linked to studies on the effects of the residues.
- How are pesticide residues and their metabolites transformed and transported in the environment? This is one of the most intriguing, baffling, and important questions about pesticidal residues. Its importance stems both from the usefulness of answers to the question to help predict adverse effects of pesticides and from the general interest in transport mechanisms of many different types of chemical pollutants, such as heavy metals, radionuclides and other chemicals. No monitoring programs have yet been linked to studies on the transport of residues.

Monitoring efforts could be designed to help give information about the source of injection of pesticides into the environment. Four general types of operations could lead to the introduction of the chemicals:

- Manufacture. Some monitoring systems do give information that can be linked to spills of material from pesticide manufacturing plants.
- Transit. Accidents during transportation are discussed in Chapter 5. No monitoring programs are designed to track this source.
- Use. All monitoring systems track use in a very general way in that the level of pesticides found is in some way proportional to the amounts of material used. Use, however, is for specific purposes, such as on a particular crop, or for a particular disease organism. There are no monitoring programs designed to track residues from specific types of use.
- Disposal. Pesticides are frequently disposed of either intentionally or inadvertently. Batches of poor quality at the manufacturing plant, unused chemicals at the end of an agricultural season, and small amounts left in "empty" pesticide containers may all end up in some type of dump from which they travel into the general environment. No monitoring programs follow pesticides from this source.

The final type of monitoring program is to have the activity linked directly to a program for regulating the concentration of the chemical. Such programs now constitute a large part of our total monitoring system because of the efforts of the FDA and the USDA in monitoring residues in food and feed products. These programs are discussed in Chapter 8.

THE DEVELOPMENT OF CONTEMPORARY MONITORING PROGRAMS

With the exception of monitoring programs of the FDA for residues of pesticides in food and feed, regular monitoring schemes did not originate until after 1964. The development of monitoring systems for media other than food and feed was closely tied to the development of the Federal Working Group on Pest Management.

The first efforts to coordinate pest control activities among different federal departments came with the establishment of the Federal Pest Control Review Board in 1961. This Board allowed representatives of the Secretaries of Agriculture; Interior; Defense; and Health, Education, and Welfare to meet and review pest control operations conducted by their respective departments. The President's Science Advisory Committee called for a strengthening and broadening of this Board's role in its report, *The Use of*

Pesticides (1963). As a result, the Board was superseded in 1964 by the Federal Committee on Pest Control, which contained representatives from the same four departments as did the Federal Pest Control Review Board. Over a period of time, the Federal Committee established a series of subcommittees, one of which was assigned the task of coordinating monitoring efforts. The Monitoring Panel, established in 1964, was responsible for two developments that added to pesticide-monitoring capacity: (1) the formation of a National Pesticide Monitoring Program, and (2) the initiation of a new journal, *Pesticides Monitoring Journal*, for the publication of data derived from monitoring activities.

The National Pesticide Monitoring Program derived directly from the recommendations of the President's Science Advisory Committee that the government monitor pesticides in "air, water, soil, man, wildlife, and fish." The monitoring panel published a catalog of federal pesticide monitoring activities in 1968 (Federal Committee on Pest Control 1968), by which time a National Pesticide Monitoring Program had been implemented. The first issue of *Pesticides Monitoring Journal* was published in June 1967, and it has been published quarterly since that time.

The Federal Committee on Pest Control and its subcommittees experienced a number of reorganizations and name changes, but its functions remained relatively unchanged: It became the Working Group of the Subcommittee on Pesticides of the Cabinet Committee on the Environment in 1969; in 1970 the Cabinet Committee was dissolved, but the Working Group continued and was made responsible to the Council on Environmental Quality. Its membership was expanded to include the Departments of Transportation, State, and Commerce and the EPA when it began to report to the Council on Environmental Quality. Observers from the Office of Science and Technology, the Office of Management and Budget, and the Office of Intergovernmental Relations were also added at this time. In 1971, the Working Group again changed its name to the current one, Federal Working Group on Pest Management.

The first National Pesticide Monitoring Program was published in 1967 (Pesticides Monitoring Journal 1967). A revised version was published in 1971 (Pesticides Monitoring Journal 1971), and a third version is being prepared by the Environmental Protection Agency. Thus the monitoring program conducted by the federal government is a dynamic program constantly undergoing change in light of past monitoring results and new changes in pest control practices. This is as it should be, but our task is made somewhat more difficult by the constant changes: We want to examine the scope, methodology, and costs of the monitoring program, but these parameters are constantly changing. The following discussion is thus an overview of monitoring as it was in about the period 1971-1974. Details of monitoring activities will change rapidly in the future.

MONITORING PROGRAMS, 1971-1974: SCOPES, METHODOLOGIES, AND COSTS

Responsibility for different parts of the National Pesticide Monitoring Program was originally scattered across many agencies of the federal government: the Bureau of Sport Fisheries and Wildlife, the Bureau of Commercial Fisheries, the Federal Water Pollution Control Administration, and the U.S. Geological Survey in the Department of the Interior; the PHS and the FDA of the USDHEW; and the Agricultural Research Service, the Consumer and Marketing Service, and the Forest Service of the USDA. The consolidation of many environmental programs into the newly formed EPA in 1970 and the enactment of the Federal Environmental Pesticide Control Act in 1972* resulted--over a period of time--in the consolidation of the monitoring activities into the Technical Services Division of the Office of Pesticide Programs in the EPA. By early 1974, prime responsibility for monitoring pesticide residues in human tissue, soils, fresh water, and fish from estuaries and oceans were in the hands of the Technical Services Division; some of the programs are conducted with the help of other agencies, however. Prime responsibility for monitoring freshwater fish and wildlife remain in the hands of the Bureau of Sport Fisheries and Wildlife of the Interior Department. We now turn to an examination of each of these efforts.

Human Tissues

Monitoring of pesticides in human tissues is designed to follow levels and changes of the amounts of chlorinated hydrocarbons in adipose (fat) tissue of people in the continental United States. The EPA contracts with

*86 Stat. 996. Section 20 of FEPCA explicitly directs the EPA to formulate a National Monitoring Plan and conduct monitoring activities.

pathologists in 75 randomly selected Standard Metropolitan Statistical Areas (SMSA) to collect adipose tissue at time of autopsy or therapeutic surgery. Analyses for chlorinated hydrocarbons are conducted under contract by laboratories of the public health departments in the states of Colorado and Michigan.

Soils

Land is divided into three basic categories for the soil-monitoring program: cropland, rural noncropland, and urban land. The sampling intensity for cropland is one 10-acre site per 40,000 acres; for rural noncropland the intensity is one 10-acre site per 400,000 acres. Urban locations are identified by random selection of SMSA. For those SMSA selected, one site per square mile is picked within the boundaries of the core city and one site per 20 square miles in the surrounding suburbs. Samples consist of multiple soil cores that are sieved, composited, and then analyzed. Other data collected includes (1) pesticides used on the land, (2) crops grown, and (3) samples of crops if they are available. Samples are collected by the USDA and analyzed in laboratories of the EPA.

Air

The EPA has plans to monitor air for pesticide residues, but they are not operational due to technical difficulties.

Fresh Water and Freshwater Sediments

Surface water and the sediment under it from 17 major drainage basins is monitored at 162 randomly selected sites for ambient levels of pesticide residues. Water samples are depth integrated where possible and are sampled four time per year. Sediment samples are collected by a series of cores across the riverbottom and are sampled twice a year. Samples are collected by the U.S. Geological Survey and are analyzed in laboratory facilities of the EPA.

Estuarine Fish

Net-collected samples of 25 herbivorous fish and 25 carnivorous fish are composited separately from each of 107 primary estuaries twice a year. If high levels of residues are found, lower life forms such as mollusks, crustaceans, and plankton may be sampled to pinpoint the origin of the chemical. Collection of materials and analyses are both performed by EPA personnel.

Ocean Fish

The EPA began a pilot program of monitoring fish in the open ocean in 1974 in cooperation with the National Marine Fisheries Service (NMFS) of the Department of Commerce. The NMFS collects fish on regularly scheduled cruises over the continental shelf and over deeper waters. Samples are sent to laboratories of the EPA for chemical analyses. Larger fish are analyzed as individuals, but smaller specimens may be composited to yield sufficient tissue for analysis. Selection of species for anlysis is somewhat haphazard in that analyses are performed on the animals that become available.

Freshwater Fish

The Division of Fishery Services of the Bureau of Sport Fisheries and Wildlife, U.S. Department of the Interior, has responsibility for monitoring fish in fresh water. A total of 100 sites were selected on the basis of (1) availability of desired species, (2) relationship to a station in the freshwater monitoring network, (3) existence of a state-operated monitoring program or cooperative participation by a state conservation agency, and (4) availability of suitable manpower and facilities to make field collections. During the fall of each year, 3-5 uniform adults are selected in each of three species at each site. The animals within each species are composited and then analyzed. Replicate collections are made of one of the species, and a total of 400 samples are thus analyzed each year. Personnel of the Division of Fishery Services, state agencies, and commercial fisherman collect the animals. Chemical analyses are performed in the Denver Wildlife Research Center of the Department of the Interior. Approximately 10 percent of the samples (40 per year) are cross-checked for laboratory quality control by the Fish-Pesticide Laboratory of the U.S. Department of the Interior in Columbia, Missouri. Material sent to Missouri is also checked for a variety of other pesticides and other types of toxic chemicals.

Birds: Starlings and Ducks

The Division of Wildlife Services of the Bureau of Sport Fisheries and Wildlife, U.S. Department of the Interior, is responsible for monitoring starlings (*Sturnus vulgaris*) for residues of some persistent chlorinated hydrocarbons and for some heavy metals. The nation is divided into 40 blocks, 5 degrees latitude and 5 degrees longitude on each side. One to four randomly selected sampling sites are then identified in each block; a total of 128 sampling sites have been established. At each sampling site, 10 birds are trapped or shot and then composited to make one sample. Collection of birds is performed by the Division of Wildlife Services and analyses are performed under contract by the WARF Institute (chlorinated hydrocarbons) or by the Denver Wildlife Research Center (USDA) (heavy metals). The program is designed solely to follow the levels of general pesticide residues in a common bird rather than to determine the effect of the residues on the starling population.

The Patuxent Wildlife Research Center of the Bureau of Sport Fisheries and Wildlife has responsibility for monitoring pesticide residues in ducks (either mallard ducks, *Anas platyrhynchos*, or black ducks, *Anas rubripes*).* Ducks are monitored every third year as a part of the Nationwide Survey on Wildlife Productivity. Cooperating hunters mail duck wings to the Patuxent laboratory. The wings are separated by state where shot and put into pools of 25 wings. Individual pools within each state's set are selected randomly, and the number of pools selected from each state for analysis is rougly proportional to the total harvest from that state. Chemical analysis, performed under contract by the WARF Institute, is conducted on a composit of each pool of 25 wings.

Costs

It is difficult to estimate precisely the costs of monitoring, because the programs are sometimes closely connected to other programs and it is difficult to separate the costs of the different programs. Furthermore, some monitoring

*The Patuxent laboratory until 1974 listed a monitoring program for bald eagles (*Heliaeetus leucocephalus*); this program continues but is now considered strictly a research program rather than a monitoring effort.

programs rely on cooperation with state agency personnel or with private citizens; thus, the costs for this labor do not show up on the budget estimates of the federal agencies. Despite these difficulties, we estimate that at the current levels, monitoring programs cost about $1.2 million per year. If the costs of regulating the levels of residues in food and feed are ascribed to monitoring rather than to regulation, the cost of monitoring is $4.6 million per year. In either case, the Committee judges that monitoring as it now stands cannot be called a very expensive effort.

Summary Charts

The total monitoring network is complex, but its highlights are summarized in Tables 30, 31, 32, and 33. Regulatory programs in which feed and food are monitored for residues are included in the tables, but they were discussed in more detail in Chapter 8.

EVALUATION OF MONITORING PROGRAMS

It was beyond the scope of this Committee's charge to evaluate the National Monitoring Network exhaustively as it now stands. Nevertheless, our survey of monitoring networks leads us to make some general comments. Our motivation for these comments is twofold: First, there are some gaps in the current efforts to generate information about the prevalence of pesticide residues that leads us to believe we may be ignorant of some of the costs that are associated with the use of pesticides. Thus, our examination of the costs and benefits associated with the use of different pest control technologies must necessarily remain incomplete for the moment. Second, we hope that our comments on current monitoring efforts will be helpful to those trying to improve them.

Coverage of Media

Monitoring plans for soils, fresh water, and sediments under fresh water appear to be well formulated. The soils-monitoring network has been in actual operation for almost 10 years; the monitoring of fresh waters and their sediments is of more recent origin, and we urge that sufficient funds be available to ensure the operation of this program.

We are distressed by the absence of monitoring efforts in air, estuarine water, ocean water, and groundwater. First, these four media represent sinks for the ultimate disposal of pesticides and other pollutants; this condition is especially true of the atmosphere, the oceans, and some groundwater. For this reason, we must know the residue levels in these resources. Second, pesticides--like radionuclides and other chemicals--may have a dynamic, complex circulation pattern between these four media and other parts of the biosphere. Our understanding of the transport mechanisms between the different media is very fragmentary, and a monitoring system well integrated with research efforts on the transport mechanisms in the circulation of pesticides would be welcome. Third, all of these media, especially air and groundwater, are resources that are directly utilized by humans. There is no replacement for the air and no practical replacement for groundwater. We must carefully monitor these irreplaceable resources so that we can prevent their contamination.

Selection and Collection of Species

The principles by which species and individuals should be identified as useful candidates for monitoring are the following:

1. Man among all the species is the only obvious candidate for monitoring. The reason for monitoring man is clearly one of self-interest.
2. Species from a variety of taxonomic groups should be monitored. Green plants, microbial organisms, and a wide variety of animals--both vertebrates and invertebrates should be selected.
3. Species should be selected from a variety of trophic levels. Primary producers (green plants), herbivores (plant-eating animals), primary carnivores (herbivore-eating animals), secondary carnivores (animals that eat primary carnivores), omnivores (animals that eat a mixture of plant and animal materials), and decomposers.
4. Organisms should be selected from a variety of geographic locations. Cropland, rural noncropland, urban areas, wetlands, and bodies of water should all be sampled.
5. Individuals selected from a given species should include a variety of ages and both sexes (if applicable). (Should research indicate that age and sex of an individual are irrelevant to the individuals' load of residues, this requirement can be dropped.)

Program	Responsibility[a,b]	Collection[a,c]	Analysis[a,d]	Initiated[e]	Method of Site Selection	Number of Sites	Method of Sample Collection	Sampling Frequency
Human tissue	TS	Contract	Contract	1967	Random selection of SMSA. Individuals allocated by sex, age, race.	75 SMSA	Adipose tissue taken in surgery or autopsy	2,000 individuals per year
Soils								
Cropland	TS	USDA	TS	1968	One 10-acre site for every 40,000 acres	10,468	Fifty soil cores, 2" in diameter by 3" deep composited for each 10-acre site	Once every 4 years at each site
Rural, noncropland	TS	USDA	TS	1968	One 10-acre site for every 400,000 acres	3,832		
Urban	TS	USDA	TS	1969	Random selection of SMSA. One site per square mile in core city; one site per 20 square miles in suburbs.	2,000	Sixteen soil cores 2" in diameter by 3" deep composited for each site	Four times yearly
Air	TS	Not Operating Due to Technical Problems						
Fresh water and Freshwater sediments	TS	USGS	TS	1973	Nation divided into 17 major drainage basins. Sites selected randomly within basins.	162	Depth integrated where possible. Core series across riverbottoms	Four times yearly Two times yearly
Estuarine fish	TS	Contract	TS	1972	Primary estuaries	About 107	Net collection of 25 herbivorous fish and 25 carnivorous fish collected and composited separately. Mollusks, crustaceans, and plankton sampled if levels are high.	Two times yearly
Ocean fish	TS	NMFS	TS	Pilot 1974	Regularly scheduled cruises of NMFS. Deep water and continental shelf both sampled.	--	Various species. Individuals may be composited to attain sufficient weight.	--
Freshwater fish	DFS	DFS & Others	DWRC FPRL Contract	1967	(1) Availability of desired species (2) Relationship to freshwater monitoring station (3) Existence of state program (4) Availability of manpower and facilities for collection	100	Select 3-5 uniform adult fish in each of 3 species in fall. Composit members of each species with one another. Two replicas of one composite mixture at each sampling site.	Yearly

TABLE 30 (continued)

Program	Responsibility[a,b]	Collection[a,c]	Analysis[a,d]	Initiated[e]	Method of Site Selection	Number of Sites	Method of Sample Collection	Sampling Frequency
Birds								
Starlings	DWS	DWS	Contract	1967	National divided into 40 blocks--5° latitude and 5° longitude on each side. One-four randomly selected sites per block.	128	Trap or shoot 10 birds per site. Composite to make one sample.	Alternate years
Ducks	PWRC	Hunters	Contract	1965-1966	Part of nationwide survey on wildlife productivity	--	Wings separated by state where shot, put into pools of 25 wings each. Pools selected randomly but roughly proportionally to total state harvest. Composite wings in each pool for analysis.	Every third year
Food and feed								
Market basket	FDA	FDA	FDA	1964	Retail market basket samples taken from representative municipalities within broad geographic areas of the United States and based on USDA household food-consumption data	30	Simulate the 14-day diet of 15-20-year-old males in the area of collection (approximately 117 food items). Geographic differences recognized in relative amounts of food collected.	One time yearly at each site
Pesticide program	FDA	FDA	FDA	1973	Surveillance sampling of nationally important foods required within each FDA District where produced; locally important produce sampled according to District initiated plans	--	Items selected on basis of production volume in area. Samples constructed so as to yield estimates of (1) relative contamination of same item in different areas, and (2) general contamination across nation. Samples also indicate items requiring regulatory action. Imports also sampled.	Continuous
Egg products	AMS	AMS	AMS	1971	Plants producing liquid, frozen, and dried eggs	150	Four aliquots drawn from production line over 2-day period and composited for	Quarterly at each plant

poultry	slaughter house	animals arriving for slaughter: calves, cattle, swine, lambs, chickens, turkeys, ducks, geese, and horses. Total of 3,200 animals tested in 1973.	uous except for animals slaughtered at special times of year only

[a]Abbreviations: AMS (Agricultural Marketing Service, USDA); APHIS (Animal and Plant Health Inspection Service, USDA); DFS (Division of Fishery Services, Bureau of Sport Fisheries and Wildlife, U.S. Department of the Interior); DWRC (Denver Wildlife Research Center, Bureau of Sport Fisheries and Wildlife, U.S. Department of the Interior); DWS (Division of Wildlife Services, Bureau of Sport Fisheries and Wildlife, U.S. Department of the Interior); FDA (Food and Drug Administration, Department of Health, Education, and Welfare); FPRL (Fish Pesticide Research Laboratory, Bureau of Sport Fisheries and Wildlife, U.S. Department of the Interior); NMFS (National Marine Fisheries Service, U.S. Department of Commerce); PWRC (Patuxent Wildlife Research Center, Bureau of Sport Fisheries and Wildlife, U.S. Department of the Interior); TS (Technical Services Division, Office of Pesticide Programs, Environmental Protection Agency); USDA (U.S. Department of Agriculture); USGS (U.S. Geological Survey, U.S. Department of the Interior).

[b]Responsibility means the agency responsible for constructing and operating the monitoring program.

[c]Collection means the agency or person who actually collects the sample.

[d]Analysis means the agency or organization responsible for analyzing the sample for residues.

[e]Initiated means the year in which the program began in the shape it is currently operated. The 1973 initial date for the Pesticide Program reflects 1973 data on pesticides and related industrial chemicals recovered through the multiresidue methods used in current and prior programs.

TABLE 31 Production and Monitoring of Selected Herbicides, 1974*[a]

Chemical		Production, Million Pounds of Active Ingredient	Human Tissue	Soils, Fresh-water Estuarine Fish[b]	Ocean Fish	Fresh-water Fish	Star-lings	Ducks	FDA[c]		Egg Prod-ucts	Meat and Poultry
Common Name	Trade Name								Fatty Food	Nonfatty Foods		
Atrazine	AAtrex	90		+					ND	P(25)		
2,4-D[d]	Several	45		+					mb	mb		
MSMA-DSMA	Several	35										
Sodium chlorate	Several	30										
Trifluralin	Treflan	25		+					C	C		
Propachlor	Ramrod	23							ND	ND		
Chloramben	Amiben	20										
Alachlor	Lasso	20							ND	ND		
CDAA	Randox	10							ND	NR		
Bromacil	Hyvar	8										
Nitralin	Planavin	8							ND	P(60)		
Dicamba	Banvel-D	6										
Diuron	Karmex	6							ND	ND		
Propanil	Stam F34	6							ND	NR		
Butylate	Sutan	6										
2,4,5-T[d]	Several	6		+					mb	mb		
---	DEF	5		+					P(60)	C		
EPTC	Eptam	5										
Simazine	Princep	5		+					ND	NR		
Dalapon	Dowpon	5										
Fluometuron	Cotoran	4										
Propazine	Milogard	4		+					NR	P(41)		
Maleic hydrazide	Several	3										
Dinoseb	Premerge	3										
Diphenamid	Dymid, Enide	3										
Merphos	Folex	3		+					C	C		
2,4,5-TP	Silvex	3		+					mb	mb		
Picloram	Tordon	3		+								
Cacodylic acid	Several	2										
Vernolate	Vernam	2										
Naptalam	Alanap	2										
Linuron	Lorox	2										
Norea	Herban	2										
TBA	Trysben	2										

Acrolein	Aqualin	<1			
---	Ammate	<1			
Amitrol	Several	<1			
---	Bandane	<1			
Barban	Carbyne	<1			
Benefin	Balan	<1	+	C	C
Bensulide	Betasan, Prefar	<1		C	P(70)
CDEC	Vegadex	<1		C	C
Chlorbromuron	Maloran	<1			
Chlorboxuron	Tenoran	<1			
Cycloate	Ro-Neet	<1			
Diallate	Avadex	<1			
Fenuron	Dybar	<1			
Fluorodifen	Preforan	<1			
Molinate	Ordram	<1			
Monuron	Telvar	<1		ND	ND
Pebulate	Tillam	<1			
Siduron	Tupersan	<1			
Terbacil	Sinbar	<1		ND	ND
Terbutol	Azak	<1			
TCA	Several	<1			
Triallate	Avadex BW	<1			
2,4-DB	Butyrac	*e*		mb	mb
Dichlobenil	Casoron	*e*		C	P
MCPA	Several	*e*		mb	mb
Neburon		*e*		ND	ND
Nitrofen	TOK	*e*		C	C
Prometryne	Caparol	*e*		ND	P(50)
2,3,7,8-tetra-chlorodibenzo-*p*-dioxin[h]		*e*		P(70)	ND
Trichlorobenzene		*e*		P(60)	C
Trichlorobenzoic acid		*e*		mb	mb

*Footnotes follow Table 33.

TABLE 32 Production and Monitoring of Selected Insecticides and Acaricides, 1974*[a]

Chemical: Common Name	Trade Name	Production, Million Pounds of Active Ingredients	Human Tissue	Soils, Freshwater Estuarine Fish[b]	Ocean Fish	Freshwater Fish	Starlings	Ducks	FDA[c] Fatty Food	FDA[c] Nonfatty Foods	Egg Products	Meat and Poultry
Toxaphene	Several	50		+	+	+		+	C	C		+
DDT[f]	Several	45	+	+	+	+	+	+	C	C	+	+
PCB[g]	Several	40	+	+	+	+	+	+	C	C	+	+
Carbaryl	Sevin	45								pp,mb		
Methyl parathion	Several	45		+					C	C, S		+
Malathier	Cythion	35		+					ND	C, S, pp		+
Chlordane	Several	25	+	+	+	+	+	+	C	C		+
Parathion	Several	15		+					C	C, S		+
Aldrin	Several	10		+	+	+		+	C	C	+	+
Methoxychlor	Several	10		+	+			+	C	C		+
Diazinon	Spectracide	10		+					C	C, S		+
Carbofuran	Furadan	8										
Disulfoton	Di-Syston	8		+					ND	P(50), S		+
Phorate	Thimet	8		+					P(80)			
Heptachlor	Several	6	+	+	+	+	+	+	C	C	+	+
Metalkamate	Bux	6										
Chlorpyrifos	Dursban	5		+					C	C, S		
Dicofol	Kelthane	4		+	+			+	P	P		
Azinphos-methyl	Guthion	4		+					ND	S		
Fensulfothion	Dasanit	4		+					ND	S		
Methomyl	Lannate	3										
Fonofos	Dyfonate	3		+					ND	C, S		
Ethion	Nialate	3		+					C	C, S		
Ronnel	Korlan	3		+					C	C, S		+
Carbophenothion	Trithion	2		+					P(60)	C, S		
Naled	Dibrom	2							ND	NR		
Dimethoate	Cygon	2		+					ND	S		
Aldicarb	Temik	2										
Endosulfan	Thiodan	2		+	+				C	C		
Chlorobenzilate	Several	2		+	+				P(80)	P(80)		
Crufomate	Ruelene	2							ND	NR		+
Biothion	Abate	<1		+								
Coumaphos	Co-Ral	<1		+					ND	S		+
Fenthion	Baytex	<1		+					ND	P(45)		+

Dichlorvos	DDVP	<1							ND	NR		
Mevinphos	Phosdrin	<1							ND	NR		
Phosphamidon	Dimecron	<1							ND	S		
Dicrotophos	Bidrin	<1		+					ND	S		
Monocrotophos	Azodrin	<1							ND	S		
---	Aspon	<1										
TEPP	Several	<1										
Ethoprop	Mocap	<1							ND	P(55)		
Dieldrin	Several	<1	+	+	+	+	+	+	C	C	+	+
Endrin	Several	<1		+	+	+	+	+	C	C	+	+
Lindane (BHC)	Several	<1	+	+	+	+	+	+	C	C	+	+
Mirex	Dechlorane	<1	+	+	+	+	+	+	P(70)	C		+
Chloropropylate	Acarlate	<1							ND	C		
Propargite	Omite	<1										
Binapacryl	Morocide	1							P(65)	P		
Bromophos-ethyl	Nexagan	*e*							ND	C		
Chlorbenside		*e*		+	+				C	C		
Chlordecone	Kepone	*e*		+	+			+	P	P(45)		
Chlorfenvinphos		*e*							ND	NR		
Chlorinated naphthalene		*e*							P(65)	C		
Crotoxyphos	Ciodrin	*e*							ND	S		
Demeton	Systox	*e*		+					ND	S		
Dialifor	Torak	*e*							P(50)	C		
Dioctylphthalate		*e*							C	ND		
EPN		*e*		+					C	C, S		
Famphur		*e*							ND	NR		
Fenitrothion		*e*							ND	C, S		
Methidathion	Supracide	*e*							ND	P(35)		
---	Perthane	*e*							C	C		
Phenkapton		*e*							ND	C, S		
Phosalone	Zolone	*e*							C	C, S		
Phosmet	Imidan	*e*		+					ND	S		
---	Phostex	*e*							ND	P(65)		
Sulfotepp	Bladafume	*e*							ND	C		
Tetrachlorvinphos	Gardona	*e*							ND	NR		+
Tetradifon	Tedion	*e*							C	C		
Tetrasul	Animert	*e*							C	C		
Trichlorfon	Dylox	*e*		+								+
---	Zytron	*e*							ND	C		

*Footnotes follow Table 33.

TABLE 33 Production and Monitoring of Selected Fungicides and Miscellaneous Compounds, 1974[a]

Chemical		Production, Million Pounds of Active Ingredient	Human Tissue	Soils, Fresh-water Estuarine Fish[b]	Ocean Fish	Fresh-water Fish	Star-lings	Ducks	FDA[c]		Egg Prod-ucts	Meat and Poultry
Common Name	Trade Name								Fatty Foods	Nonfatty Foods		
Pentachlorophenol	Several	46								mb		
Dithio-carbamates	Several	40										
Tetrachlorophenol	Several	20										
Captan	Orthocide	18		+					ND	P(20)		
PCNB	Several	3		+					C	C		
Dodine	Cyprex	3										
Captafol	Difolaton	2							ND	P		
Folpet	Phaltan	2							P	P		
Cu-naphthenates	Several	2										
Aminobutane	Several	<1										
Benomyl	Benlate	<1										
Chloroneb	Demosan	<1							ND	P(75)		
DCNA	Botran	<1							P(50)	P(35)		
Dinocap	Karathane	<1							P(60)	P		
DMTT	Dazomet, Mylone	<1										
Glyodin	Glyodin	<1										
Tetra-CP	Several	<1										
Hg-Salts	Several	<1										
Inorganic sulfur and sulfides		150										
Inorganic Cu salts		2										
Warfarin	Several	12										
Methyl bromide	Several	22										
Dibromochloro-propane	Several	10										
Acyclic organics	Several	75										
Dichlorobenzene	Several	60										
Naphthalene	Several	2										
Phenyl phenol	Several	*e*								pp, mb		
Hexachlorobenzene		*e*	+	+	+	+		+	P(60)	C	+	+

in the environment, and dieldrin is the material detected in the analysis. Similarly, DDT is frequently detected as DDE or DDD.

[b]The freshwater monitoring covers both water and sediments.

[c]The two columns, Fatty Foods and Nonfatty Foods, refer to foods selected in both the Pesticide Program and the Market Basket Studies. The data are taken from the FDA *Pesticide Analytical Manual*, Vol. I, Table 201-A. Abbreviations: C = Complete (>80 percent recovery of residues); P = Partial (<80 percent recovery) (numbers in parentheses indicate actual percent recovery, when known); NR = Not Recovered; ND = No Data. Compounds with NR or ND should be considered "not monitored," while compounds with C or P can be considered "monitored" in the two programs.

Some chemicals in nonfatty foods are not detected by the methods described in Table 201-A, but they are detected by a special new method, the "Storherr Method." Compounds for which the Storherr Method is useful are indicated by an S. Some compounds are detected by both the Storherr Method and the methods of Table 201-A, and such is indicated.

In the Pesticide Program, grain samples are analyzed for malathion, and fresh fruit and leaf and stem vegetables are checked for carbaryl and orthophenylphenol. This special monitoring activity is indicated by a "pp." If there is a suspicion of contamination by a compound not monitored in a particular sample, an analytic procedure specific for that compound is used.

In the Market Basket Studies, chlorophenoxy acids, pentachlorophenol, and trichlorobenzoic acid are monitored in all commodities by special procedures. This monitoring is indicated by "mb." Carbaryl and orthophenylphenol are monitored by special procedures in all commodities except dairy products, meat, fish, poultry, oils, fats, and shortening. This activity is also marked by "mb."

[d]2,4-D and 2,4,5-T are monitored by special procedures in the Market Basket Studies (see footnote *c*). The regular analyses of commodities by the techniques of the *Pesticide Analytical Manual*, Vol. 1, Table 201-A, do detect the following esters (1) of 2,4-D in both fatty and nonfatty foods in both the Pesticide Program and the Market Basket Studies: butoxy ethanol, *n*-butyl, ethyl hexyl, isobutyl, isooctyl, and isopropyl. (2) of 2,4,5-T: butoxy ethanol, *n*-butyl, isooctyl, isopropyl.

[e]Production data for certain compounds were not available.

[f]Most domestic uses of DDT were prohibited as of December 31, 1972. Production of DDT is now considerably less than 45 million pounds.

[g]PCB (polychlorinated biphenyls) is not an insecticide, but, analyses for DDT are confounded by the presence of PCB. Estimated production by Monsanto Company in 1971. Monsanto is the only U.S. producer of PCB. Production in 1970 was 85 million pounds. (*Chemical and Engineering News*, December 6, 1971, p. 15)

[h]A contaminant of 2,4,5-T.

6. The distribution of residues within individuals of a given species should not be obscured by compositing of individuals; otherwise the comparison of average levels of residue across time and space may not be possible.

7. Endangered species should not be killed for monitoring purposes.

The organisms currently monitored on a regular basis include man, a variety of herbivorous and carnivorous fish from estuaries, a variety of freshwater fish not specified as to trophic level, and two omnivorous birds. A pilot program has been established for ocean fish, but no criteria for selection of species have been established. Eagles, secondary carnivores, are monitored on an informal basis. Soil contains a variety of green plant residues and decomposer organisms, but the soil-monitoring program does not identify the location of the residues as to speci of organisms.

The monitoring of pesticides in human tissues is still inadequate, and improvements are urgently needed. The goals of the program are designed to follow levels and changes of the amounts of the chlorinated hydrocarbons in tissues of people of the continental United States. There is an urgent need to expand the number of chemicals monitored in man, especially because the use of the chlorinated hydrocarbons will probably continue to decline and substitute chemicals will come into use in larger quantities. Simple analyses for urinary phenolic and dialkyl phosphate compounds are now possible; monitoring programs should incorporate them.

The method of sample selection in the human-monitoring program is not adequate. Although efforts are made to collect tissue from individuals of different ages, sexes, and races, insufficient attention has been given to the question of sample size necessary to determine accurately the distribution pattern of residues that are found. The differences found between the distribution of residues of DDT and its metabolites and the distribution of dieldri residues illustrates the problem.

Residues of DDT and its metabolites are not homogeneously distributed in the United States. Average levels of DDT and its metabolites in fat or in serum are associat with race, occupation, social class, age, and probably sex Clustering of residues occurs: Northerners have half as much as Southerners. These profound demographic and geographic characteristics of DDT residues are in striking contrast to the distribution of dieldrin residues, which is far more homogeneous. The differences between the two distribution patterns probably reflect the different sources of the two types of chemicals in the human organisms. Dieldrin residues may arise largely from eating food with residues of dieldrin, and residues of DDT and its metabolites may come both from food and from other items such as household dust.

The implications of significant differences in the distribution patterns for pesticide residues in humans are fairly straightforward. Materials that are homogeneously distributed can be accurately monitored by smaller samples than can materials showing significant differences demographically or geographically. The problem is that the very existence of significant demographic and geographic variation requires large sample sizes for its detection. Thus until we know how different chemicals behave, we must maintain large sample-size monitoring programs. The rapid changes now occurring in the use patterns of pesticides accentuate the need for larger, properly stratified sample size, because humans will be coming into contact with new types of chemicals and new mixtures of older chemicals.

There is no monitoring of mammals other than man. Two major classes of vertebrates--amphibians and reptiles--are not monitored. Invertebrates are not monitored except in estuaries when high residues are found in fish. The only green plants monitored are crop remains analyzed in conjunction with the soil-monitoring program. No decomposer organisms are monitored.

Man is the only monitored organism in which a conscious effort is made to collect tissue from individuals of different ages, sexes, and races. A deliberate effort is made to sample only adults in the freshwater fish program. Variation in age, sex, and biotype is ignored in all other programs.

Wetlands and bodies of water are well sampled by the programs on freshwater fish, estuarine fish, ocean fish, and ducks. Starlings are wide-ranging and well distributed; thus, monitoring these birds for residues is in effect a monitoring of the general terrestrial ecosystem. No programs specifically sample organisms from cropland, rural noncropland, and urban areas.

Compositing of individuals before analysis is routinely done in the programs on estuarine fish, freshwater fish, ocean fish, starlings, and ducks. Only in the human tissue program is the analysis performed on tissue from one individual at a time.

Chemicals

We found a surprising lack of integration between monitoring programs and data on the production and use of major chemicals. The intensity of monitoring should depend upon the perceived hazard; and the hazard will in turn depend both upon the properties of the chemical and its concentration. The most intensive monitoring programs tend to be carried out (Tables 31-33) for chemicals having proved risks for man and the environment, but there is little tendency for monitoring programs to be closely adjusted in space or time to take into account variations in production or use of the chemicals. Perhaps more important, the agencies responsible for monitoring have apparently made little effort to specify their needs for production and use data.

Monitoring and Research Effort

Monitoring should be treated as a special type of research activity or it will be a rote collection of unused and useless statistics. Accordingly, the structuring of monitoring programs should depend upon the types of research questions in need of answers. The problem of relating monitoring to research efforts is generally centered in selecting species to be monitored and devising protocols for how individuals are to be analyzed or in the selection of media to be sampled. The following are the major types of questions to which a monitoring program can contribute:

1. What is the trend of residues in the environment through time?
2. What are the effects of the residues on organisms?
3. What is the source of the residues?
4. What is the fate of the pesticide molecule once it is released into the environment? How is it changed chemically and how is it transported through space and time?

All monitoring programs conducted over a period of time are a research effort of the first type. A question remains, however, as to whether those programs monitoring composited organisms produce averages that can be compared across time with statistical reliability.

None of the programs, except for the monitoring of ducks is directly linked to a research effort on the effects of the pesticide residues on the monitored species. The reason for this may be largely organizational: Only the duck-monitoring program is conducted by an agency with explicit responsibility to conduct research on the effects of pesticides. The Technical Services Division, Office of Pesticide Programs--the division of EPA responsible for monitoring human tissues, estuarine fish, and ocean fish--has no such explicit mandate. Likewise, the divisions within the Department of the Interior that monitor freshwater fish and starlings have no assignment in the effects research field. The FDA no longer has responsibilities for both monitoring food and feed for pesticide residues and for conducting research on the effects of the residues.

The separation of monitoring programs from research agencies does not mean that monitoring data are not or cannot be used in research on the effects of pesticides. However, the separation of the two activities must necessarily raise the cost and effort required to link monitoring to research. Furthermore, the separation will impede the feeding of new research results into efforts to restructure monitoring networks. The net effect will thus be a decrease in the relationship between the two activities, even if efforts are made to keep personnel in touch with one another.

No monitoring programs attempt to detect the source of the injection of pesticides into the environment. Rather, they measure the aggregate of all introductions. Likewise, no programs are geared to research efforts on the chemical and physical fates of pesticide molecules. The absence of monitoring programs related to studies of the transport of pesticides in the environment is particularly distressing, because an understanding of how pesticide residues reach their destinations is of crucial importance to an evaluation of the environmental hazard posed by the chemicals. The absence of an operating program for monitoring residues in the air is of particular concern because of the studies that have suggested that air is one media in which significant transport of pesticides occurs (Woodwell et al. 1971, Cramer 1973). An air-monitoring system closely linked to studies on the transport of pesticides in the biosphere might help resolve the disputes that have arisen over the role of the atmosphere in the movement of residues (Stewart 1972). It should be emphasized that current monitoring efforts for pesticides (as for other chemicals introduced into the environment) are scattered and fragmentary to the extent that the system as a whole is of little utility in *predicting* major concentrations of chemicals in time to take corrective action. For a variety of purposes, it is important to have effective monitoring on a national basis, and at adequate scale.

RESEARCH EFFORTS ON THE EFFECTS OF PESTICIDES

Six federal agencies make substantial research expenditures to determine the effect of pesticides on man, valued crops and livestock, and other nontarget organisms. During FYs 1971-1973, these expenditures amounted to between $21 and $25 million annually (GAO, personal communication, 1974).* The significance of these expenditures in our pest control system is very similar to that of monitoring: The public must finance research on the effects of pesticides to supplement the privately financed toxicological research of the chemical industry. The cost of the research must accordingly be deducted from the benefits achieved by the pesticides in overall assessment of their costs and benefits.

Approximately one-half of the effects research is performed by agencies with research and operational responsibilities in pest control, and it is often an integral part of the latter. For example, when scientists of the Agricultural Research Service evaluate an insecticide for use against cotton insects, they collect data on its impacts on beneficial insects and on the cotton plant itself. Similarly, when the Bureau of Sport Fisheries and Wildlife (Department of the Interior) performs research on the control of unwanted wildlife species, it also investigates the effects of candidate chemicals on nontarget creatures.

The remaining agencies with major expenditures on effects research have little or no responsibility for conducting pest control operations but may have programs of research on pest control methods. The EPA and the National Institute of Environmental Health Sciences (Department of Health, Education, and Welfare) are the two agencies with the largest programs on effects research. The National Science Foundation has also financed significant research on the effects of pesticides.

---

*Expenditures on effects of pesticides were thus about 20 percent of the total funds expended by federal agencies on all research related to pesticides and pest control ($120 million) in FY 1973. The agencies involved in effects research are the EPA, USDA, USDHEW, the Department of the Interior, the Department of Defense, and the National Science Foundation.

REFERENCES

Cramer, J. (1973) Model of the circulation of DDT on earth. Atmos. Environ. 7:241-256.

Federal Committee on Pest Control (1968) Catalog of Federal Pesticide Monitoring Activities in Effect July 1967. Washington, D.C.: U.S. Government Printing Office. 131 pp.

Moriarty, F. (1973) The effects of pesticides on wildlife: Exposure and residues. Sci. Total Environ. 1:267-288.

Pesticides Monitoring Journal (1967) Volume 1, Number 1, June. Washington, D.C.: U.S. Government Printing Office.

Pesticides Monitoring Journal (1971) Volume 5, Number 1, June. Washington, D.C.: U.S. Government Printing Office.

Stewart, C. A. (1972) Atmospheric circulation of DDT. Science 177:724-725.

Woodwell, G. M., P. P. Craig, and H. A. Johnson (1971) DDT in the biosphere: Where does it go? Science 174: 1101-1107.

PART II

PROSPECTS FOR PEST CONTROL

10

FUTURE OF THE PEST CONTROL ENTERPRISE

INTRODUCTION

In this chapter, we will attempt to project changes that may affect the matrix of domestic and international institutions within which pest control operations are carried out. The commodities that are produced with the aid of pest control procedures are bought and sold on world markets; so are the compounds used in pest control, and so in turn are the petrochemical feedstocks used in their manufacture. More broadly still, the entire international monetary equilibrium affects the prices that will be paid for U.S. commodities and thus controls incentive structures that influence U.S. farm firms in making decisions about what to grow. In a similar way, the public's estimate of complex domestic political processes affect environment-- or health-related "external costs" of pesticide use--and so produce shifts in federal or state regulatory policies. Changes in crops and in the chemicals used to protect them affect the rate at which genetic resistance develops in pest organisms and thus independently alter the allocation of research effort.

Somehow, this bewildering maze of parameters will determine the technological and economic climate in which pest control will be carried out in the medium-range future--the late 1970s and early 1980s. To predict any single influence is risky enough; to forecast how they are likely to interact may seem foolish. Yet well-defined trends are already discernible in many of them; and *not* to attempt prediction would, in our view, be more foolish still. We will discuss factors that could change our demands on the pest control process, drawing for illustrative purposes from the work of our four study teams. In each case, we attempt to concentrate on a "most likely" future, stating where possible how different assumptions could yield a different scenario. The projections themselves depend, of course, on some other assumptions, e.g., that the unprecedented current inflation rate will be brought under some control.

AGRICULTURE AND THE AGRICULTURAL ECONOMY

What kind of demands are likely to be in store for agricultural commodity production in the United States, and what will be the derived demand for the technical inputs among which pest control materials and procedures are so prominent? One may begin with a recent set of "alternative future" staff projections from the USDA, summarized in a recent paper entitled "1980 and Beyond: The Impact of Zero Population Growth and Alternative Export Levels on Cropland Requirements in U.S. Agriculture." Projections anticipate the effect of expected changes in productivity (crop yield and animal feed efficiency), shifts in the composition of commodity production, growth in domestic demand (both from income and from population growth), and growth in foreign demand.

Domestically, the projections imply that, if the low population growth rate of the early 1970s continues, U.S. agriculture can support a rather substantial increase in export demand without placing undue stress on its cropland base. This projection seems sound enough, granting for argument its assumptions about population growth, provided that gains in crop yield and livestock feed efficiency do not become more difficult to attain over the next decade than they have been in the recent past. A decrease in productivity would place more intensive pressure on such techniques as double-cropping, intensification of management of presently marginal cropland, and increased emphasis on engineering approaches to land improvement.

But it is possible that even the relatively high export levels built into these projections may underestimate the external demands that will be placed on U.S. agriculture by food-aid requirements and, in particular, by the growing affluence of the major commercial importers. Many economists initially viewed the sharply rising demands for U.S. agricultural commodities, and the dramatic price increases that took place in domestic and world commodity markets, as a temporary event caused by poor harvest in South Asia, parts of Africa, and elsewhere, and by inept management of U.S. grain reserves. But it has become increaasingly clear that a major role is played by rising standards of living in Japan, western Europe, and by a realignment of currencie which has made U.S. agricultural commodities cheaper for the strong-currency importer. We now suppose that there will be a slow plateau in U.S. commodity prices; it is not even difficult to envisage a combination of liberalized international trade policies and currency relationships that would make the USDA export projections seem conservative. Such projections would presume some spillover of the energy crisis into agriculture, in the form of tightened fertilizer and fuel supplies.

Even if we assume, as a balanced view from the community of U.S. agricultural economists would suggest, that the present situation will not permanently mark a sharp upturn in commodity price structure, it is unlikely that agricultural commodity prices will resume the downward drift relative to nonagricultural prices that they showed in the 1950s and 1960s.

Reasonable prediction would seem to entail the following assumptions: (1) There will be continued pressure, for the next several years, to meet U.S. international balance-of-payments objectives with agricultural commodities; and Japan, the western European economic community, and some eastern European countries are likely to continue their emergence as competitors for U.S commodities on international markets; (2) fuel shortages, at least at a low level, are likely to be chronic for the remainder of the decade, and fertilizer production may well run behind demand over the next 3-5 years, and perhaps longer, until production capacity is built up in the newer producing areas and until hoped-for proof arrives that some of the "natural" fertilizer supply problems of the last 2 years are indeed transitory; and (3) the domestic economic situation is not likely to change so dramatically as to make resubstitution of labor-intensive agricultural practices for capital-intensive ones a likely prospect.

These assumptions foreshadow a continued, and probably an increasing, production pressure on the agricultural sector of the American economy. The farm legislation passed during 1974 will only be the first in a series of legislative efforts to provide incentives for increased production: The expressed purpose of the removal of acreage restrictions and changes in farm subsidy programs is the addition of 58 million acres to the nation's total of productive cropland. These changes, and future ones that may be anticipated if the economic incentives described above are correct, will certainly increase the demands upon our pest control technology.

CHANGES IN PATTERNS OF AGRICULTURAL PRODUCTION

For a variety of reasons the demand for specific U.S. agricultural products may be anticipated to change over the next decade. Newly available international markets emphasize different commodities because of unique national needs or dietary preferences. Anticipated shortages of particular inputs may change the profitability of growing particular crops. Finally, expected technological changes may offer strong incentives favoring a change in cropping patterns. A fully accurate prediction of these trends would require knowledge we do not possess about potential changes in world food demand and about technologies that are now well over the horizon. But certain present trends can be identified and watched carefully to see if, at some later time, they are beginning to produce the expected effect.

World Markets

Certain cropping patterns have already shown changes in response to world demand. Elevated prices for soybeans, cotton, and wheat on world markets have resulted from such demand effects and were reflected in acreage increases for the growing season of 1974. In the succeeding year, prices for cotton especially proved much less stable than predicted, and acreage reductions took place. Despite such examples of price fluctuation, there is reason to expect that soybeans, small grains, and certain commodities will continue to enjoy relatively high world demand, subject to modifications by technological advances. Part of any policy-planning exercise in pest control ought to include a detailed examination of projected world markets for the important commodities grown in the United States.

Fuel and Fertilizer Supplies

Crops differ in their demand for fertilizers and other inputs; since farmers' decisions to grow crops depend on the relationship between the marginal cost of these inputs and the marginal rate of return they generate, the choice of what to grow will be strongly affected by these cost relationships. For the next several years, we are expected to be in a condition of low fertilizer supply; hence there will be incentives to grow crops, soybeans for example, that are relatively low in nitrogen demand. Whether a crop with low fertilizer but high insecticide requirements can be substituted for one with the reverse requirements will depend upon the relative cost of the two inputs, among other things. The most significant shifts are likely to occur in the direction of crops that need relatively little fertilizer, since it usually is the highest of all the variable-cost inputs. Price changes in either area will affect the desirability of alternative methods or the intensity with which the input is applied.

Technological Breakthroughs

These are perhaps the most difficult influences of all to predict. An example provided in Volume III, *Cotton Pest Control*, may be cited by way of illustration. If a means can be found to remove the glands containing gossypol from cottonseed during processing (or if cotton cultivars lacking the glands can be developed), it is anticipated that the marketability of cottonseed oil and flour will increase dramatically for both human and livestock consumption. To date, partial reductions of gossypol have been obtained by genetic means; while the value of seed products has been increased, the new varieties have not proven preferable to competing strains that have higher lint yields. Moreover, reducing gossypol content by genetic means may increase insect vulnerability. If these difficulties are overcome, as they would be by a method for removing the glands during processing, a strong incentive for growing cotton would be added to that already provided by recent increases in the price of cotton lint on the world market. If major increases in the acreage devoted to cotton in the South resulted, total insecticide use would rise significantly there.

Institutional Changes

A variety of forces at work in contemporary American society have important impacts on pest control processes. In many cases, the magnitude of these effects is difficult to ascertain, and to predict even their direction is risky. Nevertheless, these influences are important and should be described, however imprecisely.

As described earlier in this volume, the economics of American agriculture have resulted in a consistent tendency over the past several decades toward the concentration of farm ownership into the hands of a relatively small number of growers, and thereby towards an increase in the average number of acres owned by single-farm families. Conventional wisdom maintains that increased size of holding is associated with an increase in pesticide use, i.e., that "chemistry substitutes for husbandry" when operations enlarge. The only modern study of which we are aware, on cotton in the Central Valley of California (see Volume III) does not support this thesis strongly. Nevertheless, concentration of ownership in many agricultural regions is bound to result in further increases in the continuous acreage devoted to single crops--in other words, crop diversity is very likely to be reduced. As far as we know, this phenomenon *is* likely to be associated with increases in pest problems if all other factors remain unchanged.

We may also consider whether changes in the domestic price structure of pesticides themselves are likely to change the nature of pest control practices. Continuing energy shortages will affect the availability of petrochemical feedstocks for pesticide production; these effects, in turn, will influence not only the availability but also the cost of the pesticides themselves. It is unlikely, however, that in a time of rapidly rising world food prices this is going to be a very significant influence. This is especially true where pest control practices do not constitute a large proportion of the grower's variable costs. As pointed out in Volume II, *Corn/Soybeans Pest Control*, even a doubling of the cost of herbicides for corn would not affect the use of these materials as a substitute for mechanical tillage by most farmers. We would expect similar responses from the growers of high-insecticide-requiring crops, with the possible exception of situations in which pest control is an unusually large input. Even there, the agricultural price structure we expect over the next few years should make pest control "worth it" at almost any foreseeable price. However, important effects can be expected on the number and diversity of control chemicals available and the price differentials between them. Already, the availability of feedstocks has had its influence on certain compounds: At least two domestic manufacturers have announced that during the 1975 growing season they will be forced to limit the amount of methyl parathion that they will make available to customers on the basis of a percentage of last year's allocation. Such effects are bound to restrict the freedom of choice of growers in deciding which compounds to use and ultimately will have the effect of concentrating use on particular materials.

These effects are already having serious consequences for the international availability of pesticides. An FAO conference warned early in 1974 that, because of lowered availability of petrochemical feed stocks, 1975 allotments of pesticides might be drastically reduced, especially for the developing countries where almost none of these materials are manufactured.

Finally, there is every reason to expect continued public vigilance regarding environmental-quality issues, and this will result in a continuation and extension of federal and state regulation regarding the introduction of chemical and biological materials into the environment. This has a self-renewing quality. As regulation is installed and monitoring programs become more sophisticated, a more accurate assessment of the "spillover" costs of using particular chemicals becomes possible. Improved public knowledge of these costs, in turn, increases the political pressure for further regulation. The impact of present regulation, as we argue elsewhere in this volume, may exert a negative influence on the rate of introduction of new chemical entities by increasing the cost and time required of the private firms developing them--although this effect is not evident yet. In addition, of course, regulatory constraints may result in the cancellation or suspension of compounds, or their restriction in certain applications. The net result is that we may anticipate a decrease of current-generation chemicals available for pest control.

This is not the place to consider the costs and benefits of such regulation--a matter that has not been adequately studied for pesticidal chemicals. But, even under the assumption that regulation has yielded society a *net* benefit, we still must reckon in a predictive sense with the costs it has generated within the regulated enterprise.

Biological Effects

The considerations just discussed indicate that a variety of technological and institutional factors will continue to change the structure of U.S. agriculture and the practice of pest control within it. We can anticipate continued upward pressure on the value of U.S. commodities on world markets and, in response to balance-of-payments objectives, more legislative incentives to increase agricultural production. Current trends in concentration of farm ownership and increase in average farm size, together with technological incentives for crop concentration that are difficult to evaluate, are likely to reduce still further the regional diversity of agroecosystems. At the same time, public concern over environmental quality will likely stabilize or even extend the present regulatory structure, resulting in a decrease in the diversity of third-generation chemicals for pest control. What do these things portend for the biology of pest and crop?

It is difficult to estimate how large the trend towards monocultures of increased size will be, or how much these are likely to affect the pest vulnerability of the crops. It seems to us a safe assumption, however, that this trend--added to the increased value of the crop--will provide substantial incentives for the use of more pesticides per acre on the average. Moreover, incentives to draw new land into cultivation may result in the growing of crops in places where, under acreage restriction, it was thought les desirable to grow them. In the case of some crops, the undesirable feature may well have been greater pest vulnerability. It thus seems logical to anticipate a modest increase in the average *level* of crop protection needed for unretired land.

This protection, however, will have to be supplied by a more limited arsenal of chemical weapons. The history of pest control in U.S. agriculture during the 1960s and early 1970s has already seen valuable pest control agents lost at the crop protection enterprise for various reasons. In many cases--as with DDT in the middle 1960s--the utility of a compound decreased drastically because of the onset of genetic resistance in many target pest species. The last domestic uses of DDT were eliminated by an EPA decision to cancel the registration of this compound, and restrictions on the use of other chlorinated hydrocarbon insecticides (aldrin, dieldrin) followed. At the same time, regulatory constraints have slowed the rate of introduction of new compounds to replace those removed by use restriction. We thus foresee a situation in which a larger total volume of a smaller variety of pesticidal chemicals is applied to U.S. croplands. We expect that similar, though perhaps not so marked, effects will take place in industrial and private use sectors.

The development of genetic resistance depends upon many factors, but the regime under which selection occurs is one of the most important of them. Where a diverse array of chemical control agents is used, no single compound dominates in its selectional effects and no given one is likely to be present in overwhelmingly large concentrations Concentration upon a single compound, however, creates a regime in which the selection of resistance takes place more rapidly. The effect is magnified by crop concentration. Because special insecticides tend to be associated with particular crops, broad regional monocultures are likely to expose an entire pest population to a spatially more homogenenous selection pressure and thereby produce resistance more rapidly.

Thus it seems likely that, during the remainder of this decade, intensified pest control will be undertaken with relatively fewer chemicals. Even with the present diversity of materials, genetic resistance is eroding the efficacy of these methods. With further concentration, that process may be expected to accelerate.

Our conclusion is that the contemporary chemical technology will experience increasing trouble in achieving its objectives. But of course chemical control is not the only kind we know of; a variety of alternative technologies are already in use or under development, although the magnitude of these efforts is substantially smaller than that expended on chemical technology. To what extent can these techniques relieve the increasing burdens we foresee for chemical pest control? In Chapter 11, we will make a brief evaluation of the present status of each alternative technology and deal with some of the scientific and institutional factors that affect their future prospects.

FORESTS AND THE FOREST ECONOMY

The general increase in demand, both for forest land for recreational and other noncommodity uses and for forest products, experienced during the past two decades will probably continue through the later 1970s and early 1980s. The impacts of these expanding demands on problems of forest pest management fall into three broad categories: impacts associated with changing patterns of forest land use, those associated with pressures to increase current timber harvests, and those associated with expansion of efforts to grow timber for the longer run. These three sorts of impacts show significant quantitative and qualitative differences.

Changes in Forest Land-Use Patterns

For the past 30 years, forest land use in the United States has been characterized both by rapidly expanding allocation of forest areas to park, wilderness, and other recreational uses and by intensification of the demand for productive land for use in timber growing. Over the same period, there has also been some diminution of the forest land base by conversion to suburban or transportation uses and some accretions resulting from abandonment of cropland. All of these trends except the last are likely to continue into the 1980s. As a result, the areas of forest earmarked for exclusive or predominant recreational use and for intensive timber growing are both likely to increase. At the same time losses of forest land to urban, suburban, and transportation uses will continue and may well be augmented by selective conversion of forest to cropland, if our appraisal of agricultural incentive trends is correct.

The result of these trends is likely to be a relatively small reduction in total forest acreage, of the order of 1 to 2 million acres by the 1980s, with considerable intensification of management on much of the remaining area, whether for timber growing, wilderness use, or general recreation use. The implications of intensified timber growing for forest pest control are discussed in more detail below. In the case of intensified recreational use (including designation of additional wilderness area), increases in the use of selective pesticidal chemicals and biological controls including microbial agents may be expected. In developed recreational areas, both the need and the opportunity for silvicultural treatments to minimize susceptibility to pests will encourage approaches to potential pest problems that stress avoidance of buildup of pest potentials and minimize the need for direct control of the pest organisms involved. On wilderness and other lightly used recreational areas, organisms that are regarded as pests in the timber-growing context often present only minor or negligible hazards in terms of wilderness management objectives. But biological and silvicultural controls may be justified where infestations in such areas present a hazard to adjacent forests that are being intensively managed for purposes other than wilderness.

Pressures to Increase Current Timber Harvests

The outlook for continued increases in the market demand for wood products, particularly for softwood sawtimber, has been amply documented in recent detailed analyses (USFS 1973, President's Advisory Panel 1973). The demand for domestically grown wood about 1980 is likely to be as much as 20 percent above 1970 consumption, even at lumber prices 25 percent or more above existing levels. The only potential source of softwood sawtimber to meet a demand increase of this magnitude is the considerable inventory of such material on western national forests. The existence of this substantial inventory in the face of expanding demand led the President's Advisory Panel on Timber and the Environment to recommend in mid-1973 that the annual harvest on western national forests be increased substantially. Such proposals for faster liquidation of old-growth forests have been sharply opposed by some interests; in addition, technical revisions of some national forest timber management plans have had precisely the opposite effect. It is therefore difficult to foresee how this particular supply/demand squeeze will be resolved. If harvesting of national forest old growth should be accelerated during the next 10 years, it would tend to reduce those insect and disease problems associated with old forest age classes and thus speed up the process of eliminating pest problems arising from decadence. On the other hand, if national forest harvesting is maintained at or below recent levels, increasing demand may produce additional pressures to minimize pest losses in such stands, since further losses there would be seen as a basis for curtailing the allowable cut in national forests. Increasing political pressure for use of pesticides to combat specific outbreaks in western old-growth timber might well then emerge (as has been the case with the current outbreak of the Douglas fir tussock moth in the Pacific Northwest).

Expansion of Timber-Growing Activities

Regardless of how national forest harvest rates are determined, the strong outlook for continuing expansion of demand for wood beyond the 1980s and into the very long run will almost certainly mean a substantial increase in timber-growing efforts during the next decade. Artificial regeneration of new forests has averaged close to 1.5 million acres per year for the last 20 years. Recent stumpage price increases have strengthened the economic incentives for planting, emergence of strengthened forest practice legislation in some states is placing heavier institutional pressure on achieving higher standards of restocking, and new federal legislation looking to additional timber-growing investment on public lands was adopted in the summer of 1974. All of these factors will undoubtedly stimulate continuing and relatively rapid growth in plantation forestry. Thus within the next decade the pest problems of newly established forests are likely to increase greatly in relative importance. Although silvicultural approaches to control will undoubtedly be stressed, other approaches and strategies will be needed to provide an adequate level of protection. Both the experience and the research needed to develop such new technology are limited thus far. New kinds and species of pests different from those considered important for mature trees will undoubtedly prove to have significant effects on the form and growth of young planted trees. If this occurs, pressures for direct control will be unusually heavy, because, contrary to most past forest pest incidents, the threatened forests will represent very heavy cash investments to the owners.

Anticipated increases in the intensity of timber-growing efforts also have a positive side. Substantial progress may be expected during the next decade in the application to young forests of such silvicultural measures as stocking control, salvage of mortality, improvement of species composition, and use of genetically resistant planting stock (e.g., slash pines resistant to fusiform rust in the South). Knowledge and technique already exist that will permit these silvicultural tools to be applied in ways that will reduce the susceptibility of the stands to future pest damage. Indeed, minimizing such potentials will be one of the specific objectives of the newly emerging silvicultural programs. The use of specific microbial agents and nontoxic behavioral chemicals for critical pests will also increase as their relatively high cost is justified by enlarged investments and rising expectations of yield from these "high yield" forests.

An essential prerequisite for the development and application of the new forest pest control technologies and strategies foreseen above, especially those involving the use of biotic agents, will be closer monitoring of pest populations and damage. Capabilities must be developed to anticipate and predict the occurrence and potential rates of increase of specific pest organisms. To accomplish this both additional funds and trained personnel will be needed for expanded surveillance monitoring programs.

PUBLIC HEALTH

There seems no reason to doubt that local, national, and multinational agencies will retain substantial obligations in the area of vector control over the next decade or so. Although many of the major new ventures in disease control in the developing countries have been successful enough to permit maintenance of control at lower effort, few have resulted in "eradications." Furthermore, anthropod vectors have produced an especially large number of genetic resistance problems, each of which requires research and development activity to solve. There is, finally, the disturbing interaction between agricultural development and the development of resistance in vectors as a consequence of crop-spraying programs. Such problems may be expected to multiply as high-technology agriculture is exported to developing countries.

As Volume V, *Pest Control and Public Health*, points out, vector control has never been a pest control technology on the scale of agriculture. We expect it to continue as a relatively small-scale but absolutely essential activity, making heavy use of research and of field experts and requiring a variety of materials in relatively modest amounts. It is likely to expand only locally and temporarily, as the need requires. For the foreseeable future, the problem that public health applications present to the policymaker may be described in much the same terms as we see it today: an area whose importance for the welfare of mankind far outreaches its economic leverage. It relies on a technology largely devised for, and manipulated by, other markets. We shall need continued surveillance to ensure that this application, relatively static in size yet vital in importance, is not allowed to suffer unforeseen consequences as a result of rapid growth in the rest of the enterprise.

REFERENCES

President's Advisory Panel on Timber and the Environment (1973) Final Report. Washington, D.C. 541 pp.

U.S. Forest Service (1975) The Outlook for Timber in the United States. USDA Forest Resource Report 20. Washington, D.C. 367 pp.

11

THE STATUS OF SELECTED ALTERNATIVE PEST CONTROL STRATEGIES

The National Academy of Sciences has, within the last 8 years, published three major reviews of different pest control technologies, and we would refer the interested reader to them for a comprehensive discussion of the technical aspects of many pest control strategies (NAS 1966, 1968-1970, 1972). We have selected here, however, a number of alternatives for discussion in the context of the costs and benefits of their use. Where appropriate, we have updated the information on the feasibility of the alternatives selected. Items selected were chosen either because they exhibit very promising potential for a favorable cost-benefit balance or because they are widely discussed and it was appropriate for us to evaluate their positive and negative features.

We clearly do not intend an exhaustive survey of every pest control technology; such an undertaking would be a major project in itself. We hope our method of evaluation, however, will stimulate others to make similar studies of different alternatives.

INSECT HORMONES AND PHEROMONES

The major classes of chemical compounds in our present insect control armamentarium were found to be insecticidal before their mode of action on the target organism was discovered. Indeed, the organophosphate insecticides were developed primarily for other purposes. Most contemporary insecticides have a broad spectrum of toxicity to animal groups within and outside the arthropods and some are also poisonous to various plants. It has been realized at least since the middle 1960s (Williams 1967) that the developmental endocrinology of insects--and perhaps other sensitive chemically controlled processes as well--provide favorable targets for intervention with specific chemical materials. These compounds have become known as "third-generation" insecticides, and the approach of identifying vulnerable points in the pest organism's biochemistry and intervening specifically at these points has been called the "bio-rational" approach to chemical pest control.

A major advantage claimed for the new compounds is that, potentially, they can be used to attack a much narrower spectrum of organisms. This is undeniably an environmental advantage, but greater specificity is an economic disincentive for the development of new compounds, at least in the private sector, because smaller potential markets are involved. Furthermore, potential users of the materials may also face economic disincentives; because, in many cases, more than one pest is present in the area to be treated and a single broad spectrum toxicant would be cheaper and easier to apply than several narrow-spectrum materials. Nonetheless, significant corporate commitments have been made to developments of this kind--for example, by Hoffman-Laroche and Minnesota Mining and Manufacturing Company. At least one new corporate venture of substantial size has been launched for the specific purpose of developing third-generation insecticides: Zoecon Corporation of Palo Alto, California, initially capitalized at $25 million.

Classes of Third-Generation Compounds

Insects grow and metamorphose under the influence of three major classes of hormones. A "brain hormone" stimulates the secretion by special glands of a hormone that promotes growth and differentiation. This second hormone, called "ecdysone," is a steroid compound resembling some of the hormones of man and related vertebrates. The chemistry of this class of compounds is well known; but, because of their similarity to vertebrate hormones, there is the possibility that compounds of this class might also interfere with endocrinology of nontarget organisms. A third hormone in insects is pitted against ecdysone; it is called "juvenile hormone," and when it is present in adequate amounts it promotes the retention of larval characters. Normally, juvenile hormone disappears just before metamorphosis to the adult; but when it is continuously present, the succession of growth and moult stages that characterizes the life of the caterpillar is prolonged. These features make the juvenile hormone a promising target for intervention; moreover, it is a terpenoid compound that does not resemble any known mammalian hormones. In principle, blocking the action of juvenile hormone early in development ought to promote premature metamorphosis into undersized, possibly ineffectual adults; whereas application late in development should prevent metamorphosis into the sexually reproducing stage. Compounds related to the juvenile hormone are found in a variety of organisms. Many plants manufacture large amounts of such substances; no one knows exactly why, although it is often speculated that they may actually constitute chemical defenses against insect attack evolved by the plants. A good deal of chemical R&D has been spurred by interest in these compounds, and a wide array of juvenile hormone analogs is now available for experimentation.

Insects also employ various classes of highly volatile chemical messenger substances called pheromones in sex attraction and (sometimes) other intraspecific communicative functions. Since successful reproduction often depends upon these compounds, they have attracted a substantial research effort aimed at control, both in the public sector (especially the USDA and the land grant universities) and in private corporations as well. Sex pheromones have been identified chemically and can be produced in useful quantities for the boll weevil, the southern pine beetle, the pink bollworm, the gypsy moth, and other important pests. Many of these compounds are active at extremely low concentrations. Traps utilizing such materials are effective ways of measuring the population density of relatively uncommon or difficult-to-find species. Devices for this purpose are being distributed by the U.S. Department of Agriculture and also marketed commercially by Zoecon Corporation and, until recently, by Minnesota Mining and Manufacturing Company. It is possible that classes of compounds inhibiting the responses of insects to sex pheromones could also be employed in confusion strategies (Beroza 1970).

Other special features of insect metabolism have suggested themselves as candidate targets for specific "biorational" control measures. For example, many insects have unique mechanisms for the synthesis of cuticular compounds that regulate their permeability. Interference with these biochemical processes might prove effective. Another approach is suggested by recent work on vertebrate predators: Specific food aversion can be produced by chemical stimuli that generate illness in herbivores, a phenomenon similar to "bait-shyness" in rats. Such experiences are retained for a considerable length of time even if only one aversive conditioning trial is employed. Practical tests have shown that this method shows promise for the control of a vertebrate predator, the coyote (Gustavson et al. 1973). Because the same phenomenon is shown by some plant-feeding invertebrates (Gelperin 1975), it is reasonable to guess that some economically important insects might be controlled in this way.

These examples only illustrate the range of possible bio-rational approaches to the design of new chemical pesticides. Although some testing work on juvenile hormone analogs, and considerable pheromone research and development, is now going on in the USDA, federal activity in these areas is not large. At the level of application, research and development on chemicals based upon the insect hormones is primarily located in a small number of private firms. Some of these are in the traditional agricultural chemicals sector (Stauffer), some are new and specialized companies (Zoecon), and some are pharmaceutical houses (Hoffman-Laroche).

Strategies of Bio-Rational Chemical Pest Control

The nature of developmental hormone analogs is such that they require carefully timed application. They are most likely to be effective when employed against insects that have short life cycles and are well exposed in their larval stages. Unfortunately, many of the major crop pests--including almost all of those members of the lepidoptera and coleoptera that have "hidden larvae," like the pink bollworm and the boll weevil--are not readily accessible. The basic strategy of the third-generation chemical developers has thus shifted to attacks upon insects that have the required vulnerability and exhibit some developmental synchrony across their populations. The Zoecon Corporation, for example, has invested its major developmental effort to date in a larvacide for floodwater mosquitoes based on a juvenile hormone analog. This compound, Altosid®, has been granted an experimental permit by the EPA; it is the first registration of any kind for a compound designed on the basis of its specific interference with pest metabolism. Zoecon Corporation officials estimate that an expenditure of at least $8 million was associated with the development of this new control agent.

Costs and Benefits of Third-Generation Chemicals

It is difficult to predict what the new agents are going to offer in the way of direct cost per unit of control. Most predictions hold that they will be at least as expensive as, and perhaps more expensive than, the compounds currently employed. Research and development costs--imposed in substantial part by regulatory requirements--are approximately the same as those for conventional insecticides, while at the same time the specificity of the products makes the potential market size smaller. It also seems probable that more expertise will be required to use the new compounds successfully. This means that industrial concerns may have to market services along with their goods, for which trained--and expensive--personnel will be required.

The efficacy or cost effectiveness of the new control techniques will have to be assessed in large-scale field tests in direct comparison with the competing contemporary technologies. Since no such large-scale tests have yet been made, nothing decisive can be said at the moment.

It has been argued that genetic resistance in target insect populations--a major cost of the current technology-is unlikely because the third-generation compounds are analogs of the insect's own hormones. There seems no compelling theoretical reason, however, to believe that such resistance is substantially less likely than with current compounds (Schneidermann et al. 1970). Indeed, some insects already resistant to contemporary chemicals show cross-resistance to juvenile hormone analogs (Cerf and Georghiou 1972). The rates of resistance development are likely to differ for different compounds, and explicit knowledge on this point will have to await the results of empirical tests.

In one area, however, the third-generation compounds offer unambiguous advantages over current compounds. Both in terms of their mammalian toxicity and their potential for ecological damage, the third-generation chemicals are relative innocuous. The juvenile hormone analog for which a temporary tolerance has been granted has an oral LD_{50} in the rate of 34,600 mg/kg, and it is extremely nonpersistent. The abundance of members of this family of compounds in various natural sources argues that the third-generation chemicals will, in general, avoid some of the external costs associated with the current technology. Whether the specificity of the new compounds will extend to avoiding damage to beneficial predatory insects is less clear, although in principle a high degree of selectivity ought to be possible.

Incentives and Disincentives

Collectively, the third-generation compounds represent an important and well-publicized hope for the future of insect control. The biological promise of such developments is great. What institutional factors affect the prospects for further progress?

We note the following impediments to rapid development in this field. First, effective new techniques depend heavily and directly upon basic research advances in insect biochemistry, physiology, and ecology. These are not areas that have traditionally been well supported by extramural federal grant programs. Second, many of the practical developments are being conducted in the private sector by relatively small firms. They are operating in a regulatory climate that was designed for compounds selected on the basis of their general toxicity and produced by an industry composed mainly of large firms. Finally, there are also problems--described above--involving the size of potential markets. These constraints combine with the inherent biological limitations of the compounds to make us cautious: It seems too much to expect that in the next decade this group of chemicals will substantially replace the contemporary technology for major crop or timber-producing systems.

These agents show considerable promise, however, for the control of certain arthropod vectors. Furthermore, their use in agriculture is not restricted to control programs in which they play a direct pesticidal role. Pheromones have already proven useful as attractants in the monitoring of pest populations. This detection and survey function is a critical one in many integrated pest management strategies.

CONTROL BY PATHOGENS

In principle, it should be possible to control a variety of pest organisms by introducing or encouraging organisms pathogenic to them. In practice, such methods have been applied only to insects and mites and only on a rather limited basis. (Some plant pathogens are affected by microbial organisms, but no practical applications have been found.)

History and Scope

It has been known for some time that a variety of pathogenic organisms attack insects, some of them with a high degree of species-specificity. In this country, systematic research efforts on the use of pathogenic microorganisms in such control have dated from the work of E. A. Steinhaus beginning in the 1940s. Today, research oriented toward pest control by microbial agents is carried out in a number of laboratories in the USDA, universities, and private corporations. We will consider bacterial and viral agents only. Some fungi and nematodes are pathogenic to insects, but in neither group do major developments toward control applications seem to be under way. Neither will we consider higher organisms that are more properly described as parasites than pathogens, because these are treated in another section of this volume.

Bacteria

Various bacteria have found use as practical control agent Perhaps the earliest successful example involved control of the Japanese beetle in the United States by the "milky disease," produced by *Bacillus popilliae*. Permanent control of this pest was achieved in large areas of the eastern United States during the late 1930s and early 1940s by the dissemination of *B. popilliae* spores; because the pathogen remained in sufficient numbers in natural beetle populations, the maneuver resulted in the establishment of a new equilibrium pest population density at far lower levels than obtained in the mid-1930s. More recently, research efforts have concentrated on a related microorganism, *Bacillus thuringiensis*. (Like *B. popilliae* this bacterium is a spore-former; one of the requirements for a practical microbial agent is that it must have a resistant stage in which it can be stored, formulated, and applied.)

The history of research on *B. thuringiensis* is an interesting case study in the development of alternative technologies; the following brief account is summarized from A. M. Heimpel (1970). As a result of proposals emanating from academic scientists, industrial concerns developed pilot-plant capability for producing *B. thuringiensis* spores; early efforts mistakenly concentrated on selecting for spore-production. Subsequently, it was shown that the active insecticidal principal is a polypeptide produced by the bacterium at the time of sporulation. Changes in the selectional strategy produced marked improvement in the efficacy of the commercial preparations, especially against lepidopteran pests. Currently, *B. thuringiensis* preparations are manufactured by Abbot Laboratories and Nutrilite Products, Inc. It is claimed that during 1970 about a quarter of a million acres of agricultural land were treated with them at a cost of approximately $3 million. A very active sales and marketing program is being mounted during the current year.

It does not appear, however, that the successful experience with *B. thuringiensis* has stimulated the development of more microbial pesticides. Formulation and distribution requirements demand a microorganism with a resistant stage; other spore-forming bacteria do not appear to be as promising candidates as the ones already available, and other groups of bacteria that are pathogenic to arthropods have complex life cycles that do not encourage their use in control. Finally, the cost of meeting regulatory requirements is now many times higher than it was in the late 1950s and early 1960s, when the major industrial commitments to development of *B. thuringiensis* were made.

Viruses

It is probable that virtually all insect species are attacked by a number of viruses, some of which are highly specific. Particularly promising as viral control agents are the nuclear polyhedrosis (NPV) and granulosis viruses, collectively termed the bacculoviruses; these tend to be quite host-specific and are known to cause massive epizootics in natural populations of pest insects.

As with the bacteria, basic research in the universities and in federal laboratories has led to development of manufacturing techniques in the private sector. The "manufacture" of viruses, however, poses some unique problems. Viruses will only grow and reproduce within the cells of their own specific host(s). In the preparation of viruses for human immunization programs, the development of tissue-culture techniques for virus propagation has often been an important step. Insect tissue culture, unfortunately, has been a relatively neglected field. Recently success has been reported from several laboratories in growing NPV in primary organ culture as well as in various insect cell lines. Viruses propagated in this way appear to be quantitatively as infective as those grown *in vivo* (Ignoffo et al. 1974). Since insect organs are small, primary cultures can be expected to yield only relatively small amounts of virus. Cell culture is more promising, but large-volume production has not yet been reported. One preliminary note suggests that several species of *Bacillus* may be infected by NPV (Wells and Heimpel 1970). Nevertheless, at present the production of viruses depends upon infecting living insects and later harvesting the virus particles from them. Industries active in this area include International Minerals and Chemical Corp. and Nutrilite Products Inc., but very few others. Only one insect virus has been granted an experimental use permit: a nuclear polyhedrosis virus active against *Heliothis* that was cleared in 1970 for experimental use on cotton. This preparation is sold as a powder containing infective virus particles and can be formulated in all of the ways used with chemical insecticides.

Because specific viruses often produce epizootic infections of prominent agricultural pests and because the preparation of virus particles from an infected insect is easy, it is not surprising that agriculturalists have made informal use of viral control measures for many years. For example, populations of the cabbage looper (*Trichoplusia ni*) frequently show heavy local infections with a specific nuclear polyhedrosis virus. In the San Joaquin Valley in California, where this pest often moves into cotton, growers will find a source of infected larvae, collect a number of the diseased insects and blend them in water to make a virus preparation that can be sprayed on cotton fields. This is a locally effective method of pest control that has been recommended frequently by independent pest management consultants in the area. At one time, extension agents in the northern San Joaquin Valley described the "home-extraction" methods in official publications for growers; but in 1972 these sections were deleted at the request of the EPA. Although the law prohibits the sale of viral agents that have not completed the registration requirements designed for chemical pesticides, large numbers of the same particles can be employed without penalty by any individual with rudimentary competence!

A number of other bacculoviruses have been applied successfully as control agents in an experimental setting and must therefore be regarded as promising. Target pests include the gypsy moth, the Douglas fir tussock moth, the tent caterpillar, the codling moth, the cabbage worm, and the beet army worm. A "fallout" benefit from research on insect viruses as control agents has been the realization that many natural populations of pests are under density-dependent regulation by viruses and that the existence of such control is a contra-indication for the application of chemicals under certain circumstances.

Costs and Benefits of Microbial Control

In general, microbial and viral agents ought to be relatively inexpensive in terms of direct costs to applicators, especially in cases where an infection can be maintained in the population following a single application. In such instances, the one-time character of the cost might argue for governmental subsidy of the procedure. Where continued application is required for control, the agents actually resemble chemicals, and the considerations here are similar to those described earlier for third-generation compounds.

Because microbial parasites have coevolved with their hosts and do so on a continuing basis, there are good theoretical reasons for believing that the problem of resistance development will not be as serious as it is with chemical agents. The conclusion is already foreshadowed by findings using the bollworm NPV (Ignoffo and Allen 1972).

As far as is known, no microorganism that has been proposed for insect control can infect a vertebrate. Especially in the case of the nuclear polyhedrosis and granulosis viruses, there have been abundant natural and experimental opportunities to test for infectivity to higher vertebrates. Insectivorous birds often eat diseased larvae during epizootics, and no infections have ever been observed as a result. Furthermore, numerous intensive encounters of humans with insect viruses in the laboratory and in informal or unauthorized pest control operations have not produced a single known instance of human infection.

Nevertheless, such agents are viewed with great caution. This is natural in view of their properties: chemicals, however toxic they may be, do not have the ability to reproduce in nature. On the other hand, man has had extensive exposure to some of the insect viruses over long periods, and no evidence of infectivity has been forthcoming. A WHO/FAO conference in 1972 urged intensified research programs on insect viruses and reviewed the evidence regarding their potential hazards. The safety and environmental impact status of the NPV and granulosis viruses (the bacculoviruses) was more recently reviewed at a joint conference (April 1974) sponsored by the EPA and USDA. This Committee believes that an intense continuing effort should be made to determine whether infectivity of vertebrate cells by the nuclear polyhedrosis and granulosis viruses is a realistic possibility. If the results demonstrate that it is not, new protocols for the acquisition of registration data should be devised for this group of organisms.

Incentives and Disincentives

It does not appear likely at this time that microbial control agents will soon constitute a major technological alternative to chemical control. Some agents now in use have considerable utility in certain applications, but development costs are considerable and regulatory requirements stringent. The amount of public sector research activity does not appear to be adequate to counterbalance the relative lack of interest being shown by private firms. Accelerated basic research in insect pathology would probably be quite productive of new agents, but disincentives for a subsequent development and production activity are strong.

GENETIC TECHNIQUES

Genetic methods have been employed in pest control in two very different ways. In the first instance, the crop plants themselves can be manipulated genetically so as to increase their resistance to pest attack. In the second, the insect pests may be subjected to genetic intervention by the introduction of mass-reared individuals with a selected phenotype. Both of these techniques have been effectively applied to certain types of pest control problems in the past, and both possess some potential for displacing or reducing chemical control methods in the future.

Genetic Resistance of Crop Plants

The successful development of disease-resistant crop varieties has been a major factor in increasing and maintaining high levels of crop productivity. From the standpoint of the grower, the use of disease-resistant varieties is usually the most effective, easiest to use, and most economical means of controlling plant diseases.

Currently, it is estimated that some 75 percent of U.S. cropland utilizes resistant varieties developed by plant breeders and plant pathologists during the past 50 years and that 95 percent to 98 percent of the vast acreage planted to alfalfa and small grains is planted with varieties that are resistant to one or more diseases.

Development of disease-resistant crop varieties requires a continuing and reasonably expensive research program; its value, however, has made it an excellent investment in terms of total agricultural production, since gains to the farmer probably now exceed $1 billion annually in the United States alone. The current research effort involves support for some 800 plant breeders and plant pathologists in federal research units, state experiment stations, and commercial seed companies.

The availability of disease-resistant varieties is particularly important for forage and field crops that have relatively low values per acre. On such crops, expensive chemical controls cannot be used because the profit margin is too narrow. For other kinds of diseases, such as those caused by soil-borne pathogens and certain viruses, no other economical or effective alternative control measures are available; and resistant varieties have often provided the only means of ensuring crop production. Examples include the rust- and smut-resistant varieties of cereals; fusarium-wilt-resistant varieties of tomato, cabbage, cotton, and watermelon; mosaic-resistant sugarcane; curly-top-virus-resistant sugar beets; bacterial-wilt-resistant alfalfa; mosaic-resistant beans; and cyst-nematode-resistant soybeans.

Not all efforts to obtain resistant varieties have been successful, because sources of resistance sometimes cannot be found. Efforts to develop resistant varieties of forest trees and orchard crops have been relatively slow because of the nature of the host plants. The rapid development of new, more virulent races of rust and smut fungi require a continuing breeding effort on small grains in order to achieve long-term success. The development of varieties that are resistant to rust, smut, and powdery mildew has contributed tremendously to plant disease control and to stabilization of small grain production.

The same genetic principles that determine transfer of other physiological and morphological characteristics control the inheritance of resistance to specific diseases in plants. Resistance may be controlled by a single gene pair or may be quantitatively controlled by two or more genes. Specific genes for resistance may not protect against all races of a pathogen, as in the case of several rust diseases, late blight disease of potatoes, and mildew of beans.

Some resistant varieties are produced that have high susceptibility for other pathogens. The development of *Helminthosporium victoriae* blight on oats is an excellent example. Oat varieties had been developed that had resistance to stem rust and crown rust and to loose and covered smut diseases. These varieties out-yielded most other commercial varieties of that time by a factor of 25 percent. One of the parent varieties used in breeding, the Victoria variety of oats, was highly susceptible to a previously unrecognized disease of oats. By the time the disease was recognized, a vast acreage was planted to the newly developed and closely related group of varieties. The organism built up quickly, and thousands of acres of oats were devastated by Victoria blight. Sources of resistance were quickly obtained, however, and incorporated in desirable oat varieties within a reasonably short period of time.

As an important part of the intensification of breeding programs, it will be necessary to give more emphasis to research on genetic variability of plant pathogens and to develop more understanding of the nature of the factors for virulence. Elucidation of the biochemical basis for resistance in plants and the factors that govern host specificity in plant pathogens would open new pathways in disease control and would also provide major contributions to knowledge.

The development of resistance to insects has been equally important. Many cultivars in present use are the results of selection programs designed to produce features that minimize pest damage. A celebrated case involves the strains of wheat resistant to the Hessian fly. New strains are continuously under development in order to counteract the depredations of new biotypes of the pest, which themselves have developed as a response to the earlier strains of the crop.

Experimental breeding is under way with almost every major crop in order to produce varieties resistant to insects. As described in Volume III, *Cotton Pest Control*, nectarless strains of cotton, resistant to *Lygus* and to bollworms, are in an advanced stage of development. Although the resistant phenotypes of certain crops derive their protection from substances that are toxic or discouraging to pests, many others involve less-obvious features, such as a leaf surface that discourages egg laying by the adult. An important and promising breeding strategy related to genetic resistance in cotton involves the development of short-season varieties, which are ready for harvest before late-season populations of the bollworm reach economically damaging levels.

Research on genetic pest resistance has been sponsored by the USDA and many land grant universities. In addition, such private funding organizations as the Rockefeller Foundation have taken a special interest in the subject; Rockefeller has supported a major project on host plant resistance in cotton that has led to cooperative efforts among university scientists in cotton-growing areas of the United States. Resistant varieties will play an important role in pest control methods developed by the current major project to develop pest management systems for cotton and five other major crops.*

Many crop plants, of course, possess a naturally evolved resistance to attack by herbivores. Biologists believe that many of the so-called secondary plant substances--including the alkaloids, various terpene compounds, the juvenile hormone analogs, and others--may play a role in discouraging attack by herbivores. Other groups of plants possess spines or other surface features that deter feeding by insects. What happens to such natural defenses when the plant is put into a regime of genetic selection designed to make it more palatable, to produce a higher yield, or to exhibit a better response to fertilizers? In some instances, the act of making a plant more palatable to humans in the process of cultivating it removes the natural defense against pest attack. For example, melons, cucumbers, and other members of the plant family *Cucurbitaceae* are obviously palatable to humans; yet their wild relatives taste bad. The taste, which has been eliminated in the course of cultivation, is due to triterpene compounds, which also serve to discourage some insects that otherwise attack the fruit.

It is more often the case that the genetic basis for insect resistance is complex and multiply determined. These deterrent features may in specific cases be negatively correlated with agronomic traits, and if so they will decline in effectiveness during the selection program. Plant breeders attempt to counteract this by exposing the experimental plots in which selection is taking place to

---

*The Principles, Strategies, and Tactics of Pest Population Regulation and Control in Major Crop Ecosystems. Sponsored by the National Science Foundation, the Environmental Protection Agency, and the Department of Agriculture. C. B. Huffaker, Principal Investigator.

the range of pest organisms that the plants will encounter under cultivation. This imposes some selection for resistance as well as on the agronomic features that are receiving the special attention of the breeder.

Elaborate schemes, little known outside the community of plant breeders, attempt to ensure access to germ plasm of major crops on a worldwide basis. Scientists from the USDA and State Agricultural Experiment Stations conduct plant exploration trips especially into the geographic centers of origin of crops for the purpose of collecting wild and domesticated types. These collections--together with a broad spectrum of improved strains from producing areas throughout the world--are maintained in germ plasm banks.* Plant breeders seeking particular attributes to incorporate into their breeding programs can request source material from the banks, which also have computer-stored information about the characteristics, including insect and disease reactions, of much of the known germ plasm of the world.

A parallel set of activities exists on the testing side of variety improvement. Plant breeders participate in coordinated uniform disease nurseries. Released varieties (and some prospective new varieties) are centrally assembled and then distributed to uniform testing stations in producing areas all over the world where they are evaluated especially for disease and insect resistance. More than 25 stations exist in the United States for testing wheat, and there are many more in the world network. Special precautions must be taken to minimize the risk of transferring unwanted diseases into new areas. In the case of small grains (wheat, oats, barley), a program for testing the response to new virulent and potentially destructive races of rust is located in Puerto Rico, where there is no risk of introducing the disease because these crops are totally absent in the native agriculture.

Elsewhere in this report, we examine the question of genetic vulnerability of crops in more depth; at this point, we call attention to it as one aspect of the overall strategy of achieving resistance by genetic means.

---

*The soybean germ plasm bank is at the University of Illinois, Champaign-Urbana. The small grain bank (wheat, oats, barley) is at the USDA research center, Beltsville, Maryland. The wheat bank has about 35,000 entries.

Autolethal Techniques

Autolethal genetic techniques all involve the introduction, into natural pest populations, of reared insects carrying nonfunctioning sperm or specific, usually dominant, genetic mutations that are lethal or conditionally lethal. There is a distinction between the "sterile male" technique, in which the released males produce inactive or inviable sperm, and other autocidal techniques, in which the sperm is active and fertilizations actually occur but the resulting embryos fail to develop. Dominant, lethal mutations are generally chromosome rearrangements induced either by high-energy radiation or by chemical agents. Males carrying these chromosome rearrangements are released in large numbers in the field, and the matings between these males and wild females produce no offspring.

The most celebrated success of the technique has been the use of sterile males for the control of the screwworm fly in several regions of neotropical America. The first pilot program was carried out on the island of Curacao in the early 1950s; later the same methods were used to eradicate this pest from Florida and drastically lower its numbers in the southwestern United States. It appeared for some time as though the pest had been effectively controlled in the Southwest, but a relapse was reported in 1972, when 95,000 cases of screwworm disease were reported compared to a few hundred in each of the several preceding years. This problem appeared to regress during the year 1973, and it is attributed to invasions from regions of Central America where the screwworm still occurs. Some scientists who have been prominently involved with the successful sterile male control programs feel that these regressions can be controlled if U.S.-Mexican government agreements succeed in the establishment of a barrier in southern Mexico. At present, the mass rearing and sterilization programs of the screwworm fly in Texas cost approximately $10 million per year.

The successful application of autocidal genetic techniques depends upon introducing genetically altered individuals at a higher population density than that of their wild congeners. This requirement limits the effectiveness of the technique to situations in which the natural pest population density is low or has already been reduced by some other means. Furthermore, the insect must be suitable for economical mass rearing. For this reason, most projected applications of the autocidal genetic techniques involve special circumstances, in which they are either viewed as part of integrated pest management programs incorporating a number of other techniques used previously or as safeguards against the introduction of a pest organism into a new region.

Among the current applications of this technique that fit the latter description are the use of sterile male Mexican fruit flies (*Anastrepha ludens*) to prevent the establishment of resident populations in California. Weekly releases of a few million sterilized flies in southern California are credited by many entomologists with preventing the invasion of this pest into the region. Similarly, the 600,000 acres of cotton grown in the San Joaquin Valley in California are threatened with invading populations of the pink bollworm from the Imperial Valley of southern California and in Arizona. There seems little question that this pest, if it invaded the San Joaquin Valley, would drastically worsen the pest control problems of an area that--compared with other cotton-growing regions--is *relatively* free of pest problems. Beginning in 1970, massive releases of sterile male pink bollworms were undertaken through a cooperative program between the USDA and the California Department of Agriculture.

These same features characterize plans for the future use of the genetic techniques. Pilot-sized tests on such organisms as the Caribbean fruit fly (*Anastrepha suspensa*) suggest that low-level populations of these organisms may be eliminated by the autocidal techniques. Further, the sterile male technique has been incorporated into the Boll Weevil Eradication Project now being planned. This project is discussed elsewhere in this volume; here it is sufficient to note that the characterization of this project as an *eradication* program places heavy reliance upon the ability of the sterile male technique to achieve its final stage.

Costs and Benefits of the Autocidal Techniques

It is clear that the "sterile male" technique presents a potentially favorable cost-benefit ratio. Because of the importance of the technique in dealing with low-density pest populations, it is sometimes thought of as being synonymous with "eradication," which is discussed below. Where eradication of a major pest is successful, substantial costs are incurred only once and can be amortized in subsequent years. This is also true of the use of the technique to prevent infestation where continual releases are necessary. If the region protected is at all substantial, these costs are apt to be much less than those that would be incurred if major infestations had to be combatted by conventional chemical techniques. Moreover, the reared and released insects present no environmental hazard and pose no occupational health or safety problems.

Unfortunately, although it is easy to conclude that these methods are relatively free of "spillover" costs, the benefits are more difficult to assess quantitatively than they are for other techniques. Because the screwworm fly eradication program was so successful, people have expected a great deal of the method. When it has been applied to *prevent* infestation, there is a natural tendency to attribute success to the technology even where explicit evidence is lacking. By their nature, field uses of the sterile male technique as a preventive measure against the spread of a pest are uncontrolled experiments. In no sense do we *know* that release of sterile male pink bollworms has prevented the spread of the pest into the San Joaquin Valley of California; and, although most observers are convinced that the Mexican fruit fly's failure to establish itself in southern California is due to release programs there, direct evidence is lacking.

The nature of sterile male techniques has placed the research, development, and conduct of these programs mostly within the public sector. The USDA and various state agencies have all contributed research personnel, facilities, and funds to these programs. Since compared to most pest control techniques the genetic methods have relatively specific and readily identifiable targets, it may be asked whether the distribution of costs and benefits is a proper one in this case. In California, a bale tax on cotton serves to defray a portion of the costs of sterile male releases. We believe that consideration should be given to allocating the entire cost of rearing programs--as opposed to research and development alone--to the beneficiaries of the pest protection so provided, namely, the growers.

Incentives and Disincentives

For a variety of reasons, we expect that research and development activities that seek to improve the autocidal techniques will remain in the public sector for some time. By their nature, these methods are highly specific and directed at pest problems that--however important--would constitute a small market for a private firm. They also depend heavily on large-scale rearing facilities that would make heavy initial capital investment demands.

The development of autocidal methods has depended upon, and probably will continue to benefit from, the research activities of some far-seeing and capable government scientists. Although applicable only in rather special circumstances, they have enormous potential, especially in combination with other techniques. Their use as an eradication device has probably been overemphasized, at the expense of a more balanced view of their role as management tools. We believe that government research programs are worthy of continued support, but that the research base for developing them should be widened by extramural programs designed to draw the participation of scientists in universities and other nonfederal institutions. The attractiveness of genetic techniques lies largely in their specificity and their lack of environmental and health hazards. These features, together with their unattractiveness to private investment, make them reasonable targets for public support; but we believe that in the future policymakers will have to confront the question of whether the allocation system for costs and benefits is an equitable one.

CULTURAL PRACTICES

"Cultural practices" is a commonly used term referring to a broad set of management techniques utilized by farmers and other resource managers to achieve their crop production goals. Tillage operations, fertilization, irrigation, selection of plant variety, cropping sequence, grazing management, harvesting procedures, handling of crop residues, and timing of operations represent specific cultural practices. To some extent cultural practices can be chosen so as to minimize pest problems, and hence they can be integrated into an overall strategy of pest management.

The sequence of crops grown affects weed, insect, and/or disease problems. Growth and reproduction characteristics of certain weeds allow them to survive in competition with some crops and not with others. Thus a narrow crop base (the extreme being monoculture) favors some weed species and disadvantages others. Interposing soybeans or a small grain between corn crops can eliminate the rootworm problem. Take-all (*Ophiobolus graminis*) can only survive on debris of wheat and certain other grasses, so it can be completely controlled by growing soybeans ahead of wheat in the Midwest. But insects that are highly mobile and diseases that are spread by wind or mobile insect vectors are relatively unaffected by cropping sequence because of invasion from adjacent fields.

Tillage is a powerful tool in reducing many weeds, insects, and diseases. Conventional seedbed preparation for corn and soybeans, for example, which involves moldboard plowing plus secondary tillage, severely inhibits both perennial and annual weeds. Insects and diseases that survive on plant residues are destroyed or reduced by being buried in the process of seedbed preparation. European corn borer and yellow leaf spot of corn illustrate the concept. Usually some trade-offs must be made for increasing the control of pests by cultural practices in the form of yield loss, additional energy, or in environmental impact. Additional tillage for weed control increases fuel requirements (though this is partially offset by the energy saved in not manufacturing the herbicide) and raises the likelihood of sediment pollution. Altering the cropping sequence usually reduces overall yield because a lower-yielding crop is introduced; the best adapted crop is to some extent moved out of its most favored geographic region or soil type.

Cultural practices are, of course, especially prominent in weed control. Weeds attain pest status by competing with crop plants for resources: light, water, nutrients. In substantial part, weed control involves the management of these resources by the grower so that crop plants receive as large a proportion as possible. The choice of early- or fast-growing varieties that can overtop weed competitors, the use of special irrigation schedules, and many other practices are examples of cultural control. Even where the predominant role in weed management is played by herbicides, cultural practices are often adopted in order to improve control efficiency, as where the spacing of row crops is arranged so as to optimize access for herbicide sprays.

In insect control, cultural techniques have developed hand-in-hand with other methods as a part of integrated pest management techniques for the major agricultural crops. Examples are given in Volumes II and III, *Corn/Soybeans Pest Control* and *Cotton Pest Control*, respectively. The diversity of approaches is suggested by the following examples: control of *Lygus* in cotton by the planting of a few rows of alfalfa, a crop more attractive for the insects; reduction of damage by *Heliothis*, a late-season pest, by earlier planting and the use of rapidly fruiting varieties; the use of stalk-destruction techniques to reduce populations of over-wintering pest insects, such as *Pectinophora*; and the destruction of wild alternate hosts in the case of pests having a two-part life cycle. In public health applications, "source reduction" and other environmental manipulations are the functional equivalent of cultural practices in agriculture.

Cultural practices may also greatly influence the health of a crop. Modifications in cultural practices to minimize disease development are among the oldest forms of disease control. Some diseases lend themselves readily to cultural control. The pathogen and the host plant must be associated under suitable environmental conditions for infectious disease to develop. Good disease control can be obtained by modifying the environment, the condition of the host, or the behavior of the disease agent. Cultural practices may be directed at reducing the amount of activity of inoculum through crop rotation, green manuring, deep plowing, sanitation, and roguing, or they may be used to avoid disease by use of pathogen-free planting stock, by varying the time of planting, by selection of planting sites unfavorable to disease development, by care in seedbed preparation, by varying spacing, and by nutrition.

Several cultural control methods involve biological control activity. For example, certain crop rotations may alter soil conditions so as to decrease the accumulation of reproductive units of a pathogen; the total pathogenic activity of the organism, whether it be a fungus, bacterium, nematode, or virus, is thereby reduced. Effective use of this means of cultural control depends upon knowledge of the life cycle of the pathogen involved and of the various factors that influence its existence, survival, and pathogenic behavior in a particular ecologic situation. For example, growing soybeans as a green manure crop in the fall after potato harvest and turning the crop under green before preparation of the land for spring planting effectively prevents the development of potato scab caused by the soil-borne actinomycete *Streptomyces scabies*. Similarly, serious losses from fusarium root rot of beans can be avoided by rotating the culture of beans with a barley crop.

Many techniques are used in modifying cultural methods to control plant disease. These generally seek to make it possible (1) to use pathogen-free seed and propagating materials, (2) to grow crops in an environment unfavorable for disease development, and (3) to modify cultural practices to avoid increase of plant pathogens and the development of disease.

In the main, cultural practices are undertaken individually by growers or adopted as part of integrated approaches to pest management. Because only changes in practice, and not a product, are involved, research in cultural techniques is carried out almost entirely in the public sector, largely through the USDA-land grant university complex. Advice from Cooperative Extension Service personnel is heavily relied upon for the diffusion of new findings. A special feature of the cultural techniques in agriculture is the degree to which they depend upon institutional features of the enterprise. Many of the methods, such as the destruction of alternate hosts, depend upon joint action by the growers of a region. Others, like the stalk destruction programs, depend critically upon the cooperation of all; 90 percent participation may, in some cases, be no better than no participation at all. Special arrangements are required to ensure the unification of cultural practices over a wide geographic area and thus minimize pest problems in the region. In turn, these may demand mechanisms for employing the legal power of the state to arrange sanctions for noncompliance.

Pest Control Districts

Pest control districts are examples of devices for accomplishing this objective. They are modeled in part after the mechanisms, at the state and federal levels, that allow the growers of a region to set uniform policies regarding research, promotion, and the control of production. For example, California's Agricultural Marketing Act permits the growers of a particular commodity to establish, by vote, marketing orders that require all growers within the region to comply with a production procedure. The planting of particular varieties and the "green drop" of fruit in a year of overproduction are examples of the kinds of practices that may be compelled under such orders. Similarly, pest control districts may require growers to destroy plant debris at the end of a growing season, to plant or harvest a crop before a certain date, or to participate in other group efforts aimed at the cultural control of a particular pest organism. The districts, like marketing orders, are devices that allow a group pursuing some common interest to produce a self-imposed regulation and then to employ the legal power of the state in its enforcement.

Districts of this kind exist in most of the agricultural states. In principle, it should be possible to give the appropriate federal agency equivalent powers to enforce procedures that would be properly applied to a particular crop over a range that extended across several states. Such an arrangement might considerably shorten the time required for the adoption of new cultural practices. Although equally sweeping powers are given to federal agencies in other areas of pest control (e.g., the plant and animal quarantine laws), no proposals of which we are aware would do the same thing for cultural pest control.

Organic Farming

A special instance of the use of cultural pest control is represented by the diverse group of people who practice a variety of farming known collectively as "organic farming." Their importance lies in two areas. First, they produce food that commands a portion of consumer expenditures; thus they are an economic reality. Second, because they are a physical embodiment of alternative technologies of farming, they prompt discussion about the validity of those alternatives.

Both producers and consumers of organic food have a variety of motivations for their beliefs and actions. The farmers themselves may believe that their methods are the best way to avoid polluting the environment, to improve the soil, and to produce superior food; the consumers may believe that organic food tastes better or is more nutritious. Many are concerned about the possible health hazards of pesticide residues. Some share the concern of many organic farmers for the encouragement of more "ecologic" farming practices.

Whatever the validity of the claims of the organic food producers and consumers, it appears likely that consumer demand for organic produce will remain a small but significant factor in the food-marketing business. It is difficult to quantify the demand for organic food alone, as it is generally classed under "health foods" (a term as yet unjustified by evidence of specific health advantage, which includes "natural" foods, usually defined as foods that are prepared without the use of additives; "organic" foods, grown without fertilizers, pesticides, growth hormones, etc.; and specialized diet products The most popular health foods are grains, flours, cereals, and juices.

Annual sales of health foods have grown from $200 million in 1971 to around $400 million in 1972, and were predicted to reach $500 million in 1973. There are roughly 3,000 health food stores in the country, including 1,300 in Los Angeles County alone (Small Business Reporter 1973). In addition, many supermarkets, drugstores, and department stores have added health food departments. Many of the products originally stocked did not sell well after an initial spurt. Such departments reached a peak in the spring of 1972 (Epstein 1973). Although the market for health foods is growing, it still represents less than 1 percent of total U.S. food sales (Spaeth 1973). *Supermarket News* (Feb. 21, 1972) has predicted that health foods will make up 5 percent of the total food market in the next several years. Other industry spokesmen, however, consider that prediction overstated.

The organic food market has experienced some problems with the false labeling of foods by retailers who want to take advantage of the generally higher prices of organic products. The problem has been taken seriously enough that a number of states are moving toward the regulation of advertising and sales of products. Oregon, Massachusetts, and New Hampshire have already passed laws setting forth definitions and labeling requirements. Such laws are also under consideration in Arizona, California, Maine, New York, Pennsylvania, Texas, and Vermont. Two federal laws have also been introduced, but no action is expected on either in the near future. Although we do not make a judgment here on the merits of "organically grown" produce, we endorse the rights of consumers to expect that foods so labeled will have been grown by understood procedures.

Organic farmers have been among those experimenting with and using nonchemical methods of pest control during the past 30 years. No single pest control practice is totally within the province of organic farmers, however, and many of their practices have counterpart practices among conventional farming techniques. The following briefly describes the major pest control activities of organic farmers:

1. The use of diversified plantings and the avoidance of large-scale monocultures. Diversity may stabilize an agroecosystem, in some cases, by providing "diversionary" host plants, a source of natural enemies, physical refuges for natural enemies, or barriers to dispersal of pest species. The continued existence of organic farms with diversified crops suggests that they have remained

economically viable by supplying a specialized market despite their avoidance of the commonly used pesticides. (They may, however, use such naturally occurring pesticides as pyrethrins and rotenone.)

2. The use of "biological control." Many organic farmers and gardeners have experimented with the introduction of parasites and predators. A common example of a commercially available predator is the ladybug, which feeds on numerous insect pests, especially aphids and greenbugs. Ladybugs have been sold extensively to gardeners, and they have been used on a larger scale on grain sorghum in Kansas. The lacewing fly (*Chrysopa* sp.) is another common predator; it has been found to be quite effective against pests of tomatoes (fruitworms and armyworms) and cotton (bollworms and budworms). Two kinds of parasitic wasps have been used to combat the alfalfa weevil and the codling moth (a pest of apple orchards). All of these, and other more specialized organisms, are available through commercial insectaries. The markets for these beneficial insects have been sustained in part through the demand generated by organic gardeners and farmers.

3. "Companion" plants. Certain plants repel insects because of their odors, colors, or other properties. If these species are interplanted with vulnerable crop plants, some protection may be given to the crop; but many of the recommendations made in the organic farming literature concerning companion planting are based on simple observations rather than controlled experiments. In this category is the evidence that the orange color of nasturtium blossoms repels some insects and that members of the aster family (especially marigolds and chrysanthemums) appear to repel insects because of their odor.

4. The main fertility problem of organic farming and gardening is lack of available nitrogen. Three practices are widely followed in order to overcome this deficiency: (a) growing a nitrogen-fixing leguminous crop (alfalfa, clover, vetch, lespedeza) every second or third year; (b) keeping livestock to supply animal manure; or (c) composting grass clippings, leaves, and plant residues. Dried municipal sludge is also used where available. Crop rotation, practiced mainly to supply nitrogen, incidentally partially or completely controls some insects and diseases and reduces the need for certain pesticides. Some pests are, however, worse in diverse cropping systems. The advent of adequate supplies of low-cost nitrogen fertilizer gave commercial farmers more flexibility in selecting cropping systems, and they often found it most economical to specialize in one or a few crops.

5. Resistant crop varieties. Varieties resistant to various pests, especially to diseases, have been widely used in conventional agriculture. In general, however, if satisfactory pesticides exist to control the most important pests, conventional farmers are unlikely to sacrifice productivity for properties of resistance in choosing their varieties. Organic farmers are more likely to place a higher priority on pest resistance.

The practice of not planting crops vulnerable to pests in a given geographical region is related to the use of resistant varieties. The alternative crops planted may provide lower profits, but the need for chemical inputs is avoided.

In general, the beliefs of organic farmers have not been carefully and scientifically tested. There is little question that it is possible to produce crops by organic technologies, but under present conditions it seems unlikely that we could raise as much food by organic methods as we currently do. There is no indication that the number of either organic farmers or organic consumers will change significantly in the future. Many claims of organic farmers must be viewed with skepticism, but their mere unorthodoxy should not be used to reject them automatically; their lore contains useful knowledge. Since the creation of options will be the key to successful pest control of the future, no source of potential alternatives should be summarily dismissed.

FORMULATION AND APPLICATION

Problems Related to Formulation and Application*

A variety of problems in pest control are potentially solvable through development and adoption of new techniques in formulating and application. They can be summarized very briefly in seven major categories.

---

*The Committee appointed a special Panel on Formulation and Application Technology, under the Chairmanship of Boysie E. Day, consisting of James L. Brann, E. Dale Threadgill, Wade Van Valkenburg, and Wesley Yates. This section draws heavily on the proceedings of a meeting of this Panel on May 9-10, 1974.

Inefficient Application

Inefficient application of a pest control material results because (1) some of the material drifts away from the target area or (2) the target is not covered evenly. Drift from the target area can occur either with ground application equipment or aerial equipment. The drifting material can be either a dust or a spray, but dusts are somewhat more prone to drift than sprays. A major cause of drifts is the production of particles that are very small (less than 50 microns in diameter); while small droplet size may help spreading at low dosages, drift is objectionable because the toxic material often affects an unintended organism. For instance, 2,4-D drifting into cotton or methyl parathion drifting into inhabited areas can create significant toxicity problems for the cotton plants and humans, respectively.

Uneven application is particularly a problem when the target is at some distance from the applicating machine. For example, a sprayer must be adjusted very carefully to cover evenly a large fruit tree or many rows of a row crop. High variation in the amount of pest control material applied to the target organism means either pests are not controlled in the lightly treated areas or that heavily treated areas are being overdosed.

Residues

Residues of toxic material in food and feed are subject to legal regulation because of toxicity hazards to humans and livestock. Additives and adjuvants in a pest control formulation can significantly alter the residues remaining on a crop at harvest time; similarly, they can change the persistence of toxic residues in the environment. Lengthening of the persistence rate is in some cases desirable: For example, Zoecon's juvenile hormone mosquito larvicide had to be made more persistent before it could be registered as an effective material. In other cases, a shortening of persistence would be a rational goal.

Worker Safety

Workers in formulating plants, applicators, and workers reentering areas that have been treated with pesticides are all subject to poisoning by the active ingredient. Both the way in which the material is formulated and the way in which it is applied can significantly alter the hazard posed to people in any of these situations.

Social Pressures

Health and environmental considerations constitute a strong impetus for decreasing the amount of toxic material used in pest control operations. Certain ways of applying or formulating pest control materials offer some promise of helping to achieve this objective while maintaining present levels of control.

Improper Use of Current Equipment

There was a consensus among members of the panel that many applicators may be casual or even sloppy in the way they calibrate their equipment and maintain it over one or more spraying seasons. The result is that an excess of toxic material is actually used in pest control operations. A reduction in this wastage would be desirable environmentally, and it may represent a savings in cost for the applicator. Furthermore, because the application equipment currently in the field represents a substantial capital investment for the applicators, it is likely to be around for several years. Thus its proper use is important, even if new application devices and formulations are developed in the very near future.

Costs

The rising costs of inputs in farming operations have been a major concern to farmers in recent years. It is difficult to generalize on the relationship between new formulation and application technologies and cost, but such innovations as ultra low volume (ULV) spraying can potentially reduce the costs of application by reducing the amount of toxic material applied. Some types of new innovations, however, such as microencapsulation of pest control materials, may increase the cost of pest control materials.

Need to Integrate with Other Technologies

Attempts at systematic improvements in pest control technologies, such as the development of integrated pest management concepts, have been done without due consideration for optimal formulation of the pest control material and efficient delivery to the target. Technologies of formulation and application cannot be ignored in the future, and policy decisions are needed to ensure proper support of R&D.

State of the Art in Formulation and Application Technologies

Many alternatives to contemporary practices are available, but each differs in its potential to solve the problems discussed above. The following paragraphs point out the strengths and weaknesses of the most common suggestions.

Aircraft

Aerial application of pesticides has revolutionized pest control on a worldwide basis. In California, for example, such crops as rice and cotton could not be grown economically without aircraft. As much as 80 percent of the commercial applications of pest control materials in California is by aircraft, and over 50 percent of the total pest control materials nationally are applied by air. In the evolution of this industry, there have been three generations of agricultural aircraft:

1. Surplus military aircraft. Early uses, before World War II, included dusting operations in cotton; one of the most popular aircraft was the Stearman biplane. Large-scale applications on forests and rangelands have utilized TBMs and others, ranging in size up to the B-17.
2. The "Ag-1." This airplane, developed by Fred Weick in the 1950s, was the first aircraft designed specifically for agricultural applications. It was small, lightweight, and had many safety features. Weick eventually joined Piper and produced the Piper Pawnee, the most commonly used aerial applicating machine in the world. Cessna, Grumman, Rockwell International, and others are now involved in the production of agricultural aircraft. Their capacities range from 100 to 450 gallons.
3. Turbine-powered aircraft, particularly helicopters. Common examples of these machines are Bell's Jet Rangers and Hughes' model 500. Even larger turbine-powered helicopters such as the Bell 205A, which can carry up to 500 gallons of spray material, have recently been utilized for a variety of agricultural applications. In New

Zealand, two Fletcher aircraft have been fitted with gas turbine engines.

The design of agricultural aircraft has been viewed as basically a transportation project, and the major emphasis has been on minimizing operational costs. For example, in New Zealand prices for aerial application are based on the ton-mile, and the prices in the United States are generally based on the gallonage applied per acre. As a result of improved efficiency of aircraft operations and keen competition, the price of aerial applications has not substantially risen during the past 20 years.

Insufficient attention, however, has been given to the quality and safety of the spray operation. The design of aircraft and distribution equipment has not been integrated so as to provide optimum coverage of plant foliage with a minimum drift out of the target area. For instance, aircraft manufacturers do not consider the design and use of spray mechanisms; and most hydraulic nozzles have been designed for ground application equipment with little or no air shear. There has been little competition in that industry, and therefore little incentive to improve nozzle characteristics. Manufacturers are not involved in evaluating or recommending types of nozzles for specific pest control situations.

The total aerial applicating system should be considered in the design of new vehicles. Attention should be given to the control of the vortex strength and its effect on atomization and distribution of the spray. Further research is needed to relate the required drop size spectrum, vortex strength, and distribution pattern for different combinations of crops, pests, and chemicals.

Air-Cushion Vehicles

A few commercial air-cushion machines appear to have some potential for agricultural applications. Sealand Hovercraft Ltd. of England manufactures a unit that will carry a payload of 1,000 pounds and clear obstacles up to 18 inches. Because of restricted clearance, however, their potential will be limited to flat, relatively smooth terrain, such as for weed control in rice. It is unlikely that they will come into use during the next 5 years, but after that some may be introduced.

Ground Applicators

Special ground machines have been developed for various types of crops. Basic types include machines for orchards, row crops, and broadcast application.

Airblast Machines During the last 25 years, the use of airblast equipment for applications to row crops, field crops, and fruit has steadily increased and is one of the most popular methods for foliar treatments. For example, more than 90 percent of the commercial fruit plantings in New York State are sprayed with airblast equipment.

The airblast provides the energy for breaking the liquid into droplets and distributing them throughout the crop. The concentration of toxicant used in the spray liquid and the volume of liquid per acre may be varied, depending on equipment design and the problem to be treated. In "dilute" applications, the toxicant concentration is low and the plants are sprayed to the dripping point with 200-800 gallons per acre. The same machines may be used to make "semiconcentrate" applications in which the toxicant is used at concentrations two to six times that used in a dilute spray and is applied at one-half to one-sixth the total volume of liquid per acre. This results in a significant saving in time and labor. True "concentrate" sprays are generally applied with specially designed equipment. In this case, the toxicant is put in the tank at 8 to 39 times the concentration used in dilute spraying and applied at one-tenth to one-fortieth the total volume of liquid per acre. This latter process--when correctly used--may save up to 25 percent of the toxicant required, due for the post part to a lack of runoff. However, precise control of the operation must be maintained, as slight errors are magnified by the high concentration.

There are several problems associated with airblast spraying:

1. Air velocity decreases rapidly with distance from the source. Total energy remains approximately the same, but the velocity is reduced in proportion to the volume of ambient air entrained.
2. Air is a poor carrier of liquid droplets. As the velocity decreases, larger droplets fall out of the airstream so that only small droplets are carried some distance from the outlet.
3. The deposition rate of droplets on the target is a function of the mass of the individual droplets and the

velocity at which they are traveling. Large droplets impinge at relatively low velocities, whereas high velocities are required for the impingement of small droplets.

4. Interfacial tension between the drops and the target surface influences the rate of deposition. The lower the interfacial tension, the higher the deposition rate.

5. The evaporation of water from droplets is correlated directly with the wet bulb depression: The deposition rate and the efficiency of the spraying operation decreases as the relative humidity becomes lower. In an airblast spraying operation, as much as 75 percent of the water may be lost by evaporation within 36 feet when the mass median diameter of the droplets is 125 microns and the relative humidity is 30 percent.

6. The size of the spray is an important factor in airblast applications. Large droplets tend to fall short of the target, while the smallest droplets fail to impinge and end up as undesirable drift. At the present moment, there is no practical system for controlling the drop size spectrum within the limits necessary for optimum deposition and distribution of spray material.

These few selected factors make it clear that airblast spraying is a complicated process. In the past, too little attention has been paid to these factors as new machines have been developed, and little is known about their effect on the efficiency of pesticide application.

At the present time, growers obtain very uneven distribution of pesticide on the target area with airblast equipment. This is due both to inadequate equipment design and to inefficient operation in the field. Improvements in design can only be accomplished with further research and development. Effective R&D will require a team of scientists, including physicists, engineers, chemists, plant pathologists, entomologists, and others. In addition, an extension educational program is needed to train growers and others making pesticide applications in the proper way to adjust, calibrate, and operate their equipment for greatest efficiency.

Row Crop Machines Special row crop sprayers have been developed for a variety of specialized crop requirements. Systems have been integrated to incorporate pesticide applications with several different field operations. These include spray and granular systems operated in conjunction with planting and cultivating operations. Also, high-clearance vehicles provide access for pesticide treatments in tall growing crops such as corn or cotton. Because of the close proximity to the target area, problems of distribution, coverage, and drift are not as serious as with aerial applications. For instance, soil applications with vertical knife injectors or subterranean horizontal "layering" blades can effectively provide controlled soil applications with a minimum of loss to nontarget areas. However, foliage applications with fine atomization required for certain pest problems or utilized for some ULV applications may present serious drift hazards. The basic problem is that present atomization technology produces a small percentage of drops with a negligible gravitational settling rate and hence a high drift; there is also frequent improper use of present equipment. Thorough training of the operator should include instruction on safe handling procedures, calibration, adjustment of equipment, and basic factors affecting spray efficiency as well as potential hazards.

<u>Broadcast Sprayers</u> Larger broadcast sprayers are available that offer rapid treatment of large acreages and may well compete with aircraft for certain jobs. For example, large "floater" rigs equipped with low-pressure tires that result in minimum soil compaction can carry more than 1,000 gallons and operate a 60-foot boom at around 15 mph.

Current practices in ground application involve mixing toxicant and diluent before spraying in a single container. Spray is dispersed without compensation for variation in the rate of speed of the machine. Four general problems result: (1) Excess mixture must be disposed of, (2) personnel have a high probability of contacting toxicants during mixing, (3) mixing may be inaccurate or nonuniform, and (4) there is no correction in application rate when wheel slippage varies during spray applications.

Recent research in the public sector has developed machines in which the toxicant remains in a separate container from the diluent until mixing occurs as the sprayer is operated. Metering devices accurately measure small flow rates of toxicant (either liquid or powder) into the diluent in a mixing chamber. The rate at which the toxicant is metered is based upon the actual ground speed of the application equipment.

Practical systems utilizing these mixing and metering devices have been successfully operated under field conditions. They reduce or eliminate the probability of worker contact with the toxicant. Furthermore, they provide the proper rate of toxicant at all times. Capital and maintenance costs to the farmer for the new devices would probably be somewhat higher than for contemporary machines. Increases would be at least partially balanced by savings resulting from less toxicant applied and less wastage from unused toxicant contaminated with diluent. Further savings to the grower might result from the introduction and use of returnable, reusable pesticide containers that would fit directly onto the spray machine. The latter innovation could also reduce the environmental hazards associated with the disposal of "empty" pesticide containers, most of which contain significant amounts of toxicant.

Uniform Droplet Size

The basic nozzle-type atomizer most commonly used today was developed several decades ago and produces a wide range of drop sizes. The large drops are too large for effective or efficient pest control and the small ones are very susceptible to drift. Although considerable research has been devoted to improving these atomizers, significant improvements in the uniformity of size of drops produced have not been made.

Drops smaller than 50 microns in diameter are very susceptible to drift under average micrometeorological conditions. The range of 80-150 microns drop diameter is optimal for general spray applications. Drops in this size range have very low susceptibility to drift and appear to have a satisfactory efficacy for a large percentage of pest control problems. They require a minimum of volume of diluent to be transported and used during application and thus significantly reduce the current cost of most spray applications. Furthermore, they allow a reduction in toxic material required to achieve pest control, because less material would drift out of the target area and more uniform, efficient coverage of the target would be likely.

Recent research has developed several laboratory devices capable of producing sprays of uniform size drops in the range of 80-150 microns. However, efforts to utilize these devices or the principles of drop production upon which they are based for practical field equipment have not been successful. The major factor that hampers success is that the orifice size required to produce these drops is so small (approximately one-half the diameter of the drop) that it is easily clogged.

In addition, some of the devices are too delicate for satisfactory operation under normal field conditions. One or two practical devices for producing sprays of relatively uniform size drops have been developed, but they produce large drops in the range of 600-800 microns and are useful only for special pest control problems and specific transport devices. A significant breakthrough in development of uniform size drop generators for general use has yet to be achieved.

Various rotary atomizers produce a narrower range of drop sizes than most nozzle atomizers, but a high percentage of these drops are so small that they are too susceptible to drift.

Atomization is achieved by applying liquid at low pressure to the center of a rotating cage, disk, or cup. Under laboratory conditions and at very low flow rates, the rotary disc unit can produce a band of uniform size drops along with some small satellite drops. Some research machines have been developed to separate the satellite drops. Because of the laboratory capabilities, it is commonly thought that rotary atomizers produce a more uniform drop size spectrum. However, commercial machines operated under field conditions with much higher flow rates produce drop size spectra similar to other types of atomizers. One of the basic advantages of rotary-type atomizers is that very fine atomization can be readily obtained with high-speed motor units. However, maintenance of high-speed units presents some operational proble

The potential benefits of reduced costs and environmental contamination and increased efficacy that would result from the development of a uniform size drop generator are very high. It would find use on both aerial and ground-based equipment. Support for research on atomizers should be increased and should include the use of improved filtration systems, better methods for mixing diluent and toxicant, and introduction of new engineering principles to equipment design.

Low Volume (LV) and Ultra Low Volume (ULV)

LV and ULV application rates utilize lesser amounts of toxicant to achieve control and hence offer savings to the farmer and reduced environmental costs. There is good equipment for and trend toward the adoption of LV application. ULV has been a success in control operations over very large areas, for example with malathion used for grasshopper control on rangeland. The problem with ULV application is that the spray drifts over a very wide area. ULV will not be practical for application in small areas until devices are built to produce uniform drop sizes.

Electrostatic Charging

It has been suggested that sprays or dusts given an electrostatic charge as they leave the application machine would adhere better to the target and hence lead to a reduction in amount of toxicant needed to achieve control. It is our judgment that there is little potential for the electrostatic charging of dusts. On the other hand, it might be useful to continue research efforts on electrostatic charging of sprays. The mechanism by which electrostatic charging works is still unclear. How is the charge given to the spray particle? What kind of charge is it? What is the relative charge of the plant? Perfection of electrostatic charging of sprays might allow the development of ULV application techniques even in the absence of uniform drop sizes.

Wet Dusting

Water can be mixed with dust in the air before the material impinges on the target area; the wetted dust may stick better to the target and improve the efficiency of the toxicant. We judge that it would be better to use and improve sprays rather than attempt to improve the qualities of dusts.

Formulations

There are a variety of alternative formulations that may be substituted for contemporary methods. A discussion of the potential of the most prominent suggestions follows:

1. *Foam generation*. True foams and air emulsions involve materials whose function is to create a frothy mass that can be more precisely applied than a spray. A true foam is stable and may have up to as much as 40x expansion ratio. Air emulsions are unstable and have up to about 2x expansion ratio. True foams have not been used in pest control. Air emulsions have been used, mostly for applications of herbicides, cotton defoliants, insecticides

for control of corn earworm, and growth regulators for tobacco sucker control. In some cases, the adjuvants used to create air emulsions have been judged ineffective. We judge that air emulsions have less general potential than other innovations in formulation and application, but research for specialized uses should continue.

2. *Granular formulations*. Granular formulations consist of solid particles between 200 and 2,000 microns in diameter, and they can be screened to size. They are useful for depositing material on water or on the ground. One of their most interesting applications is for weed control on rangeland. Highly concentrated material, such as fenuron or tordon, can be applied in large widely disposed granules. "Microgranules" (100-300 microns in diameter) have recently been developed by Sumitomo Chemical Co. in Japan. Granules of this size will stick to a plant rather than fall to the ground or water, but the coverage may not be as efficient as that achieved in a good spray operation. Two properties of granules--resistance to drift and relative safety--indicate that research on new potential uses should continue.

3. *Microencapsulation*. Microencapsulation in the past has meant the production of a particle with two phases: an external phase without toxicant and an internal phase with active ingredient. The term is now used more casuall also to refer to solutions of toxic material in some type of matrix or to the binding of toxic material by covalent, ionic, or hydrogen bonds. The basic rationale for microencapsulation is to produce something more persistent and thus decrease the number of applications necessary to achieve control.

Encapsulation appears necessary to ensure the excess of some materials, such as Zoecon's juvenile hormone for mosquito larvae. This situation may underline the general rule that the role of the technique will be to protect the chemical. A special type of encapsulation is represented by Temik, which is bound to clay for purposes of reducing the acute toxicity hazard to humans. Even if an encapsulated material is used, there may still be a problem of getting the chemical to the target organism once it is released to the environment.

Research has only begun on the applications of microencapsulation. Insufficient attention has been given to the basic release mechanism. Further work should be conducted to more fully define the situations in which microencapsulation may play a useful role.

4. *Systemic formulations*. Systemic pesticides are absorbed by and translocated through the plant or animal

to which they are applied. The property of being systemic actually belongs to the active ingredient, but we consider it here for convenience. Systemic pesticides, especially insecticides and fungicides, show considerable promise. For example, the fungicide Benomyl shows great promise as a protectant of elm trees from Dutch elm disease. With any systemic material, careful scrutiny will be given to the residues remaining within the plant at harvest time. The regulation of tolerances may thus be a barrier to the development and introduction of more systemic materials. Research on the limitations and promises of the materials should nevertheless continue.

5. *Combinations.* It is possible to combine pesticides with pesticides, pesticides with fertilizers, and so forth. In some cases, combinations of one type of material with another may be required to create truly effective pest control agents. For example, it may be necessary to combine juvenile hormone with Abate to create a fully effective mosquito larvicide. Combinations, however, are not proprietary materials and therefore are not strongly favored by industrial organizations. In general, it is not possible to patent a combination unless it does something not predictable from the properties of its separate ingredients taken by themselves.

A significant danger attends the development of combinations: It is difficult to create a combination in which both active ingredients are applied at optimal rates and times. In some cases, in fact, people may tend to apply combinations of two or more ingredients when in fact they only need one. Thus, combinations add to the potential for considerable overuse of active ingredients.

6. *Additives and adjuvants.* We judge that there is considerable potential for improving the efficiencies of chemicals through the use of a wide range of additives and adjuvants. There may be little incentive, however, for industry to do the research and development to produce these auxiliary materials. The current use of the additives and adjuvants is certainly more of an art than a science, and there is no well-developed theory that describes the interaction of the auxiliary materials with the active ingredient. It is beyond the scope of this Committee to give exhaustive suggestions on the development of a scientific exploitation of additives and adjuvants; but for purposes of illustration, we will discuss some possibilities for improving the efficiency of herbicides with surface-active agents.

A critical problem in herbicide efficacy involves the efficiency of utilization of the applied toxicant. In both foliar and preemergent applications of herbicidal liquid formulations, only 10-25 percent of the toxicant is actually utilized in control of the undesired species of plant. A number of factors affect biological efficacy, but the most important rate-controlling steps involve (1) the penetration of a lipid membrane and (2) interaction with the plant at an active site. The relationship between the concentration of active ingredient needed to achieve a given effect and the ability of the compound to penetrate lipids is expressed by

$$\log 1/C = a\pi^2 + b\pi + \rho\sigma + d,$$

where $\pi = \log P - \log P°$; P = octanol/water partition coefficient of the active ingredient; $P°$ = partition coefficient of a reference compound; σ = Hammett sigma constant; C = a concentration to yield a given biological effect; and a, b, ρ, and d are appropriate constants.

Experiments with a homologous series of herbicidal compounds indicates that two parameters, the partition coefficient and a kinetic constant, can be optimized for a given plant and herbicide. The optimum parameters vary from plant to plant and thus indicate species diversity. This specificity in the partition coefficient parameter is related to the variable lipid molecular compositions in different species of plants. In addition, it would be anticipated that as the growing season advances and the lipid composition varies, the optimum partition coefficient would also vary.

Surface-active agents come in a wide range of solubilities and are a cosolvent with water that may yield a partition coefficient (octanol/water-surfactant) either more favorable or less favorable for lipid penetration than the previous system partition coefficient (octanol/water). Thus, judicious addition of a surfactant may increase penetration through the lipid layer of a plant or seed, with the net result that the efficiency of utilization of the pesticide is increased.

A great deal of research remains to be done on the effect of surfactants on herbicidal activity. A promising approach to this problem might be to establish the most favorable partition coefficients for control of each weed species. This might be ascertained using a homologous series of hydrocarbons with a gradation of octanol/water partition coefficients as standards. Then surface-active agents can be used to alter the partition coefficients and thus increase the efficiency of utilization of the herbicide. There is not, of course, a simple correlation between herbicidal activity and the partition coefficient, but at least one of the rate controlling variables can be improved. In addition, the variability of the most favorable partition coefficient for a given weed will vary with the season of the year.

In essence, such a program would involve custom formulations resembling those made by the petroleum industry for automobile gasoline used in various areas of the country. A successful improvement in the efficiency of herbicidal utilization might reduce the amount of active ingredients required for a given control operation by up to 50 percent.

Conclusions

In four general areas, increased research and development activities might yield a particularly high return. We recommend priority attention be given to the support of these areas. We further recommend that support for innovation in formulation and application be considered an integral part of the development of integrated pest management techniques.

1. *Develop methods to control particle size.* Sprays, dusts, and granules are all subject to nonuniform particle size. Very fine particles drift, no matter what the basic nature of the formulation. Development of methods to reduce the number of fine particles would significantly reduce drift and decrease unnecessary environmental contamination. There are, however, cases in which the use of fine particles reduces the amount of pesticide needed for control.
2. *Improve efficiency of application machines under field conditions.* It has been recognized for some time that they way a pesticide is applied to the target area is as important as the chemical itself if efficient use of the material is to be achieved. From the end of World War II until 1960, several projects were under way to improve and innovate on application methods--including programs funded by the USDA, states, and others in California, Washington, Texas, Michigan, Ohio, Mississippi, and New York. During the 1960s, most of these projects were drastically curtailed or terminated due to the pressure for monies to support research in other areas. Recently, however, pest management groups working with a variety of crops have indicated that inefficiency in pesticide application is a limiting factor in reducing

the amount of chemicals needed for adequate pest control. Other groups studying the hazardous effects of pesticides in the environment point to the need for depositing and retaining these materials in the target area. Thus efficient pesticide application has again become recognized as an important factor in crop production.

We recommend that projects be established in appropriat areas of the United States to study and develop new method techniques, and equipment for the application of pesticide To be most effective, these projects should support a team of researchers, including engineers, physicists, chemists, entomologists, plant pathologists, and possibly others with biological expertise. Members of such teams should be supported at a level to ensure that problems of pesticide application would be their major responsibility.

3. *Develop a surfactant research program to put the use of surfactants on a scientific basis.* Surfactants offer considerable potential for altering the properties of pesticides in desired ways. A theoretical understandin of how surfactants work will allow potential innovations in this area to be exploited more systematically and efficiently.

4. *Develop education programs to improve the use of the currently available machines, formulations, and technologies.* There will be no immediate replacement of application machines, regardless of what innovations come along in the near future. The environment and the society would thus be well served by getting the most efficient use out of the currently available equipment.

INTEGRATED, NATURAL, AND BIOLOGICAL CONTROL

Integrated Control

Increasingly, biologists concerned with pest management have come to believe that overall strategies of emphasizing the containment of pest populations by a multiplicity of methods are preferable to over-reliance on *any* single method. Supporters of integrated programs of control point out that fixation on particular strategies leads to evolutionary adaptation on the part of the target pest at the level of the ecosystem, and to wasteful competition for resources at the level of the human institutions involved. The integrated approach emphasizes the use of natural regulatory forces inherent in the ecosystem and the location-specific character of pest problems and solutions.

Use of the phrase "integrated control," in its pest control sense, dates from the mid-1950s, but credit for the origins of the integrated control concept may belong to such early entomologists as Stephen A. Forbes and C. W. Woodworth. Since these beginnings, the definition of integrated control has broadened considerably. In 1967, an FAO panel on integrated pest control defined it as "a pest management system that . . . utilizes all suitable techniques and methods in as compatible a manner as possible and maintains the pest population at levels below those causing economic damage." Implicit in this definition is also that imposed control measures, notably conventional pesticides, should be used only where economic injury thresholds would otherwise be exceeded. When control techniques are implemented, it is of primary importance that they be minimally disruptive to the ecosystem.

"Successful development of integrated control depends on adequate understanding of the population dynamics of pests and of potential pests, of the ecology and economics of the cropping systems, and of possible harmful effects to the general environment." This expanded definition holds that maximum reliance is to be placed on natural pest population controls, along with a combination of techniques that contribute to suppression--cultural methods, pest-specific diseases, resistant crop varieties, sterile insects, attractants, augmentation of parasites or predators, or chemical pesticides as needed.

Numerous phrases with meanings that roughly parallel the integrated control concept have been suggested. Among these are "harmonious control," "protective population management," "ecological pest control," and "pest management." It has been suggested that pest management and integrated control are synonymous, and certainly the two are often used interchangeably. There are, however, real differences. Integrated control is the approach to the management of pest populations defined above. Pest management, by contrast, includes all approaches--from single-component techniques to extremely complex control systems. It may utilize such techniques as simply not growing a crop in a certain high-risk area, the establishment of quarantines against introduction of a pest species, or attempted eradication of certain species.

Natural Control

Natural control may be defined as "the maintenance of population numbers (or biomass) within certain upper and lower limits by the action of the whole environment, necessarily including an element that is density-induced, i.e., regulating, in relation to the conditions of the environment and the properties of the species." All classes of pest organisms are subject to control by elements in the environment, but the importance of natural control has been most forcefully demonstrated with insect pests by its disruption with insecticides. Most of our comments therefore focus on insects.

Biological Control: Naturally Occurring

Most insect outbreaks reflect disturbances in natural or man-made ecosystems; in general, the more artificial (managed) and/or exotic the crop or forest management system, the greater are the chances of economic losses due to insects. Although the potential destructiveness of unleashed phytophagous insect populations is large in the United States and other temperate as well as tropical regions, there are countless species of such insects that rarely, if ever, manifest severe population outbreaks. In the western coniferous forest and chaparral biomes, occasional population explosions of insects are severe enough to defoliate and at times kill patches of plants. It would be an exaggeration to suggest that the entire homeostatic nature of these biomes results from natural biological control, but it appears that natural enemies are normally responsible for regulating and keeping under control many phytophagous species that occasionally attain outbreak proportions in wooded areas. In the United States natural enemies of phytophagous insects occur in the most diverse and thus usually the most stable communities, such as the forests, and also in the most unstable kinds of ecosystems, annual agricultural crops.

The coniferous forest is probably the oldest natural community that persists in the western United States, and it is perhaps not surprising that widespread devastating insect outbreaks are relatively rare there. An example of naturally occurring biological control in this biome is that of the Douglas fir tussock moth, *Orgyia pseudotsugata* McD. In the West there appear to have been scattered outbreaks of this species about every 10 years. The outbreak phase and collapse of the 1962-1965 outbreak in California were studied, and the decline of the population was so complete that no living larvae could be found in 1966, and only 25-30 egg masses were found in one study area. A nuclear polyhedrosis virus was considered to be the primary cause of the collapse, but in addition three predaceous insects and nine parasites were found associated with the larval and pupal stages. In another area, parasitization was an important factor in the collapse of the population, and the females that emerged did not oviposit. The lack of oviposition was believed to be due to sublethal virus infections. A temporary clearance to use DDT in the control of this insect was granted in early 1974, but many forest entomologists believe that natural mechanisms of control were adequate to keep the pest below its economic threshold density without the chemical.

A different example is that of a recent outbreak of a scale insect, *Chionaspis* (*Phenacaspis*) *pinifoliae* Fitch, at South Lake Tahoe, California. The area of infestation coincided precisely with an area that had been treated with the organophosphate malathion in an attempt to control the adult mosquito, *Culex tarsalis* Coquillett. The foliage of the pine trees in the infestation area were so heavily infested with the scale that the trees appeared white. Fogging was discontinued in 1969, and by 1972 the scale population had declined considerably in all study plots. Predation was considered to be responsible for the immediate population decline, and a high level of parasitism was noted in the remaining low-density scale population. The insecticide had probably caused an initial disruption of the parasite-scale interaction and, as the scale population increased, apparently also inhibited the effectiveness of the predators. In this instance the predators feed only at high density and thus reduce the scale population to low enough levels for the parasite complex to resume regulation.

Paradoxically, some of the best evidence indicating the importance of natural biological control comes from agroecosystems. One of the important pests on grapes in California was the grape leafhopper, *Erythroneura elegantula* Osborne. A myamaid egg parasite, *Anagus epos* Girault, was found to be effective in reducing its numbers; but its occurrence in the vineyards was highly variable. It was discovered that the parasite overwintered on a noneconomic leafhopper that occurs on native and cultivated species of blackberry (*Rubus* sp.) and that the distribution of the blackberries near some vineyards and not others explained the parasites' presence or absence. The parasite is now present and effective and can be maintained throughout the year by providing nearby overwintering refuges of *Rubus*. It is claimed that the adoption of the procedure has resulted in a net saving of $2.8 million in pest control costs over the 8-year period between 1961 and 1969 (Hawthorne 1970).

Biological Control: Classical

It has been emphasized elsewhere in this report that the components of an agroecosystem are often drawn from diverse geographical regions. Often, crops are grown most successfully far from their centers of origin; their assemblage of pest species thus contains some organisms that came with them from their original locality and some that became associated with the crop plant in its new area of cultivation. In many cases, imported pests prosper because they have escaped the control of predators and parasites that had coevolved with them in their area of origin Under these circumstances, the "natural" control mechanisms discussed in the preceding section are substantially disrupted. "Biological control" is the term commonly given to strategies that employ natural biological agents in the control of pest populations.

Some biological control of plant diseases occurs naturally, particularly in root diseases, as a result of parasitic or antagonistic action of certain microorganisms toward others. Efforts to employ such techniques, however, in planned disease control programs have met with limited success. Most research has been directed toward control of soil-borne diseases that are difficult to control by other procedures. Since the work of Weindling in the 1930s, it has been known that *Trichoderma* sp. influence the development of certain root diseases. Currently, *Trichoderma* is being mass-produced in the Soviet Union for use in biological control.

Over half of the major insect pests on U.S. agricultural crops are accidental importations. A similar percentage probably applies to crops around the world; it is thus not surprising that in over a hundred instances successful biological control has been established by the importation of specific parasites and predators to control alien pests. The first--and one of the most successful--instance of a successful biological control operation, conducted in California before the turn of the century, virtually eliminated the cottony cushion scale (*Icerya purshasi* Maskell) as a pest on citrus. More recent successes include Klamath weed, the citrus white fly, and others (Messenger and van den Bosch 1971).

The research allocation to biological control nationally has been very small compared to that devoted to chemical control. As summarized elsewhere in this report, the USDA maintains facilities for importing, rearing, and doing research on predators and parasites of alien pests. In addition, some states, notably California, Hawaii, and, more recently, Florida, maintain substantial research efforts on biological control. Finally, a few firms are beginning to rear and market insects for this purpose.

Biological control efforts may be of different sorts, varying in the level of intensity of intervention. Often, especially in cases where most of the components of the agroecosystem are native coevolved species, biological control may consist primarily of the augmentation or encouragement of organisms already present in the ecosystem. Such efforts may, for example, involve cultural techniques (see below) that enhance the reproduction of a parasite or a predator. Second, parasites or predators normally present in the ecosystem may be mass-cultured and periodically released. These procedures involve substantial intervention in the density of ecosystem components but do not involve the introduction of novel species. Finally, there is the technique of importing and colonizing predators or parasites not present in the agroecosystem. It is this latter method, employed against alien pests, that the term "biological control" conjures up in the minds of most people; it is called "classical biological control" by many practitioners. This method, it should be pointed out, is not applicable to the arthropod vectors of public health importance, since they are virtually all "native species" where they occur. Neither is it applicable to the majority of agricultural pests. In these cases, biological control is a matter of amplifying and augmenting natural control.

In many respects, biological control methods overlap with other technologies along a single continuum. For example, planting to provide cover for the alternate host of a parasite is at once cultural, natural, and biological control, and there is little point in distinguishing between the methodologies except for convenience. Similarly, microbial and viral control technologies involve parasitic organisms; yet we and most other separate them out for independent discussion. In at least one respect the similarity between microbial and metazoan control agents is unfortunate, since it has led to confusion and ambiguity regarding federal regulation. FEPCA of 1972 continues the basic definition of a "pesticide" as was first made in 1947: It defines a pesticide as "a substance or mixture of substances. . . ." This definition, as currently interpreted, sweeps up living viruses and bacteria along with nonliving chemicals. But this difficulty with definition also is a problem in the area of biological control: If bacilli are considered as "substances or mixtures of substances," why not predatory beetles and mites? Should regulatory restrictions similar to those applied to chemicals eventually be imposed upon the importation or periodi colonization by higher organisms, it would have serious consequences for biological control. On the other hand, there is reason to believe that the release of large numbers of organisms, or their importation to new geographical areas, ought to be under some kind of regulatory control. We regard the design of appropriate controls to be beyond the scope of this committee, but we firmly recommend their early consideration, or else--as is now the case--many of our activities are likely to come under inappropriate regulation.

For at least some kinds of biological control, an appropriate technology and adequate facilities for the mass-rearing of various insects is necessary. Few such large-scale facilities now exist in the United States, nor have we developed a wide range of the necessary technologies. There seems little doubt that new systems of biological control will require both. The requirement is likely to become even more pressing if current research efforts develop successful programs for specific agroecosystems. It seems likely that in the reasonably near future facilities and techniques for the rearing of insects and parasites at the required scale will be a major research need. It is difficult to see any incentive for such developments in the private sector, and the task will likely fall to the public sector. We see this as an area in which the relevant agencies, including in particular the USDA and the EPA, should pool their efforts to support the development of such facilities.

Common to all the ecologically based control strategies is a dependence upon multiple methodologies and upon a base of research data on the specific agroecosystem involved. Integrated control, in particular, is a dynamic strategy, relying upon continuous input of knowledge about the local situation. The lack of a single product with broad market ability makes this strategy an unlikely candidate for development within the private sector. Successful application of integrated control depends upon research and advisory personnel and upon technical inputs that are unlikely to be provided by any source *other* than a federal agency. We believe the promise offered by this general approach makes it an appropriate candidate for high-priority attention by federal funding agencies.

ERADICATION AS A CONTROL STRATEGY

Certain types of control measures are especially attractive because they can, in principle, be undertaken once to yield a continuing benefit. The elimination of an alternate host species for a fungus, the introduction of a parasite that can maintain itself indefinitely in its host, and the complete eradication of a pest species by a combination of techniques are all examples of this class of control measures. Because continuing control costs are not incurred if the method works, much higher initial expenditures may be justified for such programs than for continuous-control projects in which no amortization is possible. Highly successful ventures of this sort have been carried out on various geographically isolated insect populations using autosterilization techniques.

Eradication is an especially attractive objective where insect pests on major crops are displaying increased genetic resistance. Chemical control over the boll weevil is now hazardous in many parts of its range in the Cotton Belt of the South and Southwest because of secondary pest problems created by the bollworm and tobacco budworm. Current losses due to the activity of the boll weevil are estimated at over a quarter of a billion dollars annually. The boll weevil is the most important agricultural pest insect in the United States, in terms of dollar value of losses and--not surprisingly--also in terms of the poundage of chemicals used against it.

A project to eradicate the boll weevil over its entire range in the United States has been proposed by the National Cotton Council (NCC), with the backing of prominent USDA scientists but without the official sanction of the USDA. The program's cost has been estimated on behalf of the Council by Stanford Research Institute at $871 million, revised from an earlier estimate of $1.6 billion. These cost estimates are for the *minimal* program deemed necessary by proponents to achieve the eradication objective. Representatives of the Council, in testimony before the House and Senate subcommittees on agricultural appropriation, requested $850,000 in Civilian Conservation Corps funds for the establishment of expanded boll weevil management programs in the affected states, and $5 million from the same source for the federal share in the eradication program during the first half of 1975.

The original program envisioned a 3-year sequence of procedures to be applied in each region where the pest occurs. During the first year, intensive chemical control will be directed against the reproducing spring and summer populations; then reproduction diapause control procedures will follow. In both, organophosphate insecticides will be used. Stalk destruction and other methods will be combined with trap surveys to further reduce weevil populations. In the second year, sterile male releases and pheromone traps will be used, along with treatment of local infestations with organophosphate insecticides. In the third and final year, sterile male releases will be made on an estimated 20-40 percent of the acreage, as needed, to complete the eradication. The original concept was that this program would be applied to eight cotton-growing regions, starting in 1975 in the Texas high plains and moving in an easterly sequence. The proponents of the program view this serial feature as providing the opportunity to perfect the methodology and obtain adequate financing from the government and growers as the project proceeded.

Recently, controversy surrounding the plan (including skepticism on the part of some growers and failure to obtain endorsement from the Texas Agricultural Experiment Station) has caused its proponents to shift to a sequence that allows a 3-year, essentially pilot-test project centered in North Carolina with some coverage in Virginia--originally District 8 in the national plan. During this period methods will be tested and evaluations of success made; NCC and APNPS personnel propose that a beltwide eradication program pend the outcome of these evaluations.

The practicability of a larger project must thus be evaluated on the basis of data from a regionally restricted test. An earlier test of the efficacy of the methods to be employed was made in the Pilot Boll Weevil Eradication Experiment, carried out in 1971-1973 in an area of southern Mississippi and adjoining parts of Louisiana and Alabama. The results of this experiment have been viewed in very different ways by entomologists who have analyzed the data. There was disagreement about the efficacy of different trapping techniques used to ascertain the presence of weevils at the end of the experiment, about the place of origin of the weevils found, and about the significance for a massive eradication program of results obtained in one limited area of the Cotton Belt. Entomologists also do not agree about the migratory flight distances of which weevils are capable, nor about the possibility that small populations of insects may remain on undetected small acreages of cotton or on three alternate host plant species that occur in the Cotton Belt.

Two committees evaluated the results of the pilot project. The Technical Guidance Committee to the National

Cotton Council described it as demonstrating that "it is technically and operationally feasible to eliminate the boll weevil as an economic pest in the United States." The Review Committee established by the Entomological Society of America reported that "the Committee is divided as to whether or not technical feasibility of eradication of boll weevil populations has been demonstrated, but unanimously expressed reservations concerning any massive eradication undertaking without further research to refine suppressive techniques." Concerns with the methods at their present stage of development center around the following problems: (1) Chemosterilization and mass-rearing techniques currently produce a low yield of sterile weevils, and some are apparently not authentically sterile. The weevils released in the pilot project carried a genetic marker, and some F_1 progeny from supposedly sterile parents have been collected in the experimental area. (2) The program envisions intensive mandatory treatment of *all cotton acreage* with organophosphate insecticides during the early stages of the project; this may have the undesirable effect of encouraging the development of resistance to these compounds.

The difficulties in interpretation that proved troublesome in the case of the Mississippi pilot project may recur when the next pilot program is evaluated.

It is fair to summarize the situation at this writing with the statement that the adequacy of present technology to ensure eradication of the boll weevil is in some doubt. Even the report of the Technical Guidance Committee for the Pilot Project reflects this uncertainty, because it redefines the concept of *eradication* to that of "*elimination* as an economic pest." The difference between these two terms is not merely academic, because it relates to an essential feature of the concept of eradication, namely, the guarantee that continued control costs will not be required. It is this latter guarantee that makes major eradication programs so attractive; but to make the guarantee reasonable it is necessary to show that elimination, and not merely dramatic suppression, of the pest has been achieved. That is why so much significance is attached to understanding the origin of the very small number of weevils left in the aftermath of the Mississippi Pilot Project. Furthermore, testimony given before Congressman Whitten's Committee indicated that the cost of the Mississippi Pilot Project was $270/acre. This figure is much higher than NCC estimates for the entire project and would be unacceptably high compared to the benefits of suppression in many regions of the Cotton Belt.

Our Committee is concerned at what may result if the prospects and costs for any such eradication project are not accurately defined and then clearly stated. The cotton-growing community now perceives the project as one that will eliminate a pest totally and with it the need for future continuous control measures. Growers and others also see the project as a testing ground for new pest control technologies, including the autolethal genetic techniques, third-generation chemicals, and various cultural practices. Temporary suppression, even if it produces considerable savings over a short period, may not be viewed as a success in view of the advance billing given the project. We are concerned lest even minor failures be viewed as evidence that new methods and integrated approaches to pest management are at fault; it would be a real tragedy for American agriculture if alternative strategies of pest control were to become the political victim of overadvertising.

To return to the language of the Technical Guidance Committee for the pilot eradication project, what is meant by "*elimination* as an economic pest"? In particular, does it signify "without continuing control costs"? Even the most celebrated success in the history of eradication efforts could not meet this criterion, though it was originally thought to have done so. In 1966 the USDA announced that the screwworm had been "eradicated from the United States . . .," yet currently over $10 million annually is being expended on continuing suppression measures. In the case of the boll weevil, even a total success might require some expenditures afterward for continuous control, but proponents of the project have not published estimates of these. It is difficult to avoid the conclusion that anything *less* than a total success would be followed by a return, in a relatively few years, to control costs roughly equivalent to those currently characteristic of the region.

The Committee recommends that the USDA, as it moves toward policy decisions on this subject and toward implementation of the second pilot program, carefully consider the political as well as the other impacts of the possible failure of eradication. It is particularly important that such programs not be seen as testing grounds for integrated pest management or for any particular alternative control strategy. Further, prospective beneficiaries of the project should be given the most objective assessment possible of the chances of failure and the consequences should it occur.

ASSISTING TECHNOLOGY TRANSFER

All of the innovations discussed in this section have, in one situation or another, been extraordinarily effective at dealing with a pest problem; all show promise for the future. Yet even if the institutional and scientific problems that attend their further development are solved, there will remain the question of whether the actual users of pest control technologies--the farmer, the structural pest control operator, the householder, the public health official--will adopt the new methods. One of the issues our Committee found it difficult to approach was whether technology diffusion--compared with rate of scientific progress--had been the critical obstacle to the use of new methods. A number of experienced scientists in the USDA-CES system believe that methods that have been perfected but not adopted could produce large savings over present ones; they express the view that growers, in particular, will not adopt relatively untested technologies because they perceive the risk to be too high. It is a fact, however, that many growers in different regions of the United States have cooperated extensively with state extension and AES personnel in testing new methods, and that many of these have been adopted even before they were fully tested. We find no evidence that U.S. farmers are inherently overconscious of risk; indeed, it is widely recognized that agriculture is more tightly locked in to the adoption of new technological advances than most sectors of the economy.

On the other hand, as we have documented elsewhere (see especially Volume III, *Cotton Pest Control*), there are substantial incentives for a grower to remain committed to chemical pest control. A large investment in advertising and in the provision of technical advice is made by the agricultural chemicals industry. By and large, the contemporary technology is successful--that is, the advice and the materials work most of the time, and the marginal return on the input is favorable. If alternatives are to be adopted, convincing evidence will have to be brought that they offer more protection per unit cost than the technology for which they are being proposed as a substitute.

Technological substitutions that involve investment by the society as a whole raise difficult questions. Before discussing devices for accomplishing the development and diffusion of new methods, we will briefly discuss two of these: Under what circumstances should society provide a subsidy for the substitution? What effort should be made to determine the distribution of benefits from such a subsidy?

Society might wish to subsidize a new technology under any of several circumstances: if it resulted in increased production, and so met some national goal; if it spared some resource important to conserve for some other social purpose; or if it reduced an external cost that society was bearing on account of the existing technology. Some policymakers have held that the latter motivation properly applies to pest control. In the case, for example, of chlorinated hydrocarbon insecticides there is adequate evidence to convince us that society is paying some external costs, in reduced environmental quality and in health hazards; but no measurement of that cost has been made and, indeed, none seems possible at this time. This does not mean that a transfer subsidy is not justified; it only means that we do not know how large it should be. But many other compounds used as pest control agents are not known to be harmful to the environment, or to pose health risks; their costs lie within the enterprise in which they are used. Hence one of the first two arguments would be required in order to justify a subsidy. Of these, the goal in increased production is the one traditionally invoked in favor of various kinds of government support of research designed to benefit agriculture.

It has been argued that the social expenditures on behalf of agricultural research have actually yielded benefits primarily to large growers, in effect enabling them to become larger; and that these distributional inequities have contributed heavily to the displacement of labor from the farm sector to other sectors of the economy, with unfortunate results. That charge is complex, and difficult either to support or refute. It is true that concentration of farm ownership has occurred during a period in which the USDA-land grant university system has contributed important technological advances to agriculture. But private sector innovation has contributed importantly as well; and other social factors having nothing to do with federal subsidy were at work during the same period. We do not find the available evidence adequate to take a position on this issue, but we do believe that wherever a new subsidy is considered, efforts should be bent to determine the distribution of the expected benefits.

We do not extend this caution to a recommendation that government efforts to transfer new technology into agricultural practice be curtailed. Beyond the inadequacy of the evidence regarding the impact of distribution problems on agriculture, there are good reasons for not discouraging such innovation. First, we already have made major commitments to a high-technology agriculture, and to withhold new advances in certain areas would hardly cure any inequities that may exist. Second, it is not at all clear that efforts to channel benefits to marginal farmers might not result in even more serious inequities for marginal consumers--a group in the society that has enjoyed conspicuous benefits from the increased efficiency of U.S. agriculture. Finally, the dissemination of technology is a crude tool to use in redistribution schemes. To the extent that inequities exist, they will be better solved by transfers made directly than by the manipulation of programs designed primarily for other purposes.

In summary, we should be prepared to recommend not only research efforts in the public sector to improve agricultural practice--with as much attention as can be paid to where their benefits fall--but also to the institutions and mechanisms that govern their adoption into practice.

Pest Control and Insurance*

It is possible to see the task of pest control not merely as one of protecting some valuable commodity, but also as one of protecting the economic viability of a farm firm. If institutions can be designed that permit the farmer to adopt a new pest control practice and yet still remain in business with his normal income after a devastating loss, then at least one of the goals of pest control activities would have been met. It has been suggested that programs of insurance or indemnification for losses incurred would accomplish this goal of protecting the farm firm. Such programs could serve as an incentive for farmers and other pest controllers to adopt alternative control strategies in which some risk is perceived and thereby reduce important delays in the diffusion of new technologies.

The farmer's desire to maintain yields at a high and uniform level is one reason for the heavy use of pesticides in agriculture. Farmers often appear to be unable or unwilling to base their pest control decisions on complex calculations of the relationships between pest infestations, yield losses, and pesticide applications. Consequently, they adopt pest control programs that tend to require regular and heavy use of pesticides. Such

---

*This section draws heavily from Jones (1973).

practices may be viewed as insurance that yields, and incomes will not be reduced or made highly variable from year to year by pest damage. Farmers might more readily adopt different pest management techniques if they had insurance protection against any potential decrease in earnings.

The Federal Crop Insurance Corporation (FCIC) of the USDA has been collecting data on crop yields for its all-risk crop insurance since 1938. Pest damage is included as one of the multiple perils of agriculture, but the Crop Insurance Act states that losses due to "failure of the insured to follow established good farming practices" shall not be covered. Further, the law limits coverage to production costs no higher than 75 percent of the average value of production and thus requires the farmer to bear a portion of the loss. The net result is considerable incentive for the insured farmer to continue pesticide applications on schedule. Clearly, the current FCIC policies offer little encouragement for the adoption of new pest control practices. In addition, the historical data collected by FCIC may be of little use in the establishment of actuarial tables for losses due to pests.

The theory of conventional insurance schemes can be briefly summarized. Insurance programs spread the economi risk of losses of a few individuals over a large group. Each insured individual pays a relatively small premium to protect himself against an infrequent but substantial loss. The system requires a risk-handling organization that (1) performs necessary administrative functions, (2) estimates the mathematical probability of loss faced by the insured group and its subclasses, and (3) identifies and measures losses as they occur. The feasibility of an insurance program in this conventional sense depends upon the characteristics of the hazard insured against, the person and property insured, and the availability of adequate pertinent data. A successful program requires a large number of participants and reasonable operating costs.

There are a number of important principles or character istics that should be applicable. First, the peril causin the loss should be of a random or fortuitous nature. The insured peril should be determinable and measurable. In addition, the person insured should be indemnified only for the actual losses. Finally, it is necessary that the mathematical probability of the loss be estimated for the universe with some degree of accuracy.

The incidence of pest damage to crops may be of a random or fortuitous nature, but pest control methods employe at the discretion of the farmer can affect the magnitude of the damage. Consequently, a conventional insuring organization would be likely to take some combination of the following three steps: (1) identify and eliminate from the covered those who would be adverse risks, (2) increase premium rates to be commensurate with the risk, and (3) establish a level of coverage less than the full value of the crop. Steps (1) and (2) would likely reduce the number of participants in the program; step (3) would put part of the risk on the insured and thus reduce the potential incentive for him to alter his customary pest control practices.

Determination and measurement of pest damage is also difficult. Some perils, such as hailstorms, occur at specific times and leave damage that is easily recognizable and measurable. Pest damage, on the other hand, generally extends throughout the growing season. There would thus be serious conceptual and administrative problems in determining both the occurrence and magnitude of crop losses due to pest damage. Even if such problems were solved, indemnifying the insured only for the actual losses would require separating and measuring crop losses caused by the changed pest control practices from losses that might have occurred under alternative or previous pest control practices.

In order to spread losses equitably over all those insured, it would be necessary that the mathematical probability of loss for the universe be estimated with some degree of certainty. The estimation would have to be done for the various subclasses, because loss probability would vary with the region, crop, soil type, and climatic condition. In a conventional insurance program, premium rates would be established on the basis of these loss probabilities. Estimation of loss probabilities generally requires knowledge or actual experience over time. Our current knowledge of losses due to pests can only be described as uncertain and inaccurate for most pest situations. Hence considerable research on loss estimations would be necessary before conventional insurance schemes could be instituted.

The basic idea of insurance is to spread losses or risks among a large number of participants. It is uncertain whether a large number of farmers would voluntarily insure against pest losses if a suitable program were available. Only about 11 percent of eligible growers participate in FCIC programs, even though the government pays the operating costs of FCIC. Low premium rates would probably attract participants but would probably require tax-based support. Projected level of participation will have to receive careful study in any serious proposals for pest insurance. Elements of flexibility such as mandatory participation or prohibition of the use of certain pesticides, might be necessary for success of the program.

There are clearly a number of difficult problems associated with the development of conventional insurance programs to cover pest damage. It might be feasible, however, to tailor an insurance scheme to an individual pest on a particular crop. The limitation of coverage to a single pest and crop might reduce the problems involved in determining the occurrence of a loss and measuring its magnitude. Also, estimation of loss probability might be made easier when the peril is limited to a specific pest.

Indemnification Programs

Government-sponsored indemnification programs might be developed to encourage the adoption of alternative pest control practices or to compensate for losses due to restrictions on the uses of pesticides. Such a program might have several distinct characteristics.

The program would require some sort of registration or "sign-up" procedure prior to the start of a growing season. The registration might require the participant to submit an acceptable or alternative pest management plan. Alternatively, participation might require producer agreement to follow the advice of certain pest control consultants (public or private). A claim for indemnification would require some type of certification that the approved plan of advice was followed.

Unfortunately, most of the conceptual and administrative problems associated with conventional insurance schemes in measuring the net magnitude of a loss would remain. An administering agency would still be responsible for determination of a loss and settlement of claims due to pest damage. Adequate schedules might be based on a farm or farmer's historical yields or production costs. Depending on the alternative pest control practice, however, an indemnification program might require considerable flexibility. The simplest alternative practice involving precise timing of pesticide applications could have a comparatively straightforward indemnification schedule based on historical yields and production costs. However, as alternative practices become more complex and involve crop rotations or crop dispersion, indemnification might require a complicated schedule based on a rate of return on the farmer's complete operation.

An additional problem would arise with the length of time a farmer would be eligible for indemnification. If the objective of the program was to provide an incentive for adopting new pest control practices, then presumably the incentive should be removed sometime after the practice has been successfully adopted.

Trained Personnel

One feature of many of the new pest control technologies is that they rely more heavily on information than on a specific product. This has always been true of biological and cultural control, where advice from Extension Service personnel and other sources on procedures to follow is the primary element. To a lesser degree, both the third-generation pesticides and the microbial control techniques show the same dependence: More special skills may be required for their effective use. One important potential barrier to their adoption, therefore, is the availability of trained advisers to assist growers in employing them.

Some of these specialists will undoubtedly be employed in the private sector, where microbial agents and new chemicals are likely to be marketed by the firms developing them as a combination of goods and services. Volume III, *Cotton Pest Control*, has documented an increase in the numbers of independent pest management consultants, especially in California, who contract their services on a fee-per-acre basis. Such private firms may increase to fill the need as integrated pest management schemes gain more favor, and they may actually function effectively to gain acceptance for these methods. But the largest share of advice for the other technologies, which are being developed largely in the public sector, will have to come through the existing system of the USDA, the land grant universities, and the CES. In most of the intensive agricultural areas of the United States, Extension Service personnel are outnumbered by advisers for the private firms engaged in contemporary control technology. Quite aside from the implications of this imbalance for the adoption of new methods, it suggests that a major expansion in the number of trained pest management specialists will be needed if the newer service rather than product-oriented methods eventually predominate.

This prospect has been foreseen in some quarters, and at least one Master's degree taining program in pest management is now supported by the federal government. To our knowledge, however, no effort has yet been made to estimate the numbers of new personnel that would be needed to provide adequate support for a service-based pest control philosophy. We recommend that the agencies concerned especially the USDA and EPA, attempt to make such estimates and to specify what educational programs need to be mounted to meet the demand for trained personnel.

Summary

Establishment of insurance or indemnification programs wil be difficult, and we do not foresee their creation within the next few years. The potential for such plans to help alter contemporary reliance on pesticides, however, makes them of considerable importance. We recommend that the USDA and EPA sponsor further research in the field.

The development of techniques for better determining and measuring losses due to pests is clearly of utmost importance in the development of any such schemes. We accordingly recommend increased research efforts on loss estimations. We also note in passing that better estimations of losses will be of considerable value to future efforts to evaluate costs and benefits of alternative pest control technologies.

REFERENCES

Beroza, M., ed. (1970) Chemicals Controlling Insect Behavior. New York: Acadmic Press.

Cerf, D. C., and G. P. Georghiou (1972) Evidence of cross-resistance to a juvenile hormone analogue in some insecticide-resistant houseflies. Nature 239:401-402.

Epstein, E. (1973) Health food report: Trends stabilize as more know-how is acquired. Chain Store Age 49(2).

Gelperin, A. (1975) Rapid food-aversion learning by a terrestrial mollusk. Science 189(4202):567-570.

Gustavson, C. R., J. Garcia, W. G. Hankins, and K. W. Rusiniak (1974) Coyote predation control by aversive conditioning. Science 184:581-583.

Hawthorne, R. (1970) Estimated damage and crop loss caused by insect/mite pests--1969. Quoted by R. L. Doutt *in* Pest Control Strategies for the Future. Washington, D.C.: National Academy of Sciences. 376 pp.

Heimpel, A. M. (1970) Insect control by microbial agents. *In* Pest Control Strategies for the Future. Washington, D.C.: National Academy of Sciences. 376 pp.

Ignoffo, C. M., and G. G. Allen (1972) Selection for resistance to a nuclear polyhedrosis virus in laboratory populations of the cotton bollworm, *Heliothis zea*. J. Invert. Pathol. 20:187-192.

Ignoffo, C. M., D. L. Hostetter, and M. Shapiro (1974) Efficacy of insect virus propagated *in vivo* and *in vitro*. J. Invert. Pathol. 24:181-187.

Jones, L. A. (1973) Insuring Crop and Livestock Losses Caused by Restricted Pesticide Use, An Appraisal. USDA-ERS No. 512. 7 pp.

Messenger, P. A., and R. van den Bosch (1971) The adaptability of introduced biological control agents. Ch. 3 *in* Biological Control, C. B. Huffaker, ed. New York: Plenem.

National Academy of Sciences (1966) Scientific Aspects of Pest Control. Washington, D.C. 470 pp.

National Academy of Sciences (1968-1970) Principles of Plant and Animal Pest Control, Volumes I-VI. Washington, D.C.

National Academy of Sciences (1972) Pest Control Strategies for the Future. Washington, D.C. 376 pp.

Schneidermann, H. A., A. Krishna Kumaran, P. J. Bryant, and F. Sehnal (1970) Endocrinological strategies in insect control. Agric. Sci. Rev. 8(213):13-25.

Small Business Reporter (1973) Volume 11, No. 2.

Spaeth, Robert F. (1973) Growth of health food sales in the supermarket. Food Prod. Dev. 7(4):28-30.

Wells, F. E., and A. M. Heimpel (1970) Replication of insect viruses in bacterial hosts. J. Invert. Pathol. 16:301-304.

Williams, C. M. (1967) Third generation pesticides. Sci. Am. 217:13-17.

PART III

SUMMARIES, CONCLUSIONS, AND RECOMMENDATIONS OF THE STUDY TEAM REPORTS

The summaries and recommendations of the four study team reports--*Corn/Soybeans Pest Control*, *Cotton Pest Control*, *Forest Pest Control*, and *Pest Control and Public Health*--are presented here for the reader's convenience. The reader is urged, however, to refer to the texts of these volumes for a more extensive treatment of the subject and a more complete understanding of how these recommendations were derived. All of the reports are available from the National Academy of Sciences.

12

CONCLUSIONS AND RECOMMENDATIONS OF VOLUME II: *CORN/SOYBEANS PEST CONTROL*

The corn and soybean sector is, in many ways, the backbone of U.S. agriculture. Sensibly protecting this sector from pest problems, while minimizing environmental, economic, and social problems, is a major challenge today and in the years ahead.

It is not difficult for members of a study team such as this to spin out dozens of recommendations for public and private action. These recommendations can be general or quite specific. They can be in conflict with each other or internally consistent depending upon the composition of the group and the specificity involved. Additional research usually can be strongly recommended on a variety of questions. However, the following general conclusions, recommendations, suggestions, and ideas are distilled as the most important from the many points discussed and debated among the team members. Full agreement was not always achieved on each of these points, and some issues are not as fully developed here as individual team members might have wished.

The study team is fully agreed, however, that pest problems and pest control strategies in the corn/soybean sector of U.S. agriculture are in a continual state of evolution and development. At this time, the immediate pest problems generally can be controlled with present methods, which emphasize herbicides, insecticides, and plant breeding for disease resistance. Yet in some cases, especially with soil insecticides, these methods are only partially effective and questionable from the environmental point of view. Consequently, this is no time for complacency and lack of concern. Unexpected insect and plant disease problems can explode at any time with disastrous results. The 1970 corn leaf blight episode provides a glimpse of the potential consequences of a major pest outbreak. Continuing sophistication in cropping methods and the changing weed population requires improved weed control and more attention to actual and potential chemical carry-over problems. Current and developing externalities (side effects) involving the target pest species, the environment, and rural society suggest the need for continued improvements in corn/soybean pest control.

On-farm pest control of weeds and insects by current, mainly chemical, methods is generally a highly profitable activity at current and probable future price and production relationships. Successful pest control typically increases output and reduces costs because (1) it permits higher average yields, and (2) it allows farmers to handle more land with the same or less labor. Because of this tendency, the substantial benefits of corn and soybean pest control are distributed widely through the economy in the form of lower product prices and larger supplies than otherwise would occur. The main benefits flow ultimately to buyers of meat, livestock products, and edible oils. Foreign buyers and food-aid recipients of U.S. farm products benefit because of lower prices and more abundant supplies than otherwise would be available. Similarly, the U.S. balance of payments is favorably affected through increased foreign exchange earnings. Pest control failures have essentially opposite results. These costs are registered mainly as higher prices and shorter supplies in the food sector.

Although the substantial benefits of modern pest control are relatively clear and direct, the persistent costs to society from corn and soybean pest control occur mainly as unmeasured externalities. These effects ripple through the environment and through time in complex and subtle ways.

On balance, successful pest control in corn and soybeans has been a vital link in the emergence of the United States as the world's lowest-cost, most-efficient, and most-reliable producer of these commodities. It should remain a high-priority item on the nation's agricultural policy agenda.

An early objective of the team was to identify possible new strategies for pest control in corn and soybeans. We feel that we have succeeded in describing some ways by which to encourage public action and research that will lead to new strategies. But we have not identified any radically different approaches ready for widescale adoption. In the near-term future greatest opportunities for improvement lie in: (1) greater reliance on pest management systems; (2) use of pesticides geared to real needs rather than insurance or prophylactic treatments;

(3) rigid adherence to the correct timing, rates, and methods of application to avoid spillover/drift and undesirable residues; (4) minimal use of persistent chemicals; and (5) better-trained applicators.

If current economic trends continue, no major policy and regulatory changes occur, and the trajectory of technological change remains steady, the study team foresees the following evolution occurring in pest control methods over the next 10-15 years.

1. Application methods and timing will continue to change but primary reliance on chemicals will remain for insect and weed control.
2. More selective, less-persistent insecticides will be available, and they will be applied by better-trained applicators.
3. Continued sophistication will occur in corn and soybean herbicides, but expanding applications are likely as total acreage increases and as mechanical tillage operations are reduced.
4. Breeding and selection for disease and insect resistance will continue as major components of the sector's pest control activity.

The assessment of current trends and their implications is not an endorsement of them. Knowledge about future developments is imperfect, yet the study team generally agrees that current methods of pest control in corn and soybeans to date have resulted in no known major environmental hazard over wide areas. Possible adverse future effects of continued use of chlorinated hydrocarbons are not fully known, but the Environmental Protection Agency recently has banned the agricultural use of aldrin and dieldrin and is considering a ban on chlordane and heptachlor. Environmental residues of these banned chemicals, therefore, will not increase in the corn/soybean sector.

Since many important pests of corn and soybeans are unpredictable in nature with regard to both time and place of population buildup and damage, a major policy priority should be to develop good predictive capabilities.

This study team, therefore, recommends that an increased public commitment be made to monitor pest populations in the field and to develop effective surveillance systems for detecting potential pest buildups, new pests, and new races of current pests.

With more effective prediction, pest control by any means can be improved. In particular, the need for routine chemical treatments can be reduced throughout the corn/soybean sector. A major research effort is needed to identify the biological, physical, and environmental factors that bear on reliable prediction techniques. Currently a substantial portion of the insecticide used in corn involves insurance or prophylactic applications by growers. The sporadic nature of infestations, relatively low price of insecticides, and the indirect nature of potential costs and consequences associated with such treatments lie behind this phenomenon. While private efforts to monitor and predict major pest problems may be helpful to individual growers and regions, such efforts probably will not be sufficient.

This study team also recommends that in addition to improved pest monitoring and prediction, the feasibility of broadened crop insurance schemes be studied, with pest insurance as an important feature. Such insurance written by public or private agencies would help achieve economic security for individual growers and thereby reduce insurance applications of chemicals. It is important to prevent prophylactic chemical applications on soybeans from becoming as widespread as they are now on corn.

It is important that pest control efforts involving corn and soybeans move further toward preventing problems rather than curing them once they have emerged. In general, we will likely be most successful if we learn to live with and attempt to "manage" most corn and soybean pest species in a rational way.

The study team therefore recommends that research and especially action programs be directed toward pest "management," broadly defined. In the corn/soybean sector, management is the key to effective and profitable pest control over the long run. Containment and suppression of well-established insects, weeds, and plant diseases at or below their economic thresholds, appropriately calculated, should be the major objective of management systems. The complex interactions of living communities make these activities very difficult, but they also make the complete eradication of well-established pests generally unworkable with present technology. However, for exotic pests in an early development stage or with new discoveries in pest control, eradication may become an appropriate management goal.

Agricultural ecosystems are "unnatural" in a biological sense. Yet, they are complex communities of plants and animals, including pest species. In dealing with pest species, the more natural and self-generating a control system is, the more likely it is to be effective and environmentally safe in the long run. Such naturalness can be fostered through biological and agricultural research. Some research activities of this kind, however, which may seem to be low-return or high-risk endeavors, are not funded when budgets are tight.

Consequently, the study team recommends that biome or ecosystem studies of the kind traditionally conducted in natural or undisturbed environments also be conducted in the disturbed or manipulated agroecosystems where corn and soybeans are produced together and constitute the major crops. The most effective principles and techniques developed and identified through previous studies, including multistate Integrated Pest Management Research Programs, should be applied to the corn and soybean sector.

Extensive base information is needed on all biological entities and their interactions in the system. Such knowledge can help in the development of alternative pest control technologies and reduce future environmental hazards. However, this knowledge must be synthesized in useful ways.

Therefore, the study team recommends that the techniques of systems analysis and modeling be applied more intensively and specifically to the evaluation of production and pest control methods in the corn and soybean sector. We encourage the development of interdisciplinary and interagency teams to achieve balance in the direction of such studies. These studies should define the interaction of production practices, pest control technologies, economics, and potential public benefits and costs. These studies might encourage a series of monographs funded through appropriate public and private agencies.

Nonchemical pest control in the corn/soybean sector is lagging behind that in other commodity sectors, such as cotton. Part of the reason for this is that it is difficult now to pinpoint major systematic environmental problems specifically associated with pest control in corn and soybeans. Although the corn/soybean sector is only one part of the total agricultural system, chemical pest control in it represents a sizeable portion of the total pesticide load introduced by man into the agricultural environment. The ability of the environment to absorb an ever-increasing load, including that from the corn/soybean sector, remains an important and worrisome unknown.

The study team recommends that research efforts be expanded in the area of alternative biological methods and strategies for pest control. These efforts combined with suggested improvements in the prediction and targeting phases of current pest management methods will help to control and possibly reduce the total pesticide load on the environment, make judicious use of the chemicals that are applied, and offset specific problems that are present. The latter include target species resistance; spillover/drift; and the buildup of unwanted residues in soil, water, and some foods.

Biological and social scientists from the nonagricultural disciplines and from nonagricultural institutions should be heavily involved in this work. The institutional and administrative barriers to professional cooperation and public funding of such research across disciplinary lines should be reduced.

The development of today's pest control technology has been largely the result of the long-standing and closely-knit interrelationship of the land grant system, the federal government, private industry, and farmers. Private universities and nonagricultural public universities traditionally have not been directly involved in agricultural pest control research and evaluation. This has occurred for a variety of historical and professional reasons. Greater interaction between the two systems as recommended above would help to avoid a more narrow and possibly shorter-range view of pest problems and control strategies than might otherwise occur.

In the corn/soybean sector, nonchemical methods other than genetic resistance are being recommended mainly for insect and plant disease control as part of mixed control strategies. At the moment, these methods mainly involve crop rotation and crop sanitation practices. More sophisticated biological methods of control probably will not involve readily marketable products or services. Consequently, publicly-funded development and implementation likely will be required. Private industry cannot be expected to provide the needed research and distribution functions. Moreover, implementation of alternative nonchemical strategies may require major changes in the traditional separation between public and private decision making. More public involvement, responsibility, and perhaps control would be needed for success of these methods.

Much of the pest control effort in the corn/soybean sector involves plant breeding for genetic resistance to disease and the extensive use of herbicides. Neither of these has generated prolonged or acute environmental problems to date. The continuing evolution of insect control technology also provides little direct evidence of acute environmental damage within the sector, although some problems have been identified. These problems mainly involve increasing resistance of some target insect species, especially corn rootworm, and the presence of soil and food residues of the persistent chemicals aldrin and dieldrin. Additional direct costs and externalities of corn and soybean pest control are caused by spillover/drift, accidental use of incorrect materials, and improper disposal of leftover chemicals.

Consequently, this study team supports the concept of increased training in the use and application of pesticides and increased public control of pesticide use as exemplified in the Federal Insecticide Fungicide and Rodenticide Act as amended in 1972 by the Federal Environmental Pesticide Control Act.

Important provisions of this 1972 Act (1) permit the classification of pesticides into general and restricted use and (2) require the certification of applicators using restricted pesticides.

In addition, the study team recommends that more and better information and educational opportunities be made available to policymakers and the general public concerning pesticide use and regulations. The study team anticipates some additional employment in or shifts of employment to rural areas stemming from these new pesticide regulations. This may be an advantage and worthy of further study.

Modern pest control is part of a technological revolution that has greatly increased farm production and yet has fostered a massive reduction in farm numbers and farm people. Many social problems in both rural and urban United States can be traced to this phenomenon.

This study team recommends that when new legislation, regulation, or public support for pest control activities are considered, specific attention should be directed to their potential employment and locational effects on rural people and rural society. Such effects may be very difficult to measure or evaluate, but the effort should be made.

It is possible that in the future serious unexpected outbreaks of insects or disease may threaten the corn/soybean sector. Currently available technology may be unable to control such outbreaks if their sources are exotic and unresearched. It is even possible that today's pest control methods may contribute, in unknown ways, to them. The capacity for such "back-up" response at present is fragmented and incomplete.

This study team, therefore, recommends that strategies for back-up systems of pest control in the corn/soybean sector be developed. This includes (1) public and private attention to building inventories of chemicals that differ from those in common use, (2) legal and administrative procedures to move quickly against pest outbreaks with currently restricted products if necessary, and (3) comprehensive and diverse inventories of plant genetic materials from which new resistance can be developed as quickly as possible.

Representing the public's interest in pest control will continue to be the responsibility of agencies such as the U.S. Department of Agriculture, the Environmental Protection Agency, and the Land Grant University System. The agricultural extension service of the land grant system has an especially important role to play. The extension service deals directly with farmers, rural people, and their organizations in an educational and advisory capacity. The ability of extension workers to respond objectively to questions and issues concerning pest control and environmental protection is more crucial now than ever before. This challenge emerges in an atmosphere of increasing public concern about the environment and about the impact of agricultural change on society as a whole. This concern has generated new legislation and regulations. It has also introduced strains upon the traditional fabric of relationships between the land grant complex, farmers, agribusiness, and the general population.

The study team recognizes that the close relationship between the agricultural chemical industry and the public agricultural establishment has been criticized, sometimes justly and sometimes unfairly. For the most part, this relationship is based upon mutual respect and responsible collaboration. Active cooperation in this field has been important in the development of many safe and effective methods of pesticide use.

In order to insure that the public's interest is protected as pest control methods continue to evolve, the study team recommends that the independence and objectivity of all publicly-supported agricultural workers who deal with pest control issues be guarded and strengthened. Although this will involve continued evaluation of the traditional ties between private industry and the public agricultural establishment, the many long-standing positive relationships between the two sectors should be sustained.

Effective and ecologically sound pest management in the corn/soybean sector for today and tomorrow depends upon (1) the development of basic knowledge, (2) the translation of this knowledge into dependable technologies and practices, (3) the implementation of these technologies and practices at the production level, and (4) the continuing evaluation of their biological, economic, and social implications. Such pest management is a complex and difficult process.

13

SUMMARY, CONCLUSIONS, AND RECOMMENDATIONS OF VOLUME III: *COTTON PEST CONTROL*

In the decade following World War II, modern synthetic pesticides and other technological developments had a great impact on cotton yields. By reducing the risk of disastrous crop losses from pests, pesticides encouraged the adoption of other highly productive practices such as fertilization, irrigation and drainage, and planting of long-growing indeterminate varieties of cotton. Profitable cotton production in the United States now depends upon the successful application of good management practices to a wide range of soil and climatic conditions. Throughout the Cotton Belt, efficient pest management practices have been and will continue to be of major importance.

The range of pest control technologies in practice toda[y] has developed, of course, in response to the changing insect, weed, and disease problems that have plagued the cotton industry. However, economic and social forces have had a considerable impact on the adoption of pest management and other cotton production practices, including government programs supporting farm incomes, registering the use of pesticides, and enforcing worker health, safety and other standards. They also include the experience and education of farmers and their access to information on changing farming practices. These forces influence regional production patterns and provide incentives for the adoption of alternative production practices, both of whic[h] affect the relative importance of the key pest problems in the major producing regions. Too often, the indirect effects of social and economic forces on pest control problems and practices have gone unrecognized by those responsible for formulating agricultural policies directing research on pest management.

A wide range of measures are used to control cotton diseases, but the most effective method of control is the incorporation of genetic resistance into commercial varieties. Except for treatment of cottonseed for seedling diseases, the application of currently available chemicals is often uneconomical, and the total acreage treated with fungicides and fumigants is relatively small. Genetic resistance to root knot nematodes has been incorporated in some varieties, but soil fumigation is more effective against nematodes.

Two potential alternatives to control of cotton diseases are planting at optimum soil temperatures for rapid emergence and development of biological control methods. Late planting, while providing disease control, often conflicts with other desirable production and pest control practices, especially insect control.

Herbicides and tillage practices are the most common weed control method. Preplanting tillage (seedbed preparation) removes germinated seedlings but does not deplete seed reserves. Postplanting tillage (cultivation) removes weeds but may prune roots and injure plants. Herbicides can be applied before planting, before emergence, or after emergence, and their use has significantly reduced the hand labor required in cotton production. The weed spectrum is changing, partly because of the widespread use of selective herbicides. In many cases, grasses are declining in importance, while some broadleaf species in the same taxonomic family as cotton are becoming more prevalent. In addition, the encroachment of perennial weed species has become more serious. To overcome these problems, the practice of rotating herbicides is gaining acceptance.

The potential alternatives to herbicides and tillage in cotton weed control involve some form of biological or other mechanical control. Development of these techniques does not appear to have reached the stage where their advantages or liabilities as practical alternatives can be evaluated. Future weed control in cotton will likely develop toward a concept of total farm vegetation management. This would include actions to reduce weeds and weed seed populations in other crops, set-aside acres, and fence rows. Crop rotations and herbicide rotations might be an integral part of future weed control programs.

In every cotton-producing region of the United States, there are one or more major insect pests that, under our present state of knowledge, usually must be controlled with insecticides if cotton is to be grown profitably. Contemporary cotton production utilizes more insecticides than any other single crop. However, the use of these insecticides can create problems more serious than the damage caused by the major pests, such as target pest resurgence, secondary pest outbreaks, and pest resistance to insecticides. In fact, most of the important arthropod pests of cotton have developed resistance to one or more insecticides.

The alternatives to continued heavy reliance on insecticides include cultural practices, biological control, and genetic manipulation techniques in combination with reduced quantities of insecticides and other chemicals. These combinations of alternative practices are designed to manipulate the environment in order to make it as unfavorable as possible for the pest species (or more favorable for the entomophagous species); to reduce the rate of pest increase and damage; or to concentrate pest numbers in small areas where direct control measures may be applied with minimum disruption to the entomophagous species.

Cultural practices, combined with chemical control, have been developed into successful programs for reducing overwintering populations of two major cotton pests, the boll weevil and the pink bollworm. These programs involve shorter growing periods and earlier harvest and disposal of crop residue. Although the application of insecticides, defoliants, and desiccants is still required, the amounts needed are greatly reduced. There is no apparent barrier to wide-scale acceptance by cotton producers, and the programs should be encouraged.

Trap cropping, with or without pheromone, can be used in cultural insect control. The weevils are then killed in the trap crop with a limited amount of insecticides. This practice shows great promise as a control measure, especially when it complements the cultural practices directed at overwintering weevils. However, the practical application of these techniques has not been developed to the point that they can be recommended to farmers.

Varietal modification of the cotton plant offers additional promise for pest suppression. Early-maturing, rapidly fruiting cotton varieties may be able to produce a crop before the population of boll weevils rises to damaging levels. In addition, they may mature so fast that harvesting and disposal of crop residues are completed before environmental conditions force boll weevils and pink bollworms into diapause.

A frego-bract cotton variety has been developed that is resistant to the boll weevil, but it is still susceptible to attack by the cotton fleahopper and *Lygus* spp. A nectarless variety moderately resistant to lygus and fleahoppers and slightly resistant to *Heliothis* spp. has been developed. Nectarless and frego-bract characters may be combined in a stock that would be resistant to the boll weevil, fleahopper, lygus, and heliothis. Resistant and short-season varieties will have their greatest utility in integrated control programs that use cultural, chemical, and biological pest suppression techniques.

The immense value of natural biological control was not fully appreciated by entomologists or cotton producers until after the introduction of synthetic organic insecticides. When the effectiveness of the approximately 600 species of predators and parasites found in cotton is disrupted, often by a single insecticide application, many secondary pests do considerable damage to the crop. Attempts at using insect natural enemies have taken the form of importation of exotic species and augmentation of native species. Importations have not been successful in controlling any major cotton insects or weeds. The programmed release of laboratory-reared insects has been successful only in experimental plots. Techniques for the large-scale rearing and release of insect natural enemies have not yet been developed. The use of pathogens, such as *Bacillus thuringiensis* and the nuclear polyhedrosis viruses, have a great potential; however, more research is needed to produce a more reliable and consistent level of control and to improve the safety, production, formulation, and application of these microbial insecticides.

Recently, sophisticated combinations of cultural practices, biological controls, genetic manipulations, and chemical control techniques have been developed into proposed programs whose objective has been the eradication of specific cotton pests. The Agricultural Act of 1973 directs the Secretary of Agriculture to eradicate the boll weevil and other insects, if feasible.

We recommend that the development of these alternative pest control practices into effective cotton pest management programs be encouraged. Although a comprehensive plan for eradicating the boll weevil from presently infested areas and to institute a containment program to prevent reentry from Mexico has been developed by USDA and state and industry scientists concerned with the boll weevil problem, the Cotton Study Team seriously questions whether technology is currently available to achieve this goal. Any consideration of such a difficult and costly program should await a critical analysis of the results of a proposed prior trial eradication program in North Carolina.

One feature of programs aimed at eradication is that the costs may be high and exceed benefits during the period that the pest is pushed toward extinction. When very low population densities are reached, benefits may be substantial. When the eradication program is relaxed, however, undetected insects may rapidly increase to repopulate the originally infested areas and costs may again exceed benefits especially in fringe areas of the pest population when large acreages are infested at population densities too low to cause significant crop losses. Proponents of the goal to eradicate the boll weevil argue, however, that although a continuing surveillance and eradicative effort may be necessary in non-isolated populations like that of the boll weevil, overall costs are likely to be lower than the benefits.

Population models on which eradication programs are based often assume that the genetic constitution of the target population will remain constant during the eradication effort. But past experience has shown that the genetic plasticity of insects is such that a pest may become resistant to intensive use of chemical insecticides or other suppressive measures employed. Migration and dispersal of the target species and the presence of alternate hosts may well defeat eradication efforts. Specific problems associated with such programs and costs of possible failure are considered in the Report of the Executive Committee. Because of these known problems and perhaps other as yet unknown problems, responsible scientists disagree about the possibility of success of the major program. Proponents argue that the important advances have been made in detection and suppression methods and that if the trial program proves successful the long-range economic benefits in relation to costs and the improvement in environmental quality would fully justify an eradication and containment effort. The Cotton Study Team, however, believes that although the individual technologies comprising the project are valuable tools for pest management programs aimed at containment, the effort to eradicate a major continental pest species is not likel[y] to achieve the successes now predicted for it by proponent[s].

Systems analysis and computer simulation modelling of the major components of the cotton agroecosystem can be used to estimate the reaction of the system to various exogenous factors; to organize and coordinate data from experiments on different parts of the system; and to clarify what additional information is needed.

We recommend that further development of these techniques be encouraged and that they be used to coordinate the research efforts of entomologists, plant pathologists, and weed scientists and to coordinate the dissemination of information to farmers for control operations.

Within the cotton industry, there are a number of developments that, on the surface, appear to be peripheral to pest management problems but that could have a significant long-run impact on future pest control problems and practices. Developments within the production sector include high-density plantings, use of growth regulators, new seed varieties, and new harvesting and ginning techniques. New techniques in the processing and utilization of cotton in the textile industry are being developed. For example, the potential for cottonseed protein in human nutrition might allow the commercial production of radically different plant types with altered pest problems.

We recommend that future pest control research planning include consideration of these developments.

For the immediate future, the combination of various pest control practices--chemical, cultural, biological, and genetic--into programs of integrated pest management offers the most promising direction for U.S. cotton production. The successful development and delivery of integrated pest management programs will require the development and the introduction of new knowledge, as well as new material inputs into the production process. In some areas, the success of a pest management program may depend on the creation of new institutions for collective action which require that participation by cotton producers be compulsory. When management programs depend upon private individuals serving as consultants, legal questions concerning licensing and liability must be considered, as well as new economic incentives for the pest control industry.

Cotton is produced under such varied conditions by so many farmers that a typical cotton farm or farmer does not exist. Even within specific cotton production regions, only broad generalizations can be made about production patterns. It should be recognized that any given pest management program will achieve varying degrees of success among different economic classes of producers. Farm operators of different educational levels and of different tenure status will vary in their ability to implement sophisticated pest management programs. The profitability of certain pest management practices may depend on the size of the farm. Although the widespread adoption of pest management programs throughout the Cotton Belt would lower aggregate cotton production costs, the magnitude of such cost reductions would not be uniform among all producers.

We believe it important that questions of economic and social equity be considered when new programs of pest management are introduced.

In addition to the general recommendation contained in this summary, the following are specific points that the study team feels need immediate attention.

1. Government farm programs based on price supports and acreage controls have in the past provided incentives for heavy use of pesticides on cotton. While not taking issue with the objective of these programs, we recommend *that if similar policies are considered in the future, consideration be given methods that will achieve the same goals without continued heavy pesticide use.*

2. The role of land tenure relationships in determinin pest control procedure in the Cotton Belt is an important area in which research is needed. We recommend *that funds be allocated for research in this area.*

3. Although examples of successful adoption of improve pest control technology have been illustrated, *stronger emphasis needs to be placed on research into the systems for delivering new pest control technologies to farmers.*

4. *A larger fraction of federal research funds allocated to pest control should be devoted to pest management*

5. It is also important to *focus more research activity on the legal structures and incentives needed to ensur the continued development of the pest management industry.*

6. *An additional effort is also needed to train individuals engaging in pest control activities, whether they are farmers or consultants.*

7. *The present level of research on the environmental impact of currently used pesticides and their potential substitutes is inadequate.*

14

SUMMARY AND RECOMMENDATIONS OF VOLUME IV: *FOREST PEST CONTROL*

The "products" provided by a forest depend upon the objectives of the managers and users of the land. In this sense, forests do not provide us with only one well-defined product, nor is there any single set of pests or any one organism that is a pest in all forest situations. Also, the future of a forest with even a brief rotation from seedlings to mature or harvestable trees cannot be decided the year it is planted, if only because of shifting and unpredictable future values of its potential products.

Thus, the existence of many different objectives for different forest lands, or for a single forest over time, creates a situation in which it is important that managers avoid irreversible pest control decisions that might have unwanted effects in the future. This philosophy also suggests that pest control should not be aimed solely at killing pest organisms and should be undertaken only when the activity of a pest can be clearly shown to interfere significantly with management objectives. This also suggests the need for careful cost/benefit analyses that ensure that those objectives will be served without undue cost or loss of other important benefits. With these general statements as a philosophical background, the Forest Study Team offers the following recommendations.

In the area of pest control research, priorities have often been set under political pressure for immediate answers, with too much regard for specific short-term problems and too little consideration for broader management objectives. Part of this problem arises from portions of the Forest Pest Control Act of 1947 that define policy, and create a division between the administrative and research arms of the U.S. Forest Service (USFS). The Act also tends to promote an "action" attitude that may inhibit solutions other than short-term, direct chemical control.

We, therefore, recommend that the Forest Pest Control Act be reexamined with a view toward repeal of its policy sections or replacement with a new act that is consistent with the management philosophy presented in this report.

Destructive harvesting techniques and harvesting without reforestation, where forest lands have been exploited for immediate profit, create "weed" problems and loss of future productivity. Reforestation laws that require that specified stocking standards be met within a certain number of years after harvest appear to be functioning effectively in Oregon and Washington and could be effective in other regions where even-age management is practiced.

We recommend adoption of similar laws by other states where such a law would be appropriate.

The impact of pests is part of the USFS inventory, but present methods are reasonably good for insects, poor for diseases, and almost nonexistent for weeds. There is a general need for better information and better inventory data on a variety of subjects related to forest management and it is particularly important that loss estimates be made in terms of the whole spectrum of social values involved and that they be consistent with all of the different management objectives involved.

We recommend that the USFS be encouraged, and provided with the necessary funds, to increase its efforts in the development of accurate methods of estimation and fuller appraisals of losses due to pests.

Detection of insect (and other pest) damage is still a major problem in forest pest management. At present, detection of insect damage is almost always based on visible damage, often by air survey. The results are often obtained so late that treatment is applied to already declining insect populations, and it is impossible in such cases to evaluate the effectiveness of control measures. With more adequate detection and more carefully controlled evaluation of control efforts, it would be possible to reduce or eliminate prophylactic treatments for insect control, which are used in many urban and suburban forests when there is no genuine evidence of significant hazard from insect pests.

We recommend that sampling procedures be improved and better methods developed; that control projects include the cost of accurate evaluation procedures, including the establishment of adequate comparisons between treated and untreated areas; and that evaluation of control methods include effects on nontarget organisms and natural enemies of the target insect.

Studies are needed on all aspects of silvicultural control, including site preparation and species selection, cultural practices after the establishment of a stand (such as thinning, including the use of herbicides), and postharvest sanitation and salvage techniques. Where ecological, silvicultural, and economic considerations permit, pest management should utilize cultural procedures. Research on silvicultural methods is especially needed in the protection of forests from bark beetle infestations, and in forest manipulation to avoid spruce budworm damage. Certain cultural techniques, such as sanitation/salvage, require political and economic coordination as well as technical skill. The case of the Southern pine beetle, for example, demonstrates the need for careful organization of procedures and development of an adequate labor force, transportation facilities, and industrial capacity for rapid salvage in the event of an insect epidemic.

We recommend expanded research on the development and use of silvicultural methods of forest pest control.

The use of more than one method to prevent damage by pests is sometimes referred to as "integrated control" or, more recently, "integrated pest management." The term also implies maximum use of natural controls of pest populations wherever feasible. The concept of integrated pest management is also relevant to vegetation management, as in the case of selective removal of unwanted trees on rights-of-way and establishment of low-growing shrub cover.

We recommend that research on forest pest control be closely coordinated with overall research on silviculture and forest management, to better achieve the objectives of integrated pest management.

The use of biotic agents to control forest pests ("biological control") is still in an early stage of research and development. With respect to insects, the greatest success has been against introduced pests, and this is also the area that shows the greatest promise. However, biological control of tree disease also shows promise. At present, the only practical use of a biotic agent to reduce the activities of pathological organisms is the inoculation of pine stumps with the fungus *Peniophora gigantea* against infection by *Fomes annosus*, but potential exists to develop biological systems antagonistic to wood decay fungi, damping-off fungi, canker and vascular disease fungi, and dwarf mistletoe.

The use of insect pheromones to attract pests to traps, or to disrupt their normal activities and thereby reduce their impact as pests, is being studied, and some pheromones have been used in preliminary control studies, but the only proven use for insect pheromones in forest pest control, to date, is as a survey tool.

Another potential form of biological control is species mixtures, in which pest problems are reduced by maintaining diversity of species. For example, there is a significantly higher incidence of *Poria weirii* root infection in pure stands of Douglas fir than in mixed stands that include red alder.

We recommend increased support of research on these and other promising methods of biological control.

Forestry has depended almost entirely upon pesticides developed for use in agriculture, because it is in the "minor use" category with respect to the pesticide market. There is little incentive for manufacturers to develop and register formulations for minor uses, and useful, relatively safe pesticides have been taken off the market when their use in agriculture declined. The result is that many pesticides that are potentially useful in forestry are unavailable, and in some cases, there have been no alternative solutions to pest problems, while in others the alternative has been more costly with more adverse environmental effects.

We recommend an examination of mechanisms that would provide a greater range of chemicals for the specific needs of minor users.

Incentive programs to improve forests on small private lands and to support sound pest management programs have had limited success, partly because the recipients of incentives have largely been landowners who would practice good forest management in any event, rather than marginal operators whose stand management is contingent upon receiving education as well as benefits.

We endorse recommendation Number 10 of the President's Advisory Panel on Timber and the Environment (April 1973), which calls for the federal government to maintain and improve incentive programs through development of income tax incentives, advice and service to forest owners and associations, and cost-sharing for intensive forest management practices.

The flow of information and recommendations about pest problems in urban, suburban, and other intensively managed areas needs to be carefully examined. Pest control decisions are frequently made by nonprofessionals, such as homeowners, on the advice of other nonprofessional or semiprofessional "experts," or pesticides are purchased and applied with no advice whatever. Cities and towns need more trained practitioners of plant welfare with practical expertise in ecology, physiology, entomology, pathology, and soils.

We recommend the training and appointment of such professionals in more urban and suburban areas and a vigorous program of public education on the use and nonuse of pesticides.

15

SUMMARY AND RECOMMENDATIONS OF VOLUME V: *PEST CONTROL AND PUBLIC HEALTH*

Vector-borne disease is now, as in the past, a major deterrent in some areas to human settlement and agricultural development. Nevertheless, the use of pesticides together with other tactics to control arthropods and other organisms of public health significance has had tremendous impact during the past quarter century on improving the health and well-being of man throughout the world. In the past decade, progress toward the elimination of disease has had setbacks in some areas, in large part because of environmental and social change. The initial successes with malaria control utilizing DDT are not being repeated in all areas. In some of the problem areas, the habits of man himself in relation to those of the vector are responsible for continued transmission. Problems are also developing with other diseases of man (see Chapter 1).

An attack on the disease vector is not the only means of reducing or eliminating an arthropod-borne disease. The other approaches are protection by vaccines and prophylaxis by suppressive chemotherapy to minimize opportunities for infection of specific vectors. However, analysis of these indicates that the majority of the important human vector-borne diseases cannot be prevented by vaccines or by chemotherapy. Of necessity, their control is based on reduction of vector sources and reduction of contacts between vector and man (see Chapter 1).

Rapid changes in density and distribution of people, in economic and social values, and in some instances in the behavior of public health pests and the characteristics of disease require continued reevaluation and study of the strategies and tactics of pest control.

The control of arthropod vectors of disease or other pests of public health should be attempted only with recognition, and insofar as possible an intimate knowledge, of the significance of the ecology and behavior of the target species and the epidemiology of the disease and with appreciation for environmental values that may be depreciated.

As every vector control problem is unique, each will require evaluation by persons trained in epidemiology, entomology, engineering, ecology, and economics to determine whether control should be undertaken, and, if so, with which tactics and strategy. The benefits of pest control, whether by application of a pesticide, drainage to reduce a vector source, or some other action, must be weighed against possible unfavorable impacts on nontarget organisms and the quality of the environment (see Chapter 4).

The very complexity of these broad ecological approaches to public health pest control and the additional burden of economic impacts create new responsibilities for pest control practitioners. Consequently, their breadth and level of training must be increased.

The study team therefore recommends that increased educational opportunities be made available in all aspects of integrated control of arthropods of public health significance.

Mosquitoes are by far the most important arthropods subject to control for general public health reasons in the United States. It is probable that 80-90 percent of the entire public health pest control effort is directed toward mosquitoes. The most important mosquito-borne diseases in the United States have largely disappeared, but their vectors remain. The United States is in what might be termed a receptive state for several important arthropod-borne diseases: all of the elements required to produce an epidemic of several arthropod-borne diseases may be present in this country at any time (see Chapter 2).

The study team recommends that to be prepared for any epidemic of arthropod-borne disease, the highly developed reporting system of the health departments of cities, counties, and states, and extensive surveillance under the general supervision of the Center for Disease Control, United States Public Health Service, be maintained.

The majority of Americans living in coastal and other areas subject to an above-average mosquito nuisance problem enjoy the benefits of organized mosquito control. The sources of mosquito annoyance in different mosquito control districts reflect geography, terrain, and degree of urbanization. Unorganized mosquito control, whether by public or private agencies, forms a small portion of all mosquito control operations. It suffers, however, from lack of technical information and regulation.

The study team recommends that, because these operations are aimed primarily at adult arthropods, and since control is achieved by use of fogs and sprays containing toxicants that may affect human beings, some form of regulation beyond general licensing should be required.

Areawide public health pest control can be accomplished by either the public or the private sector of the economy. The fact that it requires manipulation of everyone's air, water, and land, however, bestows a predominantly public element upon it. For that reason the public welfare must be served with adequate state regulation and monitoring of relevant pest control efforts, whether these are publicly or privately exerted (see Chapter 2).

The mosquito control districts of the United States appear to be in accord as to the role of insecticides now and in the near future. The organic insecticides will be useful everywhere into the foreseeable future, and they will be indispensable in some districts (see Chapter 2). As public health pest control continues to develop, pesticides with special attributes will be identified as essential in important but limited pest control programs. The size of the market for such pesticides precludes their commercial development because it is not profitable to manufacture and distribute them.

The study team recommends that some form of federal subsidy be considered to ensure the availability of critical limited-use pesticides at a reasonable cost.

The special cases of DDT and Paris green for public health use are examples of such materials. Although resistance and other problems limit the effectiveness of DDT in many areas, it remains the insecticide of choice in most malarious areas.

Steps should be taken to ensure a continuing and adequate supply of high-quality DDT formulations to meet the worldwide demand for domiciliary residual application in public health programs.

Paris green is a chemical which is relatively easy to manufacture, has a unique and effective use as a mosquito larvicide, and when used properly for this purpose, is environmentally safe (see Chapter 2).

The necessary steps should be taken to ensure the availability of Paris green at a reasonable price for mosquito control.

The reservation of certain pesticides for public health use would only rarely, if ever, be feasible economically, and in most instances, because of cross-resistance, it would not achieve the objective of avoiding the development of resistance to pesticides.

The study team does not recommend as a general practice that certain pesticides be reserved exclusively for public health purposes.

To avoid the conflicts that have arisen in some areas, the agricultural and public health authorities should attempt to coordinate the use of pesticides in their areas to serve the best interests of all concerned (see Chapters 2 and 5).

Mosquito control strategies must aim at more than ridding communities of mosquitoes effectively and efficiently. Mosquito control can no longer be a unilateral operation. Collaboration in many directions is now essential--with landowners, with agricultural scientists, with conservationists, with engineers of several sorts, with a great variety of public agencies and, most assuredly, with land-use planners. Mosquito control is now an integral part of land and water management in the public interest (see Chapter 2).

In the future, development and implementation of mosquito control through manipulation of the environment should incorporate appropriate safeguards to preserve the natural vegetation and fauna of aquatic habitats. To this end, more sophisticated forms of environmental manipulation than, for example, simple ditching should be utilized.

It is difficult to compare needs and results in vector control programs because of the many and varied methods of measuring vector densities used throughout the world.

An effort should be made to develop a universally applicable and acceptable method of population measurement for all the important disease vectors of the world.

Although complete comfort would mean no biting pests at all, complete control at the cost of insecticide resistance or environmental deterioration should yield to reasonable control (see Chapters 2 and 5).

The study team would encourage the establishment of tolerance thresholds for biting nuisances, in the belief that reduction of the pests to lower densities carries the risk of excessive external, if not internal, costs.

Managing an organized mosquito control district demands professional leadership. This leadership should not be compromised by lowered personnel standards or by restriction on exercise of professional leadership (see Chapter 2)

Organized mosquito control in the United States has always been and continues to be "integrated control," in the usual sense of the term. The term "alternative methods" is disputable if the implication is that the use of insecticides is all there is to mosquito control. There have always been alternatives, in this sense, and what is happening today is only a reordering of the various chemical and nonchemical methods. The really new "alternatives," such as genetic, biological (other than fish), and hormonal control, are still years from incorporation into the armamentarium of mosquito control.

To assist in making new alternative strategies available to public health pest control, the study team urges that support be expanded in the area of alternative strategies in vector control. Furthermore, there should be increased coordination of efforts in this area because most alternative strategies can be implemented only on a large scale.

Good economic analyses of public health programs are almost nonexistent for two interacting reasons: the general failure of the public health specialists to directly assess trade-off questions and the failure of economists to deal adequately with such issues as the value of health as opposed to illness, the value of human life, and the value of simply having fewer pest mosquitoes about. In addition, there continues to be great uncertainty regarding the technical feasibility of various kinds of control mechanisms. Even unit-cost coefficients for such standard vector control devices as source reduction, larviciding, and adulticiding vary greatly by region and will not be readily available without a major new effort in data collection (see Chapter 2).

The study team recommends that economic analyses be made of the many issues in the vector control field.

Only with this information, which will accumulate only through greater interaction among entomologists, public health specialists, and economists, will it be possible to reach any general conclusions about the economic efficiency of pesticide use versus other approaches to vector control.

Statistics on nationwide use of pesticides for public health purposes are difficult to obtain and are seldom precise when they are obtained. Hence, assessment of pesticide use in the public health field must be approached qualitatively and with historical and current trends as basic criteria (see Chapter 2).

The American Mosquito Control Association or some appropriate federal agency (e.g., the U.S. Public Health Service or the Environmental Protection Agency) should establish improved methods of assessing the levels of pesticide use for public health purposes and keep statistical data on such pesticide use and records on other methods of mosquito control.

Pest control activities for public health objectives are certain to have an impact on the environment. Much of the arthropod-borne disease problem occurs in man-made environments, domestic, urban, or rural, where it is pointless to debate the impact of vector control, if for no other reason than that simple sanitation, which can so often be the remedy, is unquestionably an improvement in environment. It is impacts on "natural" environments that are of considerable concern today, especially aquatic environments. Terrestrial environments are seldom disturbed by the control of disease vectors, with the exception of pesticide pollution or the entrance of pesticides into terrestrial life through the food chain (see Chapters 4 and 5). There appears to be no alternative to use of pesticides in the near future.

The study team recommends continuation of the current trend toward a balanced, integrated approach in pest control tactics taking into account the ecology and behavior of the pest species, the epidemiology, and environmental values. The implementation of multitactic control systems will increase research efforts; consequently, the study team recommends continued and expanded support of research on ecologically based control systems for mosquitoes and other arthropods of medical importance.

Mosquito control operations on the tidelands of the Southeast have the potential to damage the valuable shrimp fishery. However, the impact of present-day mosquito control in this area has only a very minor and very localized impact on the shrimp fishery. Shrimp authorities are more concerned about other activities of man on the tidelands (see Chapter 2).

The study team recommends that attempts to protect the tidelands from various destructive activities, such as dredging, filling, and pollution, should exempt mosquito control operations carried out with reasonable concern for environmental values and maintenance of the tidelands environment.

Until recently, the introduction of exotic (nonnative) larvivorous fish for the control of mosquitoes was considered altogether unobjectionable. Advances in ecological knowledge, particularly in aquatic systems, are now exposing areas of concern that demand that this form of biological control be practiced with more vigilance. The introduction of exotic larvivorous fish for mosquito control may pose a serious threat to native fish, both large and small (see Chapters 2 and 4).

Introduction of fish into new areas should be made with appropriate evaluation of the impact of their introduction on the endemic biota and also of the impact of the introduction of other undesirable organisms through contamination of the intended introduction.

In investigations extending over more than a half a century, a wide variety of biological controls have been studied. A few of these, e.g., the introduction of larvivorous fish, have become effective and established control procedures where mosquitoes' range is limited. A large number of other biological control agents have been identified, and, in a few instances, their characteristics and limitations described, but the full potential of these agents is largely unknown.

The developed countries may indirectly affect the supply of various insecticides for public health use in less developed countries. They may also have a more direct effect on public health programs by conducting vector control and research programs in other countries. In Chapter 3, we examine the activities of the official agencies of the U.S. government that are concerned with vector control and with vector research outside the territorial limits of the country, and we describe the major international programs for vector-borne disease control, taking note of the role the United States plays in these programs.

Pesticides are essential not only in pest control for public health purposes, but also, of course, in food production. Current practices in pesticide use are sometimes unsound from the standpoint of crop protection, human health and safety, or, in some cases, environmental quality. Improvement is needed in the pesticide management process, i.e., the proper selection, procurement, formulation, packaging, shipment, storage, marketing, application, and disposal of pesticides (see Chapter 5).

The study team recommends an agromedical team approach to the problem of poisoning, persistence, and resistance in pesticide management.

With the widespread introduction of technology from the developed northern temperate countries into the tropical areas after World War II, there has been increasing concern over the undesirable features of this transfer of technology. With this in mind, the study team has considered the impact of the introduction of new agricultural technologies on the problems of public health. Because this is such a broad area of interaction, the study team examined the impact of the changing technology of rice production in Asia (particularly Southeast Asia) on public health problems. A summary of this study is presented in Appendix A. Other examples of this type of interaction are referred to in other parts of the report. We also consider the direct (intoxication and mortality from pesticides) and indirect (induction of pesticide resistance in disease vectors) impact of pesticide use on public health. In this case, we use the situation in Central America as our primary example.

The "green revolution" has not caused sweeping changes in the way rice is grown in Asia. Still, the potential is there for far more extensive changes than have already occurred. In the future, these changes will undoubtedly occur at a slower rate than before, because of the factors limiting the spread of green revolution technology.

Where large areas have been newly irrigated for rice production, and where insecticides have been introduced for the control of rice insects, no drastic changes in patterns of disease transmission appear to have occurred. Part of the reason for this is the relatively small number of disease vectors that are associated with rice fields.

Those vectors that do use rice fields as habitats have been identified. There are only a few areas (Java, South China) in Asia where rice-field *Anopheles* are primary malaria vectors, and a few areas where they play a secondary role. The picture for filariasis is less clear, but rice-field mosquitoes appear to be relatively unimportant except on the west coast of Malaysia, and that focus has contracted sharply as a result of anti-*Anopheles* campaigns. The *Culex* vectors of Japanese encephalitis virus are closely associated with rice culture in many parts of Asia, so much so that there is some evidence, from Korea, that insecticide treatment of rice pests may dramatically reduce the incidence of the disease, at least temporarily. The snail hosts of *Schistosoma japonica* have a wide range of habitats, but in at least some areas (Philippines, China) rice culture may play a significant role in the ecology of the disease.

Where rice culture is important, any change in its production technology, or in the extent of its fields, may change the habitats of the disease vectors discussed above. In most of Asia, rice production has not been "modernized" by the introduction of agricultural chemicals, large-scale water management technology, and heavy machinery. There is still a potential for the development of sound, integrated pest control programs, which would take into account both agricultural pests and vectors of disease.

It is clear that, in addition to pesticides, a wide variety of tactics is available for use in the control of public health pests. For some of these the effectiveness and usefulness in control are well established and scientifically validated. The procedures of source reduction through environmental sanitation, the use and management of larvivorous fish and the use of screening and repellants are among such established alternatives to pesticides. Other alternatives are more limited in their current usefulness, and some are in only the earliest stages of evaluation. Some of these show great promise, and each is discussed in the report that follows. Another critical consideration will be how these several tactics will be used together in pest control or, better, in a pest management system. Identifying a new technology or tactic for pest control is not necessarily the same as identifying a functional alternative control. Such a technology must be made a part of a strategy of pest control and fitted into a management system. This adaptation of a technology into a system must include an economic analysis of that technology, including a benefit/cost study of the particular technology and of the modified system as an operational unit (see Chapter 4).

It is the view of the study team that there is in prospect no alternative to the use of pesticides in public health pest control strategy in the near future. Rather, the study team anticipates a gradual return to a more balanced, integrated approach using a variety of pest control tactics as appropriate to the ecology and behavior of the pest species, the epidemiology of the disease, and environmental values. There will be an increase in reliance on source reduction, and the goal will be containment and reduction of disease rather than eradication.

APPENDIXES

APPENDIX A: PROPERTIES OF SELECTED PESTICIDES*

Common Name: Alachlor

Trade Names: Lasso Lazo

Structure:

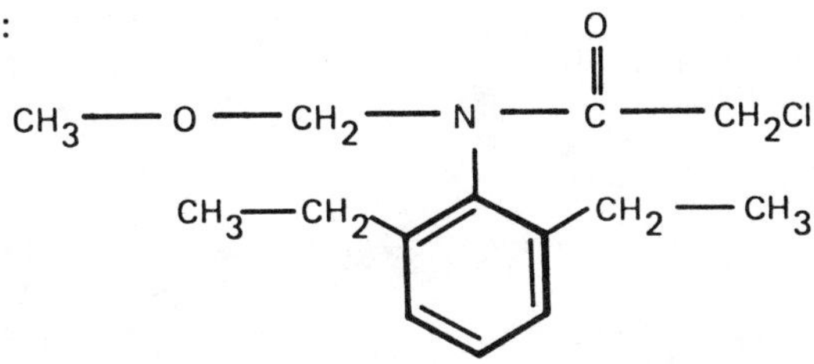

M.W., 270

Chemical Name: 2-Chloro-2',6'-diethyl-*N*-(methoxymethyl) acetanilide

Purity: 95% (2)

Solubility: Water/23°C/140 ppm (4)
Acetone, benzene, ethanol, ethyl acetate/soluble (4)
Heptane/slightly soluble (3)

Vapor Pressure, torr: 100°C/0.02 (3)

Use: Herbicide

U.S. Production, pounds: 1971/20 million (6)

Principal Manufacturer(s): Monsanto (1)

*See end of Appendix for sources and explanation of footnotes.

Market Basket Intake[a]:	Not evaluated
EPA Action[aa]:	No official action to date
Innes et al. Test[b]:	Not tested
Secretary's Commission Recommendation[c]:	None
NCI Carcinogenicity Test[d]:	Studies in progress

Common Name:	Aldrin
Trade Names:	Aldrite Drinox Seedrin Aldrosol Oxtalene
Structure:	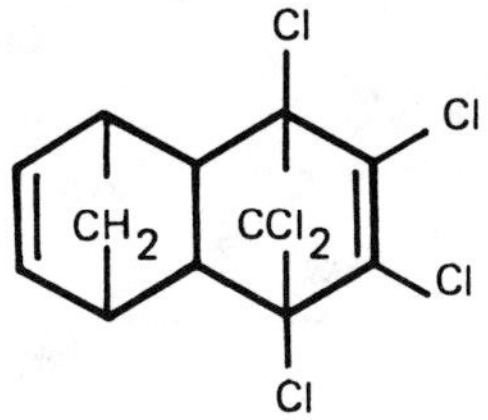 M.W., 365
Chemical Name:	1,2,3,4,10,10-Hexachloro-1,4,4a,8,8a-hexahydro-*exo*-1,4-*endo*-5,8-dimethanonaphthalene
Purity:	95% (5)
Solubility:	Water/practically insoluble (0.025 mg/1) (5) Organic solvents/very soluble in most (5) Petroleum oils/moderately soluble (4)
Vapor Pressure, torr:	20°C/2.31×10^{-5} (5)
Use:	Insecticide
U.S. Production, pounds:	1971/10 million (6)
Principal Manufacturer(s):	Shell (1,4)
Market Basket Intake[a]:	1973/<0.01 μg/kg body wt/day
EPA Action[aa]:	All uses finally cancelled in 1975 Compound prohibited in the United States
Innes et al. Test[b]:	Not tested
Secretary's Commission Recommendation[c]:	Remove from general use (Recommendation #4, #5) Carcinogenesis Panel--positive
NCI Carcinogenicity Test[d]:	Studies in progress

Common Name:	Atrazine
Trade Names:	Aatrex Gesaprim Weedex A Atranex Primatol A

Structure:

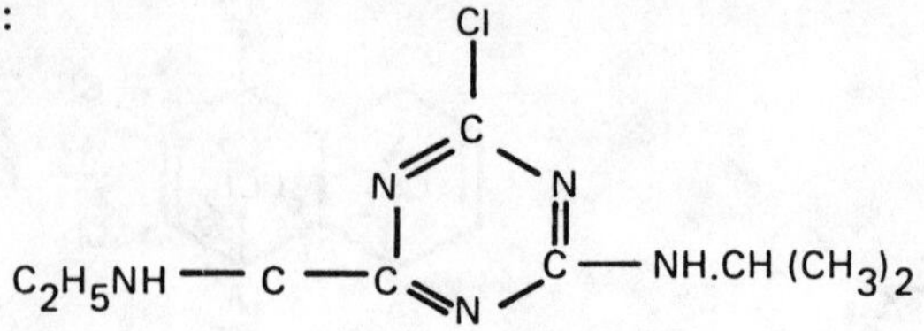

M.W., 216

Chemical Name:	2-Chloro-4-ethylamino-6-isopropylamino-*s*-triazine
Purity:	96% (2)
Solubility:	Water/20°C/33 ppm (4) Organic solvents/very soluble (4)
Vapor Pressure, torr:	20°C/3.0×10^{-7} (4)
Use:	Herbicide
U.S. Production, pounds:	1971/90 million (6)
Principal Manufacturer(s):	Ciba-Geigy (1)
Market Basket Intake[a]:	Not evaluated
EPA Action[aa]:	No official action to date
Innes et al. Test[b]:	Negative
Secretary's Commission Recommendation[c]:	None
NCI Carcinogenicity Test[d]:	Not currently being tested

Common Name:	BHC (Benzenehexachloride)
Trade Names:	Benzahex, Benzex, Dol, Dolmix, FBHC, Hexafor, Hexyclan, Soprocide
Structure:	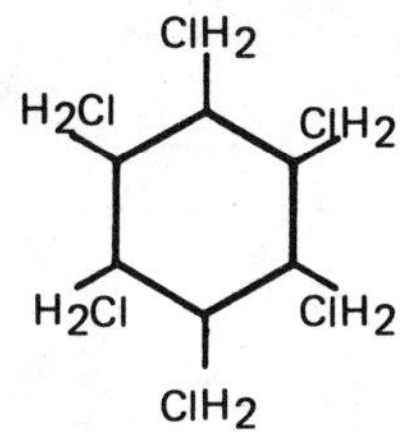 M.W., 291 (Lindane)
Chemical Name:	Mixed isomers of 1,2,3,4,5,6-Hexachlorocyclohexane
Purity:	Content of isomers depends on the manufacturer
Solubility:	α isomer: Water/practically insoluble; Chloroform and benzene/readily soluble β isomer: Water/very slightly soluble; Chloroform and benzene/slightly soluble γ isomer: Water/about 10 mg/l; Chloroform, ethanol, acetone, ether, and benzene/very soluble (5)
Vapor Pressure, torr:	α isomer: 40°C/0.06 β isomer: 40°C/0.17 γ isomer: 40°C/0.14 (5)
Use:	Insecticide
U.S. Production, pounds:	1971/<1 million (6)
Principal Manufacturer(s):	One U.S. company

Market Basket Intake[a]: 1973/0.01 μg/kg body wt/day

EPA Action[aa]: No official action to date

Innes et al. Test[b]: Not tested

Secretary's Commission Recommendation[c]: Restrict use (Recommendation #4

NCI Carcinogenicity Test[d]: Hazard indicated

Common Name:	Bis(2-chloroethyl) ether
Trade Names:	Chlorex
Structure:	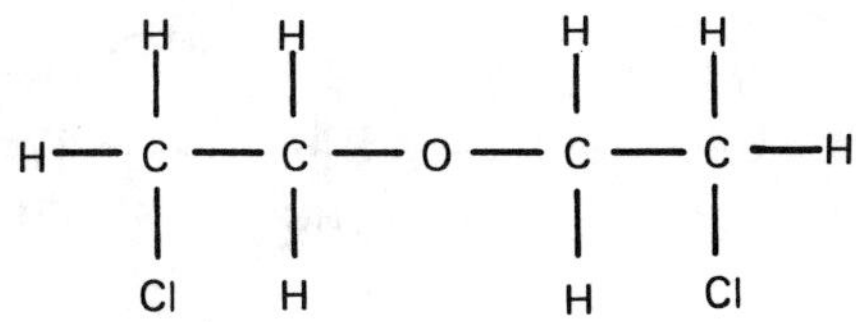
Chemical Name:	Bis(2-chloroethyl) ether
Purity:	Not available
Solubility:	Water/20°C/1.1% w/w (2) Most organic solvents/soluble (2)
Vapor Pressure, torr:	20°C/0.73 (2)
Use:	Herbicide
U.S. Production, pounds:	Not available
Principal Manufacturer(s):	Not available
Market Basket Intake[a]:	Not evaluated
EPA Action[aa]:	No official action to date
Innes et al. Test[b]:	Positive
Secretary's Commission Recommendation[c]:	Minimize human exposure (Recommendation #5) Carcinogenesis Panel--positive
NCI Carcinogenicity Test[d]:	Studies nearing completion

Common Name: Captan

Trade Names: Merpan Orthocide Vondcaptan

Structure:

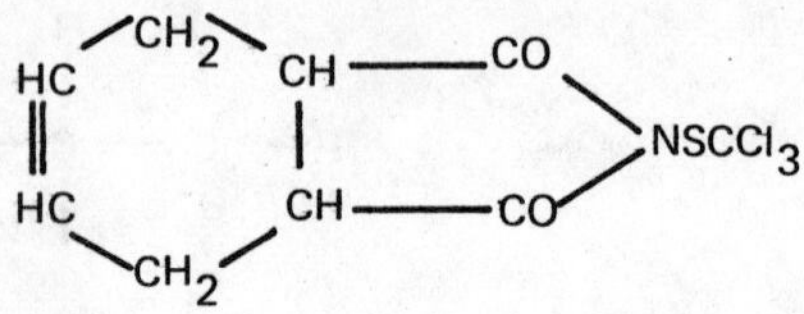

M.W., 301

Chemical Name: *N*-(Trichloromethylthio)-4-cyclohexene-1,2-dicarboximide

Purity: 90% (2)

Solubility: Water/25°C/<0.5 ppm (4)
Petroleum oils/insoluble (4)
Organic solvents/soluble (4)

Vapor Pressure, torr: 25°C/<0.01 mm Hg (4)

Use: Fungicide

U.S. Production, pounds: 1971/18 million (6)

Principal Manufacturer(s): Stauffer (1)
Chevron (1)

Market Basket Intake[a]: Not evaluated

EPA Action[aa]: No official action to date

Innes et al. Test[b]: Negative

Secretary's Commission Recommendation[c]: Minimize human exposure (Recommendation #5)
Carcinogenesis Panel--questionable carcinogen
Teratogenicity Panel--positive

NCI Carcinogenicity Test[d]: Studies in progress

Common Name: Carbaryl

Trade Names: Arylam, Hexavin, Karbaspray, Ravyon, Septene, Sevin, Tricarnam

Structure:

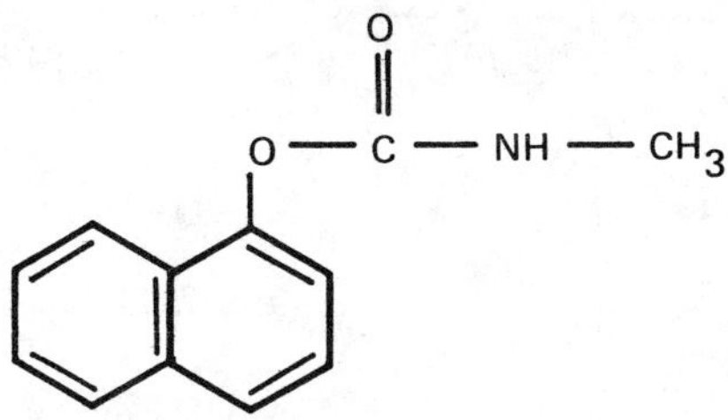

M.W., 201

Chemical Name: 1-Naphthyl *N*-methylcarbamate

Purity: 90% (4)

Solubility: Water/30°C/40 ppm
Most polar organic solvents/soluble (4)

Vapor Pressure, torr: 26°C/<0.005 (4)

Use: Insecticide

U.S. Production, pounds: 1971/45 million (6)

Principal Manufacturer(s): Union Carbide (7)

Market Basket Intake[a]: 1973/0.01 μg/kg body wt/day

EPA Action[aa]: No official action to date

Innes et al. Test[b]: Negative

Secretary's Commission Recommendation[c]: Carcinogenesis Panel--negative
Teratogenicity Panel--positive

NCI Carcinogenicity Test[d]: Studies in progress

Common Name: CDAA

Trade Names: Randox

Structure:

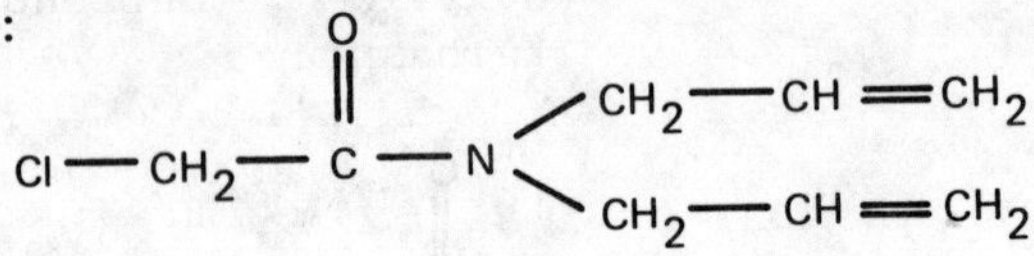

M.W., 174

Chemical Name: 2-Chloro-*N*,*N*-diallylacetamide

Purity: 95% (2)

Solubility: Water/25°C/1.97% w/w (4)
Petroleum hydrocarbons/moderately soluble (4)
Ethanol and xylene/readily soluble (4)

Vapor Pressure, torr: 20°C/9.4×10^{-3} (4)

Use: Herbicide

U.S. Production, pounds: 1971/10 million (6)

Principal Manufacturer(s): Monsanto (3)

Market Basket Intake[a]: Not evaluated

EPA Action[aa]: No official action to date

Innes et al. Test[b]: Not tested

Secretary's Commission Recommendation[c]: None

NCI Carcinogenicity Test[d]: Not currently being tested

Common Name: Chloramben

Trade Names: Amiben Amoben Verbigen

Structure:

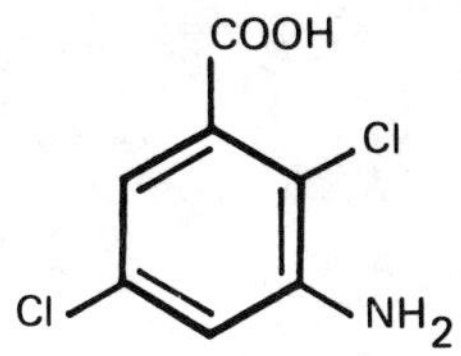

M.W., 206

Chemical Name: 3-Amino-2,5-dichlorobenzoic acid

Purity: 100% (2)

Solubility: Water/25°C/700 ppm (4)
Benzene/24°C/200 ppm (4)
Alkali salts are water-soluble (4)

Vapor Pressure, torr: 100°C/7 × 10^{-3} (4)

Use: Herbicide

U.S. Production, pounds: 1971/20 million (6)

Principal Manufacturer(s): Amchem (3)

Market Basket Intake[a]: Not evaluated

EPA Action[aa]: No official action to date

Innes et al. Test[b]: Not tested

Secretary's Commission Recommendation[c]: None

NCI Carcinogenicity Test[d]: Studies in progress

Common Name:	Chlordane
Trade Names:	Aspon, Belt, Chlor Kil, Corodane, Kypchlor, Octachlor, Ortho-Klor, Synklor, Topiclor 20, Velsicol 1068
Structure:	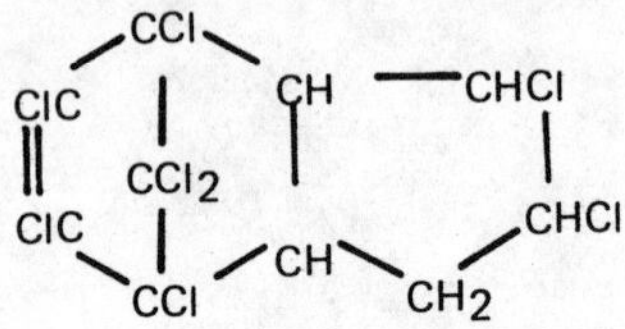 M.W., 410
Chemical Name:	1,2,4,5,6,7,8,8-Octachloro-2,3,3a,4,7,7a-hexahydro-4,7-methanoindene
Solubility:	Water/insoluble Organic solvents/soluble (4)
Vapor Pressure, torr:	25°C/0.45-1.55 $\times 10^{-4}$ (4)
Use:	Insecticide
U.S. Production, pounds:	1971/25 million (6)
Principal Manufacturer(s):	Velsicol (1)
Market Basket Intake[a]:	None detected
EPA Action[aa]:	Notice of intent to cancel issued in Nov. 1974 Notice of intent to suspend as "imminent hazard" issued Aug. 1975
Innes et al. Test[b]:	Not tested
Secretary's Commission Recommendation[c]:	Restrict use (Recommendation #4)
NCI Carcinogenicity Test[d]:	Studies in progress

Common Name:	Chlorobenzilate
Trade Names:	Acaraben Benzilan Kop-Mite Akar Folbex
Structure:	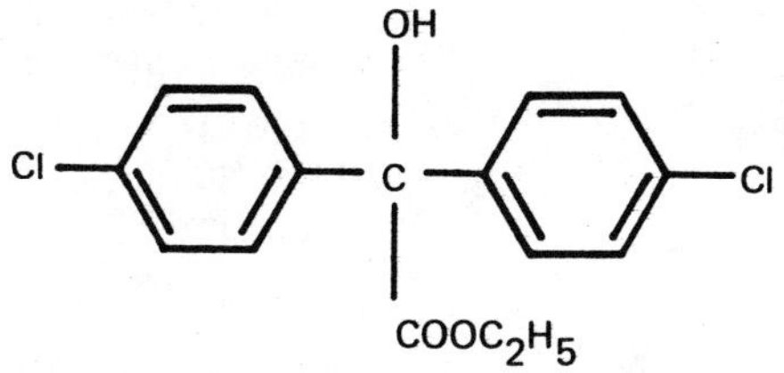
	M.W., 325
Chemical Name:	Ethyl 4,4'-dichlorobenzilate
Purity:	90% (5)
Solubility:	Water/insoluble (5) Organic solvents and petroleum oils/soluble (5)
Vapor Pressure, torr:	20°C/2.2×10^{-6} (5)
Use:	Acaricide
U.S. Production, pounds:	1971/2 million (6)
Principal Manufacturer(s):	Ciba-Geigy
Market Basket Intake[a]:	None detected
EPA Action[aa]:	No official action to date
Innes et al. Test[b]:	Positive
Secretary's Commission Recommendation[c]:	Remove from general use (Recommendation #4, #5) Carcinogenesis Panel--positive
NCI Carcinogenicity Test[d]:	Studies nearing completion

Common Name: 2,4-D (and its salts and ester

Trade Names:

Aqua-Kleen	Hedonal
Chipco Turf Herbicide "D"	Pennamine D
	Salvo
Chloroxone	Tributon
Crop Rider	Vergemaster
Dacamine	Verton 2D
Ded-Weed	Visko-Rhap Low Volatil 4L
Dormone	
DMA-4	
Esteron	Weed-Ag-Bar
Estone	Weedone
Fernesta	Weedar
Fernimine	Weedez
Feroxone	Wonder Bar

Structure:

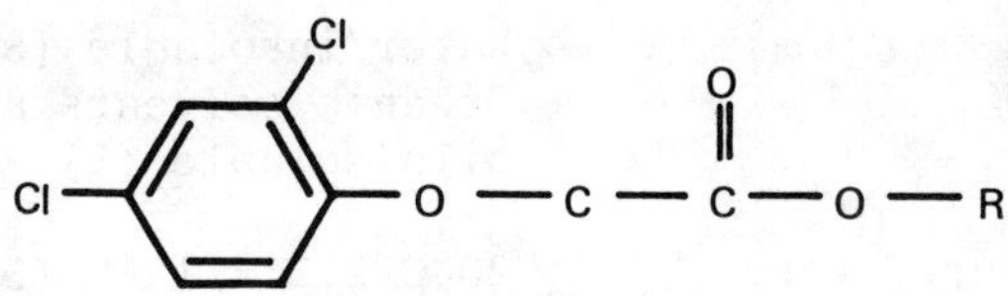

M.W., 221

(for R=H)

Chemical Name: (2,4-Dichlorophenoxy) acetic acid + esters

Purity: 90-95% (2)

Solubility: Water/25°C/620 ppm (4)
Aqueous alkali and alcohols/soluble (4)
Petroleum oils/insoluble (4)

Vapor Pressure, torr: 160°C/0.4 (3)

Use: Herbicide

U.S. Production, pounds: 1971/45 million (6)

Principal Manufacturer(s): Dow (3)
Amchem (3)

Market Basket Intake[a]: None detected

EPA Action[aa]: No official action to date

Innes et al. Test[b]:	2,4-D and its isopropyl, butyl and isooctyl esters, all negative
Secretary's Commission Recommendation[c]:	Minimize human exposure (Recommendation #5) Teratogenicity Panel--suspected teratogen
NCI Carcinogenicity Test[d]:	Studies in progress

Common Name: DBCP (Dibromochloropropane)

Trade Names: BBC 12, Nemafume, Fumazone, Nemagon

Structure:

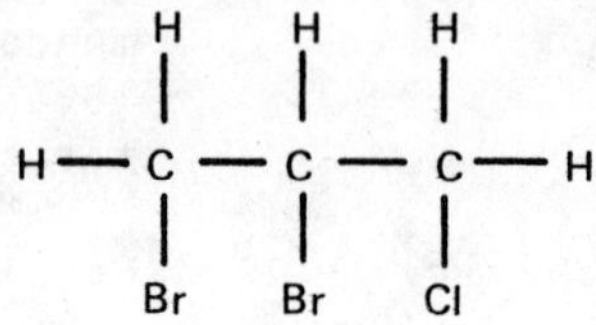

M.W., 236

(and related compounds)

Chemical Name: 1,2-Dibromo-3-chloropropane

Purity: 95% mixture of bromo- and chloro-propropane isomers (2)

Solubility: Water/slightly soluble

Vapor Pressure, torr: 20°C/0.8

Use: Soil fumigant for nematode control

U.S. Production, pounds: 1971/10 million (6)

Principal Manufacturer(s): Dow (1)
Shell (1)
Occidental (1)

Market Basket Intake[a]: Not evaluated

EPA Action[aa]: No official action to date

Innes et al. Test[b]: Not tested

Secretary's Commission Recommendation[c]: None

NCI Carcinogenicity Test[d]: Studies nearing completion

Common Name: D-D

Trade Names: D-D
Nemafene

Structure:

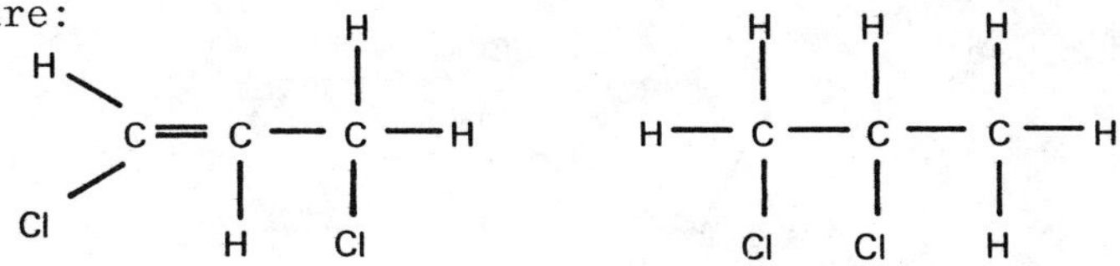

(Representatives of a mixture)

Chemical Name: Mixtures of 1,3-Dichloropropene; 3,3-Dichloropropene; 1,2-Dichloropropane; 2,3-Dichloropropene; and related C_3 chlorinated hydrocarbons (1)

Purity: 55% (4)

Solubility: Water/25°C/about 0.2% w/w (4)
Hydrocarbon and halogenated solvents, esters and ketones/soluble (4)

Vapor Pressure, torr: 20°C/35 (4)

Use: Nematocide

U.S. Production, pounds: Not available

Principal Manufacturer(s): Shell (1)

Market Basket Intake[a]: Not evaluated

EPA Action[aa]: No official action to date

Innes et al. Test[b]: Not tested

Secretary's Commission Recommendation[c]: None

NCI Carcinogenicity Test[d]: Studies in progress
3-Chloropropene and 1,3-Dichloropropene on test separately

Common Name:	DDD (Dichlorodiphenyl dichloroethane) TDE
Trade Names:	Rhothane
Structure:	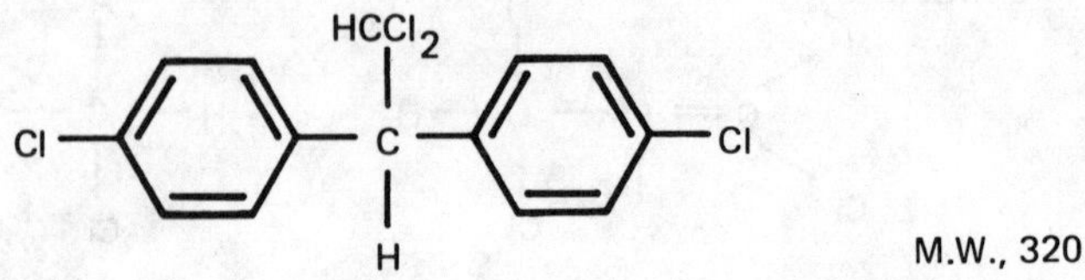
Chemical Name:	1,1-Dichloro-2,2'-bis(*p*-chlorophenyl)ethane (5)
Purity:	90% *p*,*p*' TDE. 10% *o*,*p*'-isomer (5)
Solubility:	Water/practically insoluble Fats/very soluble Most organic solvents/very soluble (5)
Vapor Pressure, torr:	Not available
Use:	Insecticide
U.S. Production, pounds:	Not available
Principal Manufacturer(s):	Two U.S. companies (5)
Market Basket Intake[a]:	1974/0.01 μg/kg body wt/day
EPA Action[aa]:	All uses prohibited in 1972, with a few minor exceptions (8)
Innes et al. Test[b]:	Requires additional evaluation
Secretary's Commission Recommendation[c]:	Eliminate use (Recommendation #3) Carcinogenesis Panel--further testing required
NCI Carcinogenicity Test[d]:	Studies in progress

Common Name:	DDE
Structure:	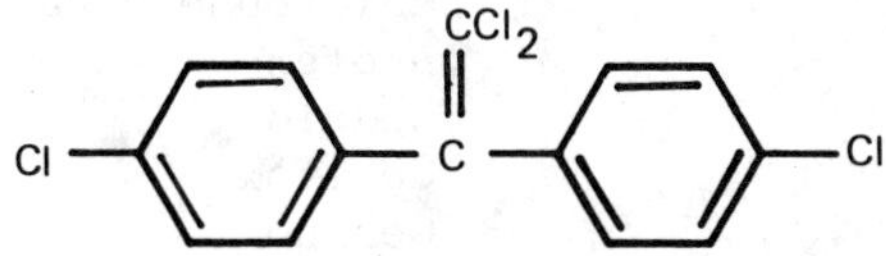 M.W., 318
Chemical Name:	1,1-Dichloro-2,2-bis(*p*-chlorophenyl)ethylene (5)
Solubility:	Water/25°C/0.12 mg/l (5) Fats/soluble (5) Most organic solvents/ soluble (5)
Vapor Pressure, torr:	Not available
Use:	DDT metabolite; not produced as a pesticide
Market Basket Intake[a]:	1973/0.07 μg/kg body wt/day
EPA Action[aa]:	See DDT, DDD
Innes et al. Test[b]:	Not tested; see DDT
Secretary's Commission Recommendation[c]:	See DDT
NCI Carcinogenicity Test[d]:	Studies in progress

Common Name: DDT

Trade Names:

Chlorophenothane	Gyron
Anofex	Ixodex
Dedelo	Kopsol
Genitox	Neocid
Gesapon	Pentachlorin
Gesarex	Rukseam
Gesarol	Zerdane

Structure:

p,p′ –DDT o,p′ –DDT

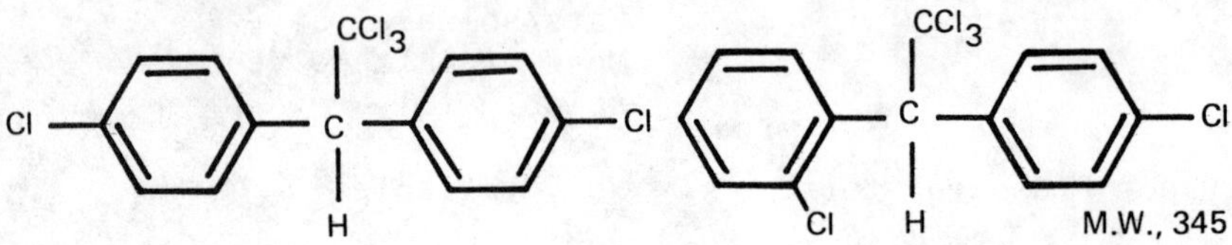

Chemical Name: 1,1,1-Trichloro-2,2-di-(4-chlorophenyl)ethane (*p,p'* DDT)

Purity: Mixture. 65-80% *p,p'*-DDT (5)

Solubility:

p,p'-DDT:	*o,p'*-DDT:
Water/practically insoluble	Water/25°C/ 0.085 mg/l
Fats/very soluble	Fats/soluble
Most organic solvents/very soluble	Most organic solvents/ soluble (5)

Vapor Pressure, torr: *p,p'*-DDT: 20°C/1.0×10^{-7}

Use: Insecticide

U.S. Production, pounds: 1971/44 million (5,6)

Principal Manufacturer(s): Montrose (1,11)

Market Basket Intake[a]: 1973/0.03 μg/kg body wt/day

EPA Action[aa]: All uses prohibited in 1972 with a few minor exceptions

Innes et al. Test[b]: Positive

Secretary's Commission Recommendation[c]: Eliminate use (Recommendation #3)
Carcinogenesis Panel--positive

NCI Carcinogenicity Test[d]: Studies nearing completion

Common Name:	Diazinon
Trade Names:	Basudin, Dazzel, Diazajet, Gardentox, Diazide, Spectracide, Diazol, Sarolex

Structure:

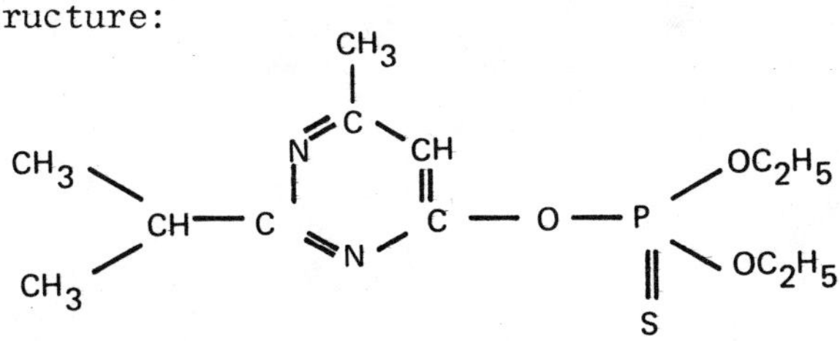

M.W., 304

Chemical Name:	*O,O*-Diethyl *O*-(2-isopropyl-4-methyl-6-pyrimidyl) phosphorothioate
Purity:	85%
Solubility:	Water/20°C/0.004% w/w (4) Ethanol, acetone, xylene/miscible (4) Petroleum oils/soluble (4)
Vapor Pressure, torr:	20°C/1.4×10^{-4} (4)
Use:	Insecticide
U.S. Production, pounds:	1971/10 million (6)
Principal Manufacturer(s):	Ciba-Geigy (1)
Market Basket Intake[a]:	1973/0.02 μg/kg body wt/day
EPA Action[aa]:	No official action to date
Innes et al. Test[b]:	Not tested
Secretary's Commission Recommendation[c]:	None
NCI Carcinogenicity Test[d]:	Not currently being tested

Common Name: Dicofol

Trade Names: Kelthane
Acarin
Mitigan

Structure:

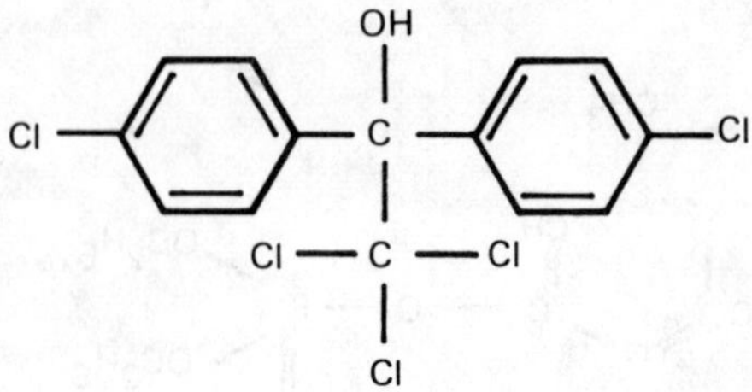

M.W., 371

Chemical Name: 1,1-Bis(*p*-chlorophenyl)-2,2,2-trichloroethanol

Purity: 80% (4)

Solubility: Water/practically insoluble (4)
Most aliphatic and aromatic solvents/soluble (4)

Vapor Pressure, torr: Not available

Use: Acaricide

U.S. Production, pounds: 1971/4 million (6)

Principal Manufacturer(s): Rohm and Haas (1)

Market Basket Intake[a]: 1973/0.01 μg/kg body wt/day

EPA Action[aa]: No official action to date

Innes et al. Test[b]: Not tested

Secretary's Commission Recommendation[c]: None

NCI Carcinogenicity Test[d]: Studies in progress

Common Name:	Dieldrin
Trade Names:	Alvit Octalox Dieldrite Panoram D-31
Structure:	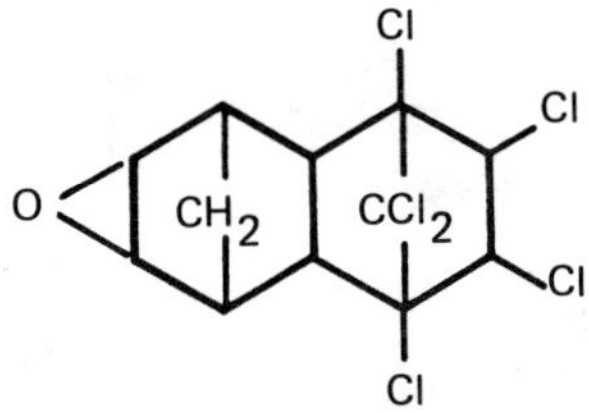 M.W., 381
Chemical Name:	1,2,3,4,10,10-Hexachloro-6,7-epoxy-1,4,4a,5,6,7,8,8a-octahydro-*exo*-1,4-*endo*-5,8-dimethanonaphthalene (5)
Purity:	85% (5)
Solubility:	Water/25-29°C/0.186 mg/l Petroleum oils/slightly soluble Acetone/moderately soluble Aromatic solvents/soluble (5)
Vapor Pressure, torr:	20°C/1.78×10^{-7} (5)
Use:	Insecticide
Principal Manufacturer(s):	Shell
Market Basket Intake[a]:	1973/0.04 μg/kg body wt/day
EPA Action[aa]:	All uses finally cancelled in 1973 Compound prohibited in the United States
Innes et al. Test[b]:	Not tested
Secretary's Commission Recommendation[c]:	Carcinogenesis Panel--positive
NCI Carcinogenicity Test[d]:	Studies in progress

Common Name:	Dithane M-45
Trade Names:	Dithane M-45
Structure:	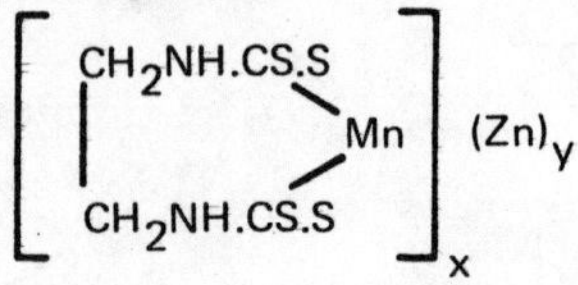
Chemical Name:	A mixture of zinc ethylenebis (dithiocarbamate) and manganese ethylenebis(dithiocarbamate)
Purity:	80% (2)
Solubility:	Water/practically insoluble Most organic solvents/practically insoluble (4)
Vapor Pressure, torr:	Not available
Use:	Fungicide
U.S. Production, pounds:	Not available
Principal Manufacturer(s):	Rohm and Haas (1)
Market Basket Intake[a]:	Not evaluated
EPA Action[aa]:	No official action to date
Innes et al. Test[b]:	None
Secretary's Commission Recommendation[c]:	None
NCI Carcinogenicity Test[d]:	Not currently being tested

Common Name: Dithiocarbamates

Trade Names: See Dithane M-45, Ferbam, Maneb (Dithane M-22), Polyram, Zineb

Structure:

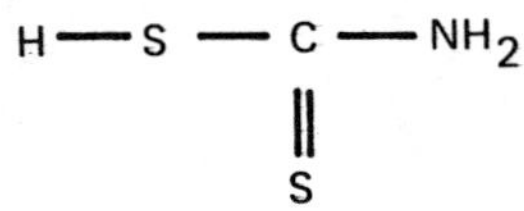

Chemical Name: Organic sulfur fungicides. Included are the metallic dimethyldithiocarbamates (such as ferric, manganous, and zinc) and ethylene bisdithiocarbamate salts (including disodium, diammonium, potassium ammonium, zinc, and manganese). Derivatives of dithiocarbamic acid. (1)

Use: Fungicide

U.S. Production, pounds: 1971/40 million (6)

Principal Manufacturer(s):
duPont
FMC
Pennwalt
Roberts Chemical Div. of Security Chemicals
Rohm and Haas (1)

Innes et al. Test[b]:
Selenium diethyldithiocarbamate--positive
Bis(2-hydroxyethyl)dithiocarbamic acid, potassium salt--positive

NCI Carcinogenicity Test[d]: Studies on lead salt in progress

Common Name: DSMA

Trade Names: Ansar DSMA Liquid, Ansar 184, Ansar 8100, Arrhenal, Arsinyl, Crab-E-Rad, Dal-E-Rad 100, Di-Tac, DMA 100, Methar, Sodar, Weed-E-Rad, Weed-Hoe

Structure:

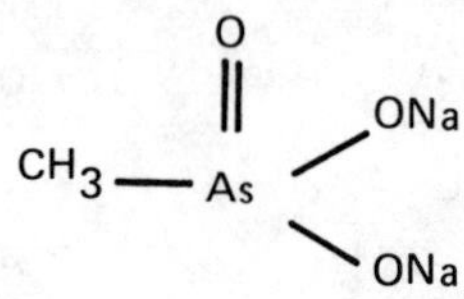

M.W., 184

Chemical Name: Disodium methanearsonate

Purity: 9% (2)

Solubility: Water/25°C/25.4% w/w
Methanol/soluble
Other organic solvents/practically insoluble (2)

Vapor Pressure, torr: Too low to measure (2)

Use: Herbicide

U.S. Production, pounds: MSMA/DSMA: 1971/35 million (6

Principal Manufacturer(s): Ansul (1)
W. A. Cleary (1)
Diamond Shamrock (1)
Vineland (1)

Market Basket Intake[a]: Not evaluated

EPA Action[aa]: No official action to date

Innes et al. Test[b]: Not tested

Secretary's Commission Recommendation[c]: See MAA

NCI Carcinogenicity Test[d]: Not currently being tested

Common Name: Endosulfan

Trade Names:

Chlorthiepin	Kop-Thiodan	Thimul
Cyclodan	Malix	Thiodan
Insectophene	Thifor	Thionex

Structure:

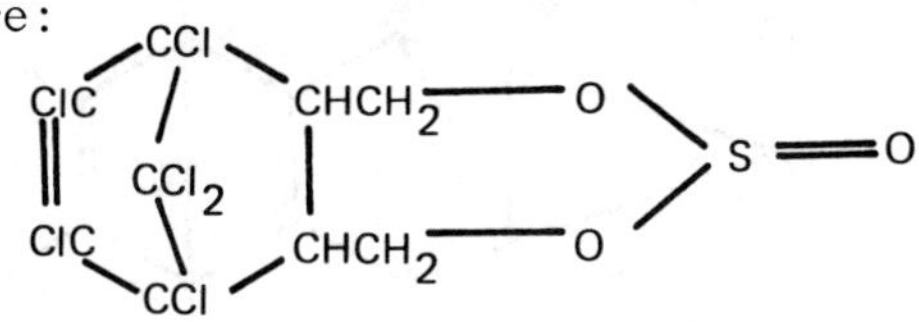

M.W., 407

Chemical Name: 6,7,8,9,10,10-Hexachloro-1,5,5a,6,9,9a-hexahydro-6,9-methano-2,4,3-benzodioxathiepin-3-oxide

Purity: 96% (2)

Solubility: Water/practically insoluble
Most organic solvents/moderately soluble (4)

Vapor Pressure, torr: 75°C/no measurable vapor pressure (4)

Use: Insecticide

U.S. Production, pounds: 1971/2 million (6)

Principal Manufacturer(s): Hooker (11)
FMC (1)

Market Basket Intake[a]: Thiodan I plus Thiodan II: 1973/0.02 μg/kg body wt/day

EPA Action[aa]: No official action to date

Innes et al. Test[b]: Negative

Secretary's Commission Recommendation[c]: Carcinogenesis Panel--further testing required

NCI Carcinogenicity Test[d]: Studies nearing completion

Common Name: Endrin

Trade Names: Hexadrin

Structure:

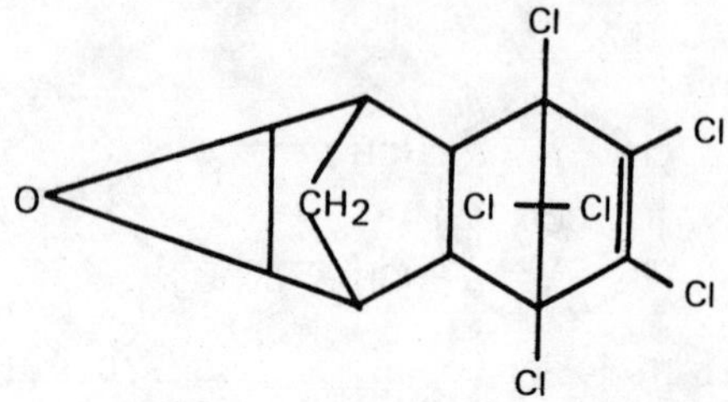

Chemical Name:	1,2,3,4,10,10-Hexachloro-6,7-eposy-1,4,4a,5,6,7,8,8a-octahydro-*endo*-*endo*-1,4,5,8-dimethanonaphthaline
Purity:	95% (5)
Solubility:	Water/practically insoluble (5 Acetone, benzene, carbon tetra chloride, hexane and xylene/ soluble (5)
Vapor Pressure, torr:	25°C/2 × 10^{-7} (5)
Use:	Insecticide
U.S. Production, pounds:	1971/<1 million (6)
Principal Manufacturer(s):	Shell (1) Velsicol (1)
Market Basket Intake[a]:	1973/<0.01 μg/kg body wt/day
EPA Action[aa]:	No official action to date
Innes et al. Test[b]:	Not tested
Secretary's Commission Recommendation[c]:	Restrict use (Recommendation #
NCI Carcinogenicity Test[d]:	Studies in progress

Common Name:	Ethion
Trade Names:	Ethodan Kwit Nialate

Structure:

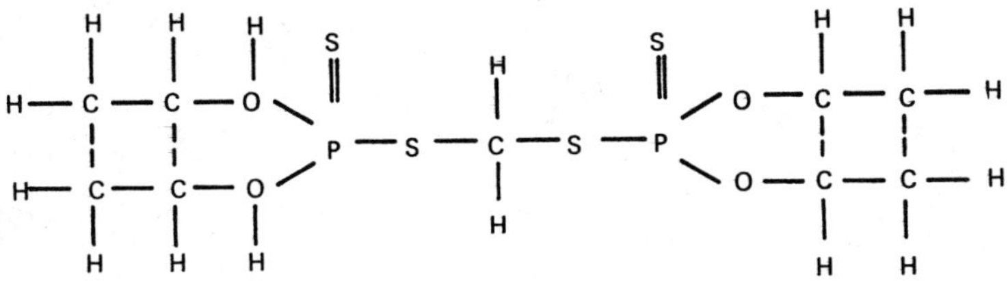

Chemical Name:	*O,O,O',O'*-tetraethyl *S,S'*-methylene bisphosphorodithioate
Purity:	95% (2)
Solubility:	Water/slightly soluble Most organic solvents including kerosene and petroleum oils/soluble (4)
Vapor Pressure, torr:	25°C/1.5×10^{-6} (4)
Use:	Insecticide and acaricide
U.S. Production, pounds:	1971/3 million (6)
Principal Manufacturer(s):	FMC (1)
Market Basket Intake[a]:	1973/0.01 μg/kg body wt/day
EPA Action[aa]:	No official action to date
Innes et al. Test[b]:	Not tested
Secretary's Commission Recommendation:	None
NCI Carcinogenicity Test[d]:	Not currently being tested

Common Name: Ferbam

Trade Names: Fermate, Ferberk, Hexaferb, Trifungol, Vancide FE-95

Structure:

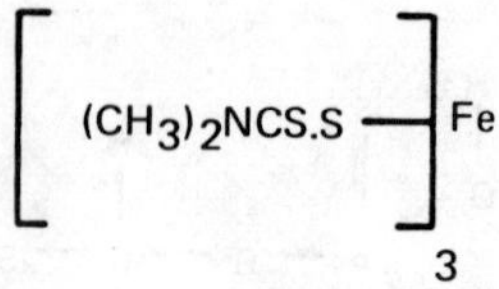

M.W., 416

Chemical Name: Ferric dimethyldithiocarbamate

Purity: Not available

Solubility: Some organic solvents such as acetonitrile, chloroform, and pyridine/soluble (4)

Vapor Pressure, torr: 25°C/negligible

Use: Fungicide

U.S. Production, pounds: Not available

Principal Manufacturer(s): FMC (1)

Market Basket Intake[a]: Not evaluated

EPA Action[aa]: No official action to date

Innes et al. Test[b]: Negative

Secretary's Commission Recommendation[c]: Carcinogenesis Panel--further testing required

NCI Carcinogenicity Test[d]: Not currently being tested

Common Name:	Heptachlor
Trade Names:	Drinox H-34 Heptamul
Structure:	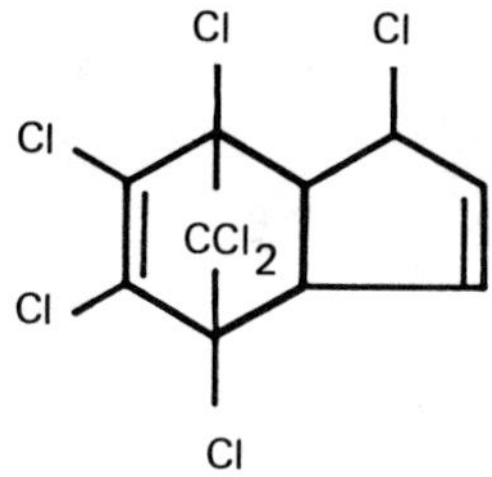 M.W., 374
Chemical Name:	1,4,5,6,7,8,8-Heptachloro-3a, 4,7,7a-tetrahydro-4,7-methano-indene
Purity:	72% (5)
Solubility:	Water/25-29°C/0.056 ml/l (5)
Vapor Pressure, torr:	25°C/3 × 10^{-4} (5)
Use:	Insecticide
U.S. Production, pounds:	1971/6 million (6)
Principal Manufacturer(s):	Velsicol (1)
Market Basket Intake[a]:	Heptachlor epoxide: 0.01 μg/kg body wt/day
EPA Action[aa]:	Notice of intent to cancel issued Nov. 1974 Notice of intent to suspend issued Aug. 1975
Innes et at. Test[b]:	Not tested
Secretary's Commission Recommendation[c]:	Restrict use; minimize human exposure (Recommendations #4, #5) Carcinogenesis Panel--positive (heptachlor epoxide)
NCI Carcinogenicity Test[d]:	Studies nearing completion

Common Name:	Hexachlorobenzene (HCB)
Trade Names:	Anticarie No Bunt Perchlorobenzene Bunt-cure Bunt-no-more
Structure:	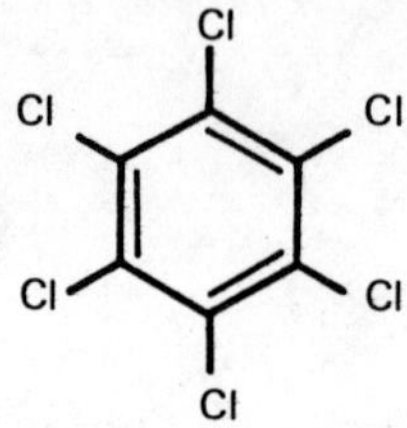 M.W., 285
Chemical Name:	Hexachlorobenzene
Purity:	Not available
Solubility:	Water/practically insoluble (4) Hot benzene/soluble (4)
Vapor Pressure, torr:	20°C/1.089 × 10^{-5} (4)
Use:	Fungicide
U.S. Production, pounds:	700,000 (2)
Principal Manufacturer(s):	Dover Stauffer (2)
Market Basket Intake[a]:	1973/0.01 μg/kg body wt/day
EPA Action[aa]:	No official action to date
Innes et al. Test[b]:	Not tested
Secretary's Commission Recommendation[c]:	None
NCI Carcinogenicity Test[d]:	Not currently being tested

Common Name:	Lindane	
Trade Names:	Gamaphex Gammalin Gammex Gammexane Lindafor	Lindagam Lintox Novigam Silvanol

Structure:

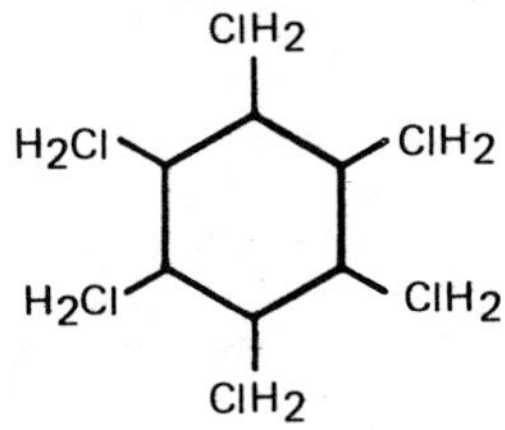

M.W., 291

Chemical Name:	γ-1,2,3,4,5,6-Hexachlorocyclohexane
Purity:	99% γ-isomer
Solubility:	Water 25°C/about 10 mg/l Chloroform, ethanol, acetone, ether, and benzene/very soluble (5)
Vapor Pressure, torr:	40°C/0.14 (5)
Use:	Insecticide
U.S. Production, pounds:	1971/<1 million (6)
Principal Manufacturer(s):	Hooker (4) Diamond Shamrock
Market Basket Intake[a]:	1973/<0.01 μg/kg body wt/day
EPA Action[aa]:	No official action to date
Innes et al. Test[b]:	Not tested
Secretary's Commission Recommendation[c]:	Restrict use (Recommendation #4)
NCI Carcinogenicity Test[d]:	Studies nearing completion

Common Name: MAA

Trade Names: Ansar
Trans-Vert

Structure:

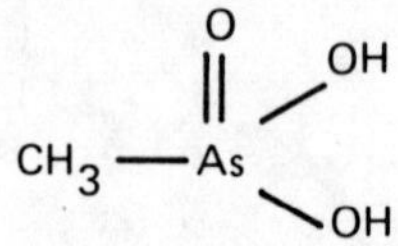

M.W., 140

Chemical Name: Methanearsonic acid

Purity: Not available

Solubility: Water/20°C/25.6% w/w (3)

Vapor Pressure, torr: Not available

Use: Herbicide

U.S. Production, pounds: MSMA/DSMA: 1971/35 million (6

Principal Manufacturer(s): Ansul
Amchem (3)

Market Basket Intake[a]: Not evaluated

EPA Action[aa]: Not a registered pesticide

Innes et al. Test[b]: Not tested

Secretary's Commission Recommendation[c]: Restrict use of pesticides containing arsenic (Recommendation #4)

NCI Carcinogenicity Test[d]: Not currently being tested

Common Name: Malathion

Trade Names:

Malathion	Kop-Thion
Cythion	Kypfos
Emmatos	Malaspray
Emmatos Extra	Malamar
Fyfanon	Zithiol
Karbofos	

Structure:

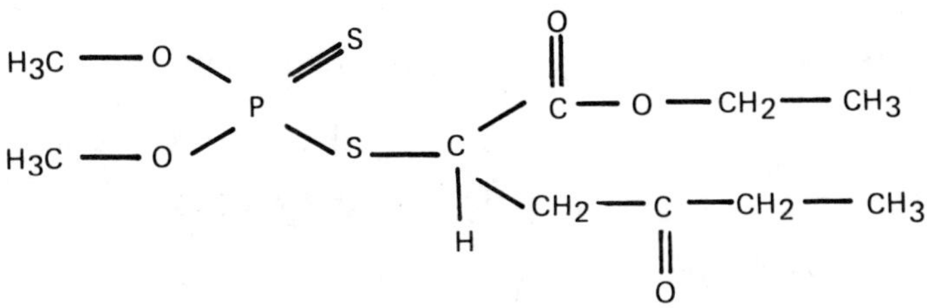

Chemical Name: *S*-(1,2-dicarbethoxyethyl) *O*,*O*-dimethyl dithiophosphate

Purity: 95% (4)

Solubility: Water/25°C/145 ppm (4)
Many organic solvents/miscible (4)
Petroleum oils/limited solubility (4)

Vapor Pressure, torr: 30°C/4 × 10^{-5}

Use: Insecticide and acaricide (4)

U.S. Production, pounds: 1971/35 million (6)

Principal Manufacturer(s): American Cyanamid (1)

Market Basket Intake[a]: 1973/0.16 μg/kg body wt/day

EPA Action[aa]: No official action to date

Innes et al. Test[b]: Not tested

Secretary's Commission Recommendation[c]: None
Subcommittee suggests should be tested for mutagenicity

NCI Carcinogenicity Test[d]: Studies nearing completion

Common Name:	Maneb	
Trade Names:	Chloroble M	Vancide
	Dithane M-22	Manesan
	Dithane M-22 Special	Manzate
		Manzate D
	Kypman 80	Sopranebe
	Maneba	Sup'R Flo
	Manebgan	Trimangol

Structure:

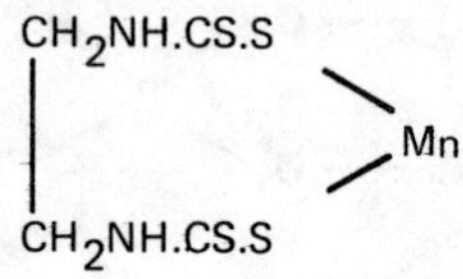

Chemical Name:	Manganese ethylenebisdithiocarbamate
Purity:	92% (2)
Solubility:	Water/slightly soluble Most organic solvents/insoluble (4)
Vapor Pressure, torr:	Not available
Use:	Fungicide
U.S. Production, pounds:	See Dithiocarbamates
Principal Manufacturer(s):	duPont (1) Rohm and Haas (1) BASE Wyandotte (1)
Market Basket Intake[a]:	Not evaluated
EPA action[aa]:	No official action to date
Innes et al. Test[b]:	Negative
Secretary's Commission Recommendation[c]:	Carcinogenesis Panel--further testing required
NCI Carcinogenicity Test[d]:	Not currently being tested

Common Name:	Methoxychlor
Trade Names:	Marlate Moxie
Structure:	

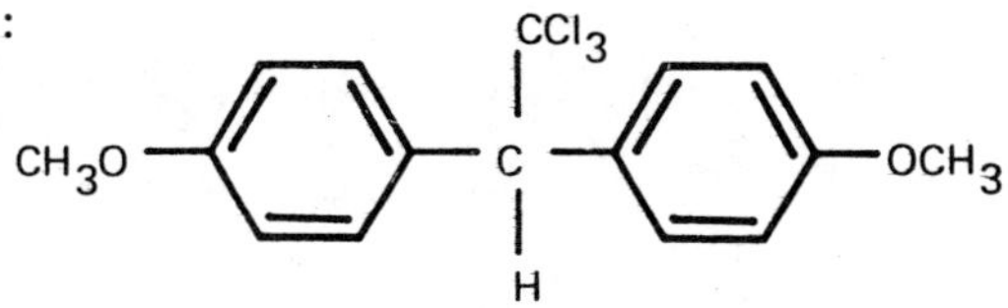

M.W., 346

Chemical Name:	2,2-Bis(*p*-methoxyphenyl)-1,1,1-trichloroethane (88%) and related compounds
Purity:	See Chemical Name above (2)
Solubility:	Water/practically insoluble Ethanol and petroleum oils/moderately soluble Most aromatic solvents/readily soluble (5)
Vapor Pressure, torr:	Not available
Use:	Insecticide
U.S. Production, pounds:	1971/9.9 million (6)
Principal Manufacturer(s):	Ansul Chemical Formulators duPont Prentiss
Market Basket Intake[a]:	1973/<0.01 μg/kg body wt/day
EPA Action[aa]:	No official action to date
Innes et al. Test[b]:	Not tested
Secretary's Commission Recommendation[c]:	Carcinogenesis Panel--additional testing needed
NCI Carcinogenicity Test[d]:	Studies nearing completion

Common Name:	Methyl bromide
Trade Names:	Brom-O-Gas, Fumigant-I, Meth-O-Gas, Pestmaster, Profume
Structure:	CH_3Br M.W., 95
Chemical Name:	Monobromomethane
Purity:	100% (2)
Solubility:	Water/20°C/17.5 g/l
Use:	Soil fumigant Insecticidal and acaricidal properties
U.S. Production, pounds:	1971/22 million (6)
Principal Manufacturer(s):	Dow, Great Lakes, Kerr-McGee, Michigan (1)
Market Basket Intake[a]:	Not evaluated
EPA Action[aa]:	No official action to date
Innes et al. Test[b]:	Not tested
Secretary's Commission Recommendation[c]:	None
NCI Carcinogenicity Test[d]:	Not currently being tested

Common Name:	Methyl parathion
Trade Names:	Dalf; Folidol M; Metron; Nitrox 80; Partron M; Tekwaisa

Structure:

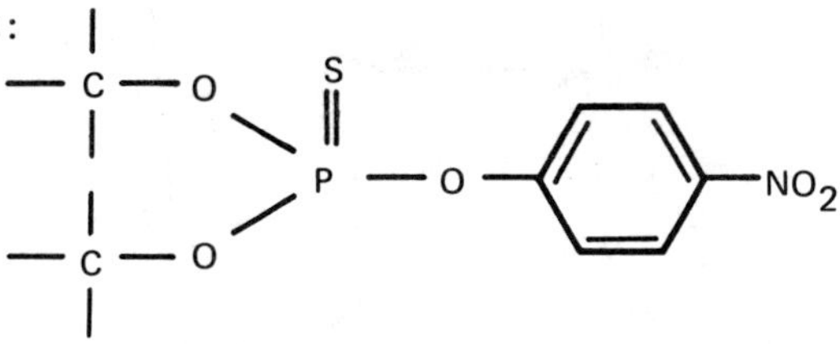

M.W., 263

Chemical Name:	*O*,*O*-Dimethyl *O*-(4-nitrophenyl) phosphorothioate
Purity:	80% (4)
Solubility:	Water/25°C/55-60 ppm (4) Light petroleum and mineral oils/slightly soluble (4) Most other organic solvents/soluble (4)
Vapor Pressure, torr:	20°C/0.97 × 10^{-5}
Use:	Insecticide with some fumigant action (4)
U.S. Production, pounds:	1971/45 million (6)
Principal Manufacturer(s):	Kerr-McGee Monsanto Stauffer (1)
Market Basket Intake[a]:	1973/<0.01 μg/kg body wt/day
EPA Action[aa]:	No official action to date
Innes et al. Test[b]:	Not tested
Secretary's Commission Recommendation[c]:	None
NCI Carcinogenicity Test[d]:	Studies in progress

Common Name:	Mirex
Trade Names:	Dechlorane
Structure:	

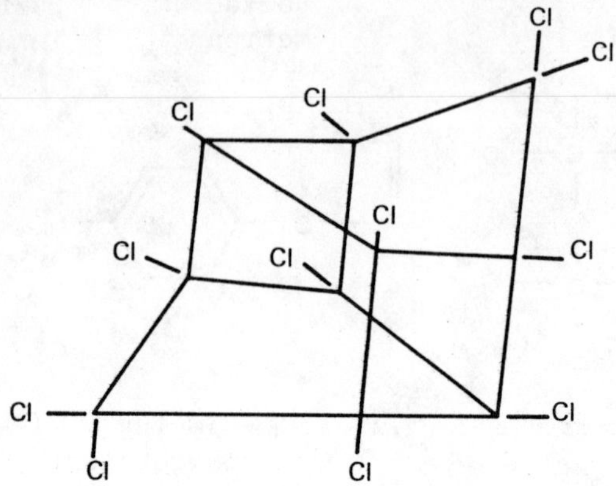

M.W., 546

Chemical Name:	Dodecachlorooctahydro-1,3,4-methano-1H-cyclobuta[cd]pentalene
Purity:	Not Available
Solubility:	Water/practically insoluble (5 Benzene, carbon tetrachloride, dioxane, methyl ethyl ketones, and xylene/soluble (5)
Vapor Pressure, torr:	Not available
Use:	Insecticide
U.S. Production, pounds:	Not available
Principal Manufacturer(s):	Allied (1)
Market Basket Intake[a]:	None detected
EPA Action[aa]:	Hearing to determine whether mirex registrations should be restricted or cancelled is currently in progress.
Innes et al. Test[b]:	Positive
Secretary's Commission Recommendation[c]:	Carcinogenesis Panel--positive
NCI Carcinogenicity Test[d]:	Studies in progress

Common Name:	MSMA
Trade Names:	Trans-Vert Ansar 170 Bueno Daconate Ansar 529 Ansar 529 H.C. Phyban H.C. Silvisar 550 Weed-E-Rad Weed-Hoe
Structure:	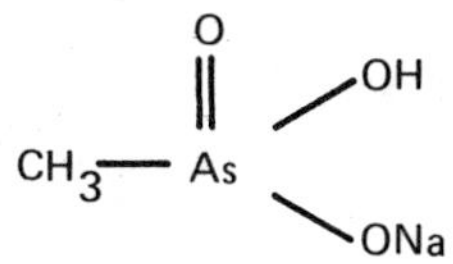
Chemical Name:	Monosodium methanearsenate
Purity:	99% (2)
Solubility:	Water/25°C/28% w/w (2) Methanol/25°C/14% w/w (2) Most other organic solvents/ insoluble (2)
Vapor Pressure, torr:	Too low to measure
Use:	Herbicide
U.S. Production, pounds:	MSMA/DSMA: 1971/35 million (6)
Principal Manufacturer(s):	Ansul (2)(1) Amchem (2) Diamond Shamrock (1) Vineland (1)
Market Basket Intake[a]:	Not evaluated
EPA Action[aa]:	No official action to date
Innes et al. Test[b]:	Not tested
Secretary's Commission Recommendation[c]:	See MAA
NCI Carcinogenicity Test[d]:	Not currently being tested

Common Name:	Para-dichlorobenzene
Trade Names:	Paracide, Santochlor, Dichloride, Paramoth
Structure:	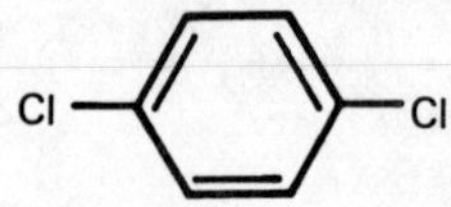 M.W., 147
Chemical Name:	1,4-Dichlorobenzene
Purity:	100% (2)
Solubility:	Water/25°C/80 ppm (4) Organic solvents/readily soluble (4)
Vapor Pressure, torr:	25°C/1.0 (4)
Use:	Insecticidal fumigant
U.S. Production, pounds:	1971/60 million (6)
Principal Manufacturer(s):	Allied (1), Monsanto (1), Dow (1), PPG Industries (1)
Market Basket Intake[a]:	Not evaluated
EPA Action[aa]:	No official action to date
Innes et al. Test[b]:	Not tested
Secretary's Commission Recommendation[c]:	None
NCI Carcinogenicity Test[d]:	Suggested for test

Common Name: Parathion

Trade Names:

Alkron	Phoskil	Panthion
Alleron	Rhodiatox	Parathene
Bladan	Etilon	Soprathion
Corothion	Folidol	Stathion
E-605	Niran	Thiophos
Parawet	Orthophos	

Structure:

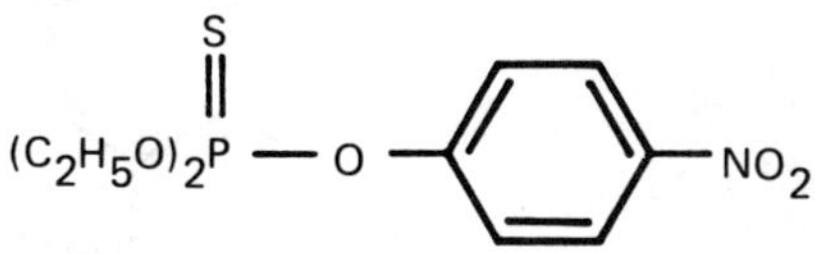

M.W., 291

Chemical Name: *O*,*O*-Diethyl *O*-*p*-nitrophenyl phosphorothioate

Purity: 95% (2)

Solubility: Water/25°C/24 ppm (4)
Petroleum oils/slightly soluble (4)
Most organic solvents/miscible (4)

Vapor Pressure, torr: 20°C/3.78×10^{-5}

Use: Insecticide and acaricide

U.S. Production, pounds: 1971/15 million (6)

Principal Manufacturer(s): Monsanto (1)

Market Basket Intake[a]: 1973/<0.01 μg/kg body wt/day

EPA Action[aa]: No official action to date

Innes et al. Test[b]: Not tested

Secretary's Commission Recommendation[c]: None

NCI Carcinogenicity Test[d]: Studies nearing completion

Common Name: PCNB (Pentachloronitrobenzene)

Trade Names:

Avicol	Folosan	Tilcarex
Botrilex	Kobutol	Tri-PCNB
Brassicol	Terraclor	Tritisan

Structure:

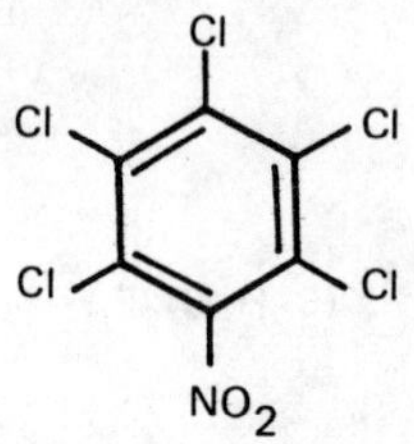

M.W., 296

Chemical Name: Pentachloronitrobenzene

Purity: 98.5% (2)

Solubility: Water/practically insoluble (4)
Ethanol/25°C/2% (4)
Benzene, carbon disulphide, and chloroform/soluble (4)

Vapor Pressure, torr: $25°C/1.33 \times 10^{-4}$ (4)

Use: Fungicide

U.S. Production, pounds: 1971/3 million (6)

Principal Manufacturer(s): Olin (1)

Market Basket Intake[a]: 1973/<0.01 μg/kg body wt/day

EPA Action[aa]: No official action to date

Innes et al. Test[b]: Positive

Secretary's Commission Recommendation[c]: Minimize human exposure (Recommendation #5)
Carcinogenesis Panel--positive
Teratogenesis Panel--positive

NCI Carcinogenicity Test[d]: Studies nearing completion

Common Name:	PCP (Pentachlorophenol)
Trade Names:	Dowicide 7 Sinituho Weedone
Structure:	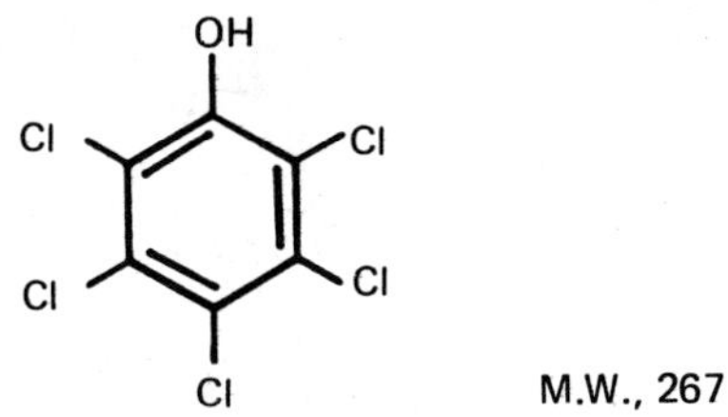
Chemical Name:	Pentachlorophenol
Purity:	86% (2) (contains polychlorodioxins)
Solubility:	Water/30°C/20 ppm (4) Most organic solvents/soluble (4) CCl_4 and paraffinic petroleum oils/limited solubility (4)
Vapor Pressure, torr:	100°C/0.12
Use:	Insecticide, fungicide, defoliant, herbicide
U.S. Production, pounds:	1971/46 million (6)
Principal Manufacturer(s):	Dow Monsanto Reichhold Sonford Vulcan (1)
Market Basket Intake[a]:	1973/0.01 μg/kg body wt/day
EPA Action[aa]:	No official action to date
Innes et al. Test[b]:	Not tested
Secretary's Commission Recommendation[c]:	None
NCI Carcinogenicity Test[d]:	Not currently being tested

Common Name. Perthane

Trade Names: Perthane

Structure:

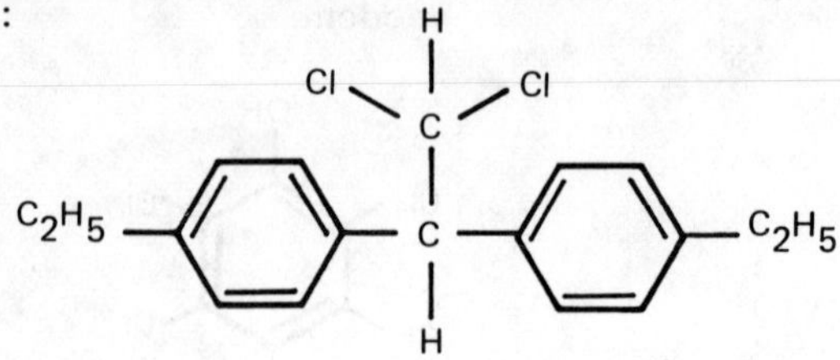

M.W., 307

Chemical Name: 1,1-Dichloro-2,2-bis(4-ethylphenyl)ethane

Purity: 88% (1)

Solubility: Water/practically insoluble
Most aromatic solvents and methylene chloride/soluble (4)

Vapor Pressure, torr: Not available

Use: Insecticide

U.S. Production, pounds: Not available

Principal Manufacturer(s): Rohm and Haas (1)

Market Basket Intake[a]: 1973/0.05 μg/kg body wt/day

EPA Action[aa]: No official action to date

Innes et al. Test[b]: Requires additional tests

Secretary's Commission Recommendation[c]: Carcinogenesis Panel--additional testing needed

NCI Carcinogenicity Test[d]: Studies in progress

Common Name:	Polyram
Trade Names:	Polyram Combi Zinc Metiram
Chemical Name:	A mixture of ethylenebis(dithio-carbamate) zinc and ethylene-bis(thiuramdisulfide)
Purity:	80% (2)
Solubility:	Water/practically insoluble Organic solvents such as acetone, benezene, or ethanol/practically insoluble Pyridine/soluble, with decomposition
Vapor Pressure, torr:	Not available
Use:	Fungicide
U.S. Production, pounds:	See Dithiocarbamates
Principal Manufacturer(s):	FMC (4)
Market Basket Intake[a]:	Not evaluated
EPA Action[aa]:	No official action to date
Innes et al. Test[b]:	Not tested
Secretary's Commission Recommendation[c]:	None
NCI Carcinogenicity Test[d]:	Not currently being tested

Common Name:	Propachlor
Trade Names:	Ramrod
Structure:	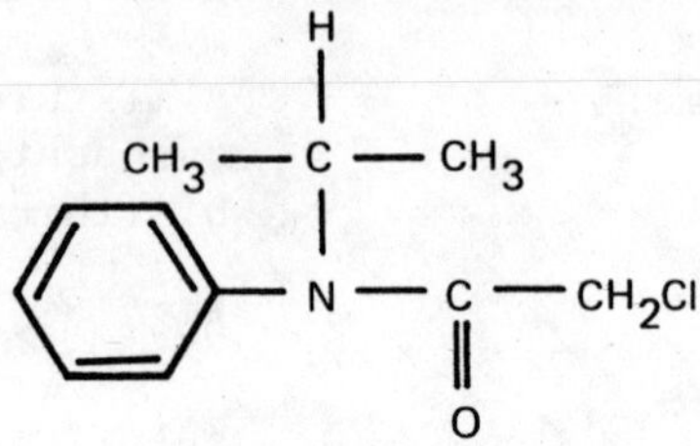
	M.W., 212
Chemical Name:	2-Chloro-*N*-isopropylacetanilide
Purity:	100%
Solubility:	Water/20°C/700 ppm Most organic solvents except aliphatic hydrocarbons/soluble (4)
Vapor Pressure, torr:	20°C/9.4 × 10^{-3} (3)
Use:	Herbicide
U.S. Production, pounds:	1971/23 million (6)
Principal Manufacturer(s):	Monsanto (1)
Market Basket Intake[a]:	Not evaluated
EPA Action[aa]:	No official action to date
Innes et al. Test[b]:	Not tested
Secretary's Commission Recommendation[c]:	None
NCI Carcinogenicity Test[d]:	Not currently being tested

Common Name:	Sodium Chlorate
Trade Names:	Atlacide, Atratol, De-Fol-Ate, Drop Leaf, Fall, Rasikal, Shed-A-Leaf, Tumbleaf
Structure:	$NaClO_3$ M.W. 107
Chemical Name:	Sodium chlorate
Purity:	99%
Solubility:	Water/100°C/230 g/100 ml (4) Ethanol and glycerol/soluble (4)
Vapor Pressure, torr:	Not available
Use:	Herbicide
U.S. Production, pounds:	1971/30 million
Principal Manufacturer(s):	Rhodia (3) Kerr-McGee (1) Occidental (1)
Market Basket Intake[a]:	Not evaluated
EPA Action[aa]:	No official action to date
Innes et al. Test[b]:	Not tested
Secretary's Commission Recommendation[c]:	None
NCI Carcinogenicity Test[d]:	Not being tested

Common Name:	Strobane
Trade Names:	Strobane
Structure:	Complex mixture (5)
Chemical Name:	Mixture of chlorinated terpene isomers (5)
Purity:	Mixture
Solubility:	Water/practically insoluble Aliphatic and aromatic hydrocarbons/miscible 95% alcohol/partially soluble (1)
Vapor Pressure, torr:	20°C/3 × 10^{-7} (2)
Use:	Insecticide
U.S. Production, pounds:	Strobane + Toxaphene = 1971/50 million (6)
Principal Manufacturer(s):	Tenneco (1)
EPA Action[aa]:	No official action to date
Innes et al. Test[b]:	Positive
Secretary's Commission Recommendation[c]:	Minimize human exposure (Recommendation #5) Carcinogenesis Panel--positive
NCI Carcinogenicity Test[d]:	See Toxaphene

Common Name:	2,4,5-T
Trade Names:	Dacamine, Ded-Weed, Esteron 245, Fence Rider, Forron, Fruitone A, Inverton 245, Line Rider, Reddon, Tormona, Tributon, Trioxone, Weedar, Weedone, Brush-Rhap

Structure:

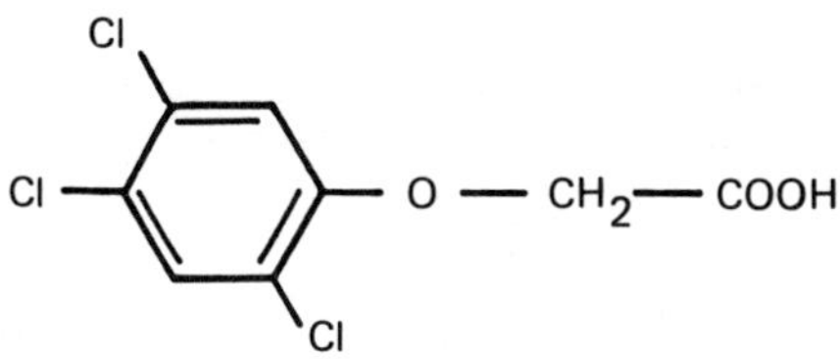

M.W., 256

Chemical Name:	(2,4,5-Trichlorophenoxy)acetic acid
Purity:	May contain traces of the highly toxic compound 2,3,7,8-tetrachlorodibenzo-*p*-dioxin as an impurity
Solubility:	Water/30°C/238 g/100 ml (3) Ethyl or isopropyl alcohol/50°C/590 g/100 ml (3) Salts: Water/soluble; petroleum oils/insoluble (3) Esters: Water/insoluble; petroleum oils/soluble (3)
Vapor Pressure, torr:	Low (3)
Use:	Herbicide
U.S. Production, pounds:	1970/12.3 million* (10)
Principal Manufacturer(s):	Dow, Transvaal, Rhodia, Thompson-Hayward
Market Basket Intake[a]:	Not evaluated

*Figure includes 2,4,5-T acid, esters, and salts.

EPA Action[aa]:	Cancellation order lifted pending outcome of special monitoring project
Innes et al. Test[b]:	Negative
Secretary's Commission Recommendation[c]:	Teratogenesis Panel--suspected teratogen
NCI Carcinogenicity Test[d]:	Not currently being tested

Common Name:	Toxaphene
Trade Names:	Chlor Chem T-590 Phenatox Motox Strobane-T Phenacide Toxakil
Structure:	Complex mixture of mono- and sesqui-terpenes (2)
Chemical Name:	Terpene polychlorinates
Purity:	Mixture
Solubility:	Water/25°C/3 ppm Organic solvents including petroleum oils/readily soluble (4)
Vapor Pressure, torr:	25°C/0.2-0.4 (4)
Use:	Insecticide
U.S. Production, pounds:	Strobane + Toxaphene: 1971/ 50 million (6)
Principal Manufacturer(s):	Hercules Sonford Tenneco (1)
EPA Action[aa]:	No official action to date
Innes et al. Test[b]:	Not tested
Secretary's Commission Recommendation[c]:	Chemical properties of toxaphene suggest that it should receive close surveillance (Recommendation #4)
NCI Carcinogenicity Test[d]:	Studies nearing completion

Common Name: Trichlorophenol

Trade Names: Dowicides 2 and 2S

Structure:

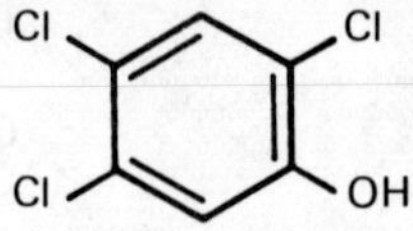

M.W., 197

Chemical Name: 2,4,5-Trichlorophenol

Purity: 95% (May contain traces of the highly toxic compound 2,3,7,8-tetrachlorodibenzo-*p*-dioxin as an impurity) (2)

Solubility: Water/25°C/<0.2 g/100 ml

Vapor Pressure, torr: Not available

Use: Fungicide

U.S. Production, pounds: 1971/20 million

Principal Manufacturer(s): Dow (1)

Market Basket Intake[a]: Not evaluated

EPA Action[aa]: No official action to date

Innes et al. Test[b]: Requires additional evaluation

Secretary's Commission Recommendation[c]: None

NCI Carcinogenicity Test[d]: Not currently being tested

Common Name:	Trifluralin
Trade Names:	Treflan Triflurex
Structure:	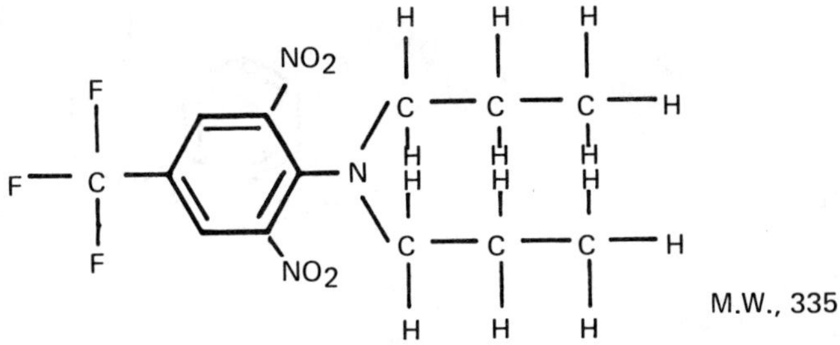
Chemical Name:	α,α,α-Trifluoro-2,6-dinitro-*N*,*N*-dipropyl-*p*-toluidine
Purity:	Not available
Solubility:	Water/27°C/<1 ppm (1) Organic solvents such as acetone, xylene, and aromatic naphthas/ readily soluble (1)
Vapor Pressure, torr:	29.5°C/1.99 × 10^{-4} (4)
Use:	Herbicide
U.S. Production, pounds:	1971/25 million
Principal Manufacturer(s):	Elanco (1)
Market Basket Intake[a]:	None detected
EPA Action[aa]:	No official action to date
Innes et al. Test[b]:	Not tested
Secretary's Commission Recommendation[c]:	None
NCI Carcinogenicity Test[d]:	Studies nearing completion

Common Name: Warfarin

Trade Names: Kypfarin

Structure:

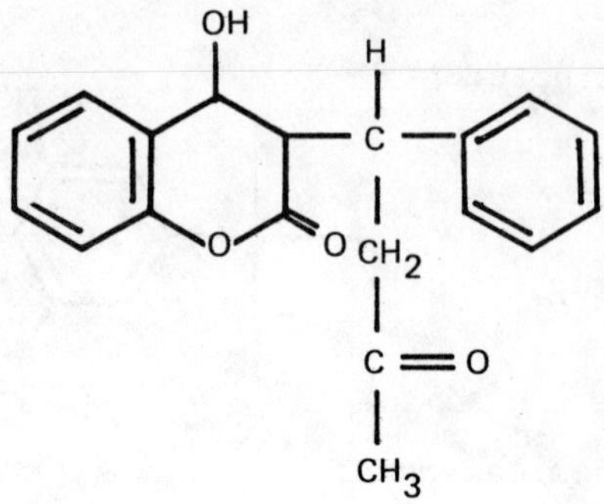

Chemical Name: 3-(α-Acetonylbenzyl)-4-hydroxycoumarin

Purity: 98%

Solubility: Water, benzene/practically insoluble
Alcohols/moderately soluble
Acetone, dioxane/readily soluble
Sodium salt: Water/soluble; organic solvents/insoluble (4)

Vapor Pressure, torr: Not available

Use: Rodenticide

U.S. Production, pounds: 1971/12 million (6)

Principal Manufacturer(s): S. B. Penick
Prentiss
Velsicol (1)

Market Basket Intake[a]: Not evaluated

EPA Action[aa]: No official action to date

Innes et al. Test[b]: Not tested

Secretary's Commission Recommendation[c]: None

NCI Carcinogenicity Test[d]: Not currently being tested

Common Name: Zineb

Trade Names:

Aspor	Parzate C
Chem Zineb	Polyram Z
Dithane Z-78	Tiezene
Hexathane	Tritoftorol
Kypzin	Zebtox
Lonacol	Zidan
Pamosol 2 Forte	Zinosan

Structure:

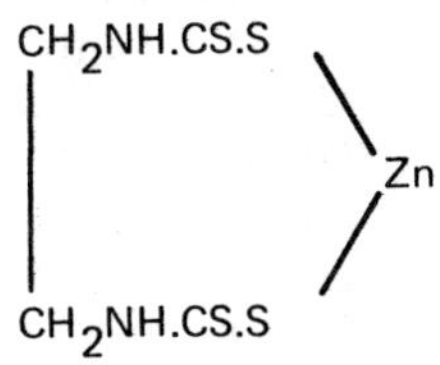

Chemical Name: Zinc ethylenebisdithiocarbamate

Purity: 90% (2)

Solubility: Water/25°C/about 10 ppm (4)
Pyridine/soluble (4)

Vapor Pressure, torr: Not available

Use: Fungicide

U.S. Production, pounds: Not available

Principal Manufacturer(s): duPont (1)
BASF Wyandotte (1)
FMC (1)

Market Basket Intake[a]: Not evaluated

EPA Action[aa]: No official action to date

Innes et al. Test[b]: Negative

Secretary's Commission Recommendation[c]: Carcinogenesis Panel--further testing required

NCI Carcinogenicity Test[d]: Not currently being tested

KEY TO APPENDIX A

Footnotes

[a]Market Basket Intake refers to dietary intake data for certain pesticide residues as derived from the total diet studies conducted by the Food and Drug Administration. Source: Office of Associate Commissioner for Compliance, USDHEW, Rockville, Maryland 20852.

[aa]EPA Action refers to regulatory status of certain pesticides within the U.S. Environmental Protection Agency under the provisions of FIFRA. Source: Office of Enforcement and General Counsel, EPA.

[b]Innes et al. Test refers to data presented in Innes, Ulland, Valerio, Petrucelli, Fishbein, Hart, Pallotta, Bates, Falk, Gart, Klein, Mitchell, and Peters (1969) Bioassay of pesticides and industrial chemicals for tumorgenicity in mice: A preliminary note. J. Nat. Cancer Inst. 42:1101-1114.

[c]Secretary's Commission Recommendation refers to the findings detailed in the Report of the Secretary's Commission on Pesticides and their Relationship to Environmental Health, U.S. Department of Health, Education, and Welfare, December 1969.

[d]NCI Carcinogenicity Test refers to laboratory analyses being conducted by and for the National Cancer Institute to determine the carcinogenicity of various compounds. Source: Dr. Elizabeth Weisburger, Carcinogen, Metabolism and Toxicology Branch, Division of Cancer Cause and Prevention, National Cancer Institute.

Sources

(1) Farm Chemicals Handbook (Farm Chemicals Magazine) (1974) Willoughby, Ohio: Meister Publishing Company.

(2) EPA pesticide registration files.

(3) Herbicide Handbook of the Weed Society of America (1970) Geneva, New York: W. F. Humphrey Press, Inc.

(4) Martin, H. (1971) Pesticide Manual (British Crop Protection Council).

(5) IARC Monographs on the Evaluation of Carcinogenic Risk of Chemicals to Man, Vol. Five (1973). Lyon, France: World Health Organization International Agency for Researc on Cancer.

(6) Lawless, E. W., R. von Rumker, and T. L. Ferguson (1972) The Pollution Potential in Pesticide Manufacturing.

Technical Studies Report No. 15-00-72-04, submitted to the Office of Water Programs, EPA.
(7) U.S. Tariff Commission (1974) Synthetic Organic Chemicals: U.S. Production and Sales of Pesticides and Related Products, 1973.
(8) U.S. Department of Agriculture, The Pesticide Review 1973. Washington, D.C.

APPENDIX B:
SELECTED LIST OF STUDIES ON PESTICIDES

NATIONAL ACADEMY OF SCIENCES

Food Protection Committee (1952) Annotated Bibliograph of Analytical Methods for Pesticides. NAS Pub. No. 241. Washington, D.C. Section One, 59 pp; Section Two, 178 pp.

Food Protection Committee (1952) Safe Use of Chemical Additives in Foods. Washington, D.C. 26 pp.
A three-part study covering safety evaluations of chemical additives for food, hazard evaluation of pesticides on foods, and recommended procedures for developing new pesticides for food crops.

Subcommittee on Animal Reservoirs and Vectors of Disease, Committee on Sanitary Engineering and Environment, Division of Medical Sciences (1954) Safe Handling of Pesticides Employed in Public Health. Washington, D.C. Mimeographed, 16 pp.
Prepared as a guide for the Air Force and other military units. Covers selected insecticides, fumigants, rodenticides, and herbicides.

Food Protection Committee, Food and Nutrition Board (1956) Safe Use of Pesticides in Food Production. NAS Pub. No. 470. Washington, D.C. 16 pp.
Report designed to replace sections dealing with pesticides in the Food Protection Committee's report of 1952, Safe Use of Chemical Additives in Foods. Report discusses need for, hazards of, and development of pesticides.

Committee on Pest Control and Wildlife Relationships, Division of Biology and Agriculture (1961) Pest Control and Wildlife Relationships, A Symposium on March 10, 1961. NAS Pub. No. 897. Washington, D.C. 25 pp.

The symposium was a way to encourage discussion of the problems eventually discussed in the three-part publication of the Committee (NAS Pub. No. 920-A, -B, -C).

Food Protection Committee (1962) New Developments and Problems in the Use of Pesticides, A Symposium, November 29, 1962. NAS Pub. No. 1082. Washington, D.C. 82 pp.

Papers cover alternate methods of pest control, metabolism of pesticides, improved analytical techniques for residue detection, contamination of water by pesticides, and protection of processed food.

Committee on Pest Control and Wildlife Relationships. Part I: Evaluation of Pesticide Wildlife Problems, 28 pp. Part II: Policy and Procedures for Pest Control, 53 pp. Part III: Research Needs, 28 pp. NAS Pub. No. 920-A, -B, -C. Washington, D.C.

Part I describes the necessity of using pesticides (agriculture, forestry, public health) and the consequent impact of use on various wildlife species.

Part II describes the general policies and procedures that should be followed in controlling pests of agriculture, public health, and forestry. The study is aimed at government pest control programs, but it is also applicable to private programs.

Part III gives a brief outline of research needed to provide alternative ways of controlling pests and to understand the effect of pesticides on wildlife.

Food Protection Committee (1965) Some Considerations in the Use of Human Subjects in Safety Evaluation of Pesticides and Food Chemicals. NAS Pub. No. 1270. Washington, D.C. 22 pp.

Paper gives recommendations of how human subjects can and should be used in the evaluation of chemical toxicity.

Pesticide Residues Committee (1965) Report on "No Residue" and "Zero Tolerance." Washington, D.C. 17 pp.

The PSAC report, "Use of Pesticides," (1963) recommended that the Secretaries of USDHEW and the USDA request the NAS-NRC to study the technical issues involved in concepts of "no residue" and "zero tolerance" as they relate to registration of pesticides, setting of tolerances for residues, enforcement of tolerances, and recommendations for use of a pesticide. This Committee recommended that "no residue" and "zero tolerance" were untenable concepts for registration and regulation of pesticides. They suggested that registration be on the basis of a published "negligible" or "permissible" residue basis.

National Research Council (1966) Scientific Aspects of Pest Control. NAS Pub. No. 1402. Washington, D.C. 470 pp.

This symposium volume has a wide range of papers on the theoretical aspects of pest populations, the impact of pesticides on the environment, research areas in pest control, the regulation of pest control by the federal government, and the social utility of pest control.

Committee on Plant and Animal Pests, Agricultural Board, National Research Council (1968-1969) Principles of Plant and Animal Pest Control. Washington, D.C.

A six-volume set designed for researchers, pest control agencies, administrators seeking guidance on allocation of resources, and general fieldworkers.

Volume 1: Plant Disease Development and Control (1968) Pub. No. 1596. 205 pp.

Volume 2: Weed Control (1968) Pub. No. 1597. 476 pp.

Volume 3: Insect-Pest Management and Control (1969) Pub. No. 1695. 508 pp.

Volume 4: Control of Plant Parasitic Nematodes (1968) Pub. No. 1696. 166 pp.

Volume 5: Vertebrate Pests--Problems and Control (1970) ISBN 309-01697-5. 153 pp.

Volume 6: Effects of Pesticides on Fruit and Vegetable Physiology (1968) Pub. No. 1698. 90 pp.

Committee on Persistent Pesticides, Division of Biology and Agriculture, National Research Council (1969) Report to U.S. Department of Agriculture. Washington, D.C. 34 pp.

This report to the USDA was intended to be a followup of the recommendation for monitoring of pesticide residues made by PSAC (The Use of Pesticides, 1963). The Committee concluded that demonstrable progress had been made in monitoring of persistent pesticide residues, but note was taken of gaps in understanding of the significance of the residues.

Panel on Monitoring Persistent Pesticides in the Marine Environment, Committee on Oceanography (1971) Chlorinated Hydrocarbons in the Marine Environment. Washington, D.C. 42 pp.

A report urging "A massive national effort . . . to effect a drastic reduction of the escape of persistent toxicants into the environment, with the ultimate aim of achieving virtual cessation in the shortest possible time." Other recommendations dealt with proposed monitoring programs and changes in laws protecting proprietary manufacturing data. (Specifically, the Committee recommended that companies like Monsanto be required to release production figures of chemicals like polychlorinated biphenyls.)

Committee on Agricultural Land Use and Wildlife Resources, Division of Biology and Agriculture, National Research Council (1970) Land Use and Wildlife Resources. Washington, D.C. 262 pp.
Chapter 6 is devoted to "Pesticides and Wildlife." This chapter was prepared by committee members Stanford N. Fertig and Edward H. Smith.

Agricultural Board, Division of Biology and Agriculture, National Research Council. Pest Control Strategies for the Future. Washington, D.C.
Published papers from symposium held April, 1971.

Committee on Genetic Vulnerability of Major Crops, Agricultural Board, Division of Biology and Agriculture, National Research Council (1972) Genetic Vulnerability of Major Crops. Washington, D.C. 307 pp.
An examination of the disease vulnerability of corn, wheat, sorghum and pearl millet, rice, potato, sugar beet, sweet potato, soybeans and other edible legumes, vegetables, and cotton. Prompted by epidemic of corn leaf blight in 1970.

U.S. GOVERNMENT

Executive Branch

President's Science Advisory Committee (1963) Use of Pesticides. Washington, D.C.: U.S. Government Printing Office. 25 pp.
This short report was one of the first broad reviews of pesticides at a high level in the federal government. The committee made several recommendations: The need for monitoring of residues in the environment and in people; that the use of persistent pesticides be reduced orderly; that alternative control measures be developed; and that toxicity testing on man and wildlife be improved.

President's Science Advisory Committee (1965) Cotton Insects. Washington, D.C.: U.S. Government Printing Office 19 pp.

Discusses the problem of increasing production costs in the cotton industry due to increasing need for chemical pest control. The committee was convinced the current chemical control practices were not adequate and recommended increase research and development of alternative control techniques.

President's Science Advisory Committee (1965) Restoring the Quality of Our Environment. Washington, D.C.: U.S. Government Printing Office. 317 pp.

Report covers pollution of air, water, and soil and the effects of pollutants on human health and other living organisms. A lengthy appendix describes alternatives to chemical control techniques. The recommendations urge an increased reliance on nonchemical control techniques, a description of the environmental effects of a pesticide before it is registered, the replacement of "routine" spraying by "when necessary" spraying, and others.

Agricultural Research Service, U.S. Department of Agriculture (1966) Pest Control by Chemical, Biological, Genetic, and Physical Means, a symposium held at the meeting of the American Association for the Advancement of Science, December 26-31, 1964. Washington, D.C.: U.S. Department of Agriculture ARS 33-110. 214 pp.

Twenty-eight papers discuss the future possibilities of pest control techniques.

President's Science Advisory Committee (1967) The World Food Problem. Washington, D.C.: U.S. Government Printing Office. Vol. I, 127 pp.; Vol. II, 772 pp.; Vol. III, 332 pp.

This massive work covers almost every facet of how to encourage improved agricultural productivity. The need to control pests to increase yield, the problem of residues of persistent pesticides, and other facets of pest control problems are discussed briefly at various places in the three volumes.

U.S. Department of Agriculture and the Association of State Universities and Land Grant Colleges (1968) Environmental Quality--Pollution in Relation to Agriculture and Forestry. Washington, D.C.: U.S. Department of Agriculture. 111 pp.

Makes recommendations for a national program of research on problems of agricultural pollution. Pesticides are a part of the report.

Secretary's Commission on Pesticides and Their Relationship to Environmental Health (1969) Report. Washington, D.C.: U.S. Government Printing Office. 677 pp.

The Mrak report. An important summary of the literature of the effect of pesticides on nontarget organisms, especially man. Effects on man include discussion of carcinogenicity, mutagenicity, and teratogenicity.

Panel on Pesticides, Arkansas-White-Red Basins. Interagency Committee (1970) The Effects of Pesticides on Water Resource Developments, a symposium, April 22, 1970. New Orleans, Louisiana.

Seven papers. Areas covered include how pesticides affect the development of water resources for agriculture, aquaculture, recreation, industry, and drinking.

Working Group, Subcommittee on Pesticides, President's Cabinet Committee on the Environment. June 30-July 1, 1970. Proceedings of National Working Conference on Pesticide Disposal. National Agricultural Library. 141 pp.

Office of Science and Technology, Executive Office of the President (1971) Ecological Effects of Pesticides on Non-Target Species. Washington, D.C.: U.S. Government Printing Office. 220 pp.

Dave Pimentel has collected and summarized all literature of the effects of pesticides on nontarget organisms. The information is presented by chemical and taxonomic class. Under Mirex, for example, he summarized eight papers describing the effects of Mirex on mammals, birds, fishes, arthropods, and plants.

President's Science Advisory Committee (1971) Report on 2,4,5-T. Washington, D.C.: U.S. Government Printing Office. 68 pp.

This report is a broad summary of the chemistry, uses, toxicology, environmental residues, and ecological effects of the phenoxy-herbicide 2,4,5-T. The recommendations are generally for more research on and monitoring of 2,4,5-T.

Advisory Committee on 2,4,5-T (1971) Report to the Administrator of the Environmental Protection Agency. Washington, D.C.: Environmental Protection Committee. 76 pp.

The majority of the Committee recommended that the registration for 2,4,5-T and its esters be restored to its original (pre-April 1970) status. A minority viewpoint recommended the continued cancellation or suspension of 2,4,5-T for use around the home, recreation areas, and on all crops intended for human food.

DDT Advisory Committee (1971) Report to Administrator, Environmental Protection Agency. Washington, D.C.: Environmental Protection Agency. 58 pp.

The Committee agreed that DDT did not present an imminent hazard to human health in terms of individual bodily functions and safety. DDT does, it was agreed, present a substantial threat to other species and was thus an imminent hazard to human welfare. The Committee also agreed that immediate suspension of DDT use would give the same results as rapid and continuous decrease in the use of DDT, which is occurring.

Advisory Committee on Predator Control (1972) Predator Control-1971. Report to the Council on Environmental Quality and the Department of Interior. Washington, D.C.: Council on Environmental Quality. 207 pp.

Recommends removal of all existing toxic chemicals from registration and use in predator control.

Mirex Advisory Committee (1972) Report to the Administrator of the Environmental Protection Agency. Washington D.C.: Environmental Protection Agency. 70 pp.

The Committee recommended that registration of products containing Mirex should be continued with labeling restrictions to minimize environmental contamination.

Aldrin/Dieldrin Advisory Committee (1972) Report to the Administrator, Environmental Protection Agency. Washington, D.C.: Environmental Protection Agency. 100 p

Committee recommended continued registration for uses in soil or items buried in soil (e.g., seed treatment, termite control). Applications by air or on aerial parts of plants, on or into water, or in structures occupied by humans or livestock were recommended for cancellation.

Legislative Branch

U.S. House, Committee on Appropriations, Subcommittee on Department of Agriculture (1965) Effects, Uses, Contro and Research of Agricultural Pesticides. 46 pp.

A report on the Surveys and Investigations Staff of the Appropriations Committee to Rep. Jamie L. Whitten,

Chairman of the Subcommittee on Agriculture. Based on over 200 interviews with scientific personnel and visits to numerous federal offices and laboratories.

U.S. House, Select Committee to Investigate the Use of Chemicals in Foods and Cosmetics (1950-1952) Chemicals in foods and cosmetics. Hearings.

The first committee at a high level in government to attempt formulation of a new public policy toward the synthetic pesticides introduced during and after World War II. Led to amendment of the Food, Drug, and Cosmetic Act.

U.S. Senate, Committee on Government Operations, Subcommittee on Reorganization and Organizations (1963-1964). Interagency coordination in environmental hazards (pesticides). Hearings. 11 volumes. Report. Senate 89:2. Number 1379. July 21, 1966.

Recommended extensive legislation for:

(1) increased regulation of production and use;

(2) improved interagency handling of pesticide matters in the government; and

(3) expanded monitoring of pesticide residues.

OTHER ORGANIZATIONS

Conference on Biological Methods of Controlling Insect Pests (1936) Belleville, Ontario. Mimeographed, 86 pp.

International Conference of Insect Pathological and Biological Control, First. Transactions. Praha. 1958. 653 pp.

American Association for the Advancement of Science (1960) Biological and Chemical Control of Plant and Animal Pests, a symposium. Washington, D.C.: AAAS Pub. No. 61. 273 pp.

American Society for Microbiology (1965) Microbial Insecticides, a symposium at the Annual Meeting of the American Society for Microbiology, Washington, D.C., May 7, 1964. Bacteriol. Rev. 29(3):364-405.

A collection of five papers discussing the possible use of bacteria and fungi to control insects and other arthropods.

International Colloquium on Insect Pathology and Microbial Control, Wageningen, 1966 (1967) Insect Pathology and Microbial Control. Amsterdam: North-Holland Publishing Company. 360 pp.

Panel on the Control of Livestock Insect Pests by the Sterile-Male Technique (1967) Vienna: International Atomic Energy Agency. 102 pp.

Panel on Radiation, Radioisotopes, and Rearing Methods in the Control of Insect Pests, Tel Aviv, 1966 (1968) Vienna: International Atomic Energy Agency. 148 pp.

Committee on Chemistry and Public Affairs, American Chemical Society (1969 and 1971) Cleaning Our Environment, The Chemical Basis for Action and A Supplement to Cleaning our Environment, The Chemical Basis for Action. Washington, D.C.: American Chemical Society. 249 and 20 pp., respectively.

Topics covered are air and water pollution, solid waste problems, and pesticidal residues in the environment. Recommendations are made for research and development of new, integrated control techniques. The Supplement updates the 1969 recommendations.

Gunther, Francis A., ed. (1969) Decontamination of Pesticide Residues in the Environment. A Symposium of the American Chemical Society, September, 1968. New York: Springer-Verlag. 213 pp. Volume 29 of the series Residue Reviews.

A collection of papers describing the degradation of pesticides in the soil, in water, and on various foodstuff

International Atomic Energy Agency (1969) Sterile-Male Technique for Eradication or Control of Harmful Insects, proceedings of a panel discussion held in Vienna, May 27-31, 1968. Vienna: International Atomic Energy Agency. 142 pp.

A symposium organized by the Food and Agriculture Organization and the International Atomic Energy Agency. The theoretical and practical details of using the sterile male control technique are discussed; examples of use of the technique are given. Recommendations for further research are made.

Kraybill, Herman (1969) Biological Effects of Pesticides in Mammalian Systems, a conference held by the

New York Academy of Sciences, May 2-5, 1967. Ann. N.Y. Acad. Sci. 160:1422.

The conference was intended to draw together the latest advances made in the epidemiologic and toxicologic research in man and other mammals with respect to chemical pesticides. Discussion includes analysis of residues, metabolism of pesticides, distribution of residues, and biochemical and pathological effects of residues.

Miller, Morton W., and George C. Berg (1969) Chemical Fallout. Springfield, Illinois: Charles C Thomas. 531 pp.

A collection of 26 papers covering the distribution and collection of pesticide residues in nature, their impact on natural populations other than man, the mechanism of toxicology and the effects in individuals, and the estimation of effect on human populations. The publication originated from a conference organized by the Department of Radiation Biology and Biophysics, School of Medicine and Dentistry, University of Rochester.

Governors' Five-State Interdisciplinary Pesticide Council. Background for Decision, a one-day educational symposium on Pesticides, December 14, 1970. Madison, Wisconsin: Governors' Five-State Interdisciplinary Pesticide Council, 801 W. Badger Road. 116 pp.

A collection of eight papers designed to give policymakers at the state level background in the toxicological and environmental problems of pesticides.

International Symposium on Pesticides in the Soil (1970) Pesticides in the Soil: Ecology, Degradation, and Movement. East Lansing, Michigan: Michigan State University Pesticide Research Center. 114 pp.

A symposium of 19 papers held in East Lansing, Michigan, February 25-27, 1970. Areas covered included biological activities in soil, the effect of pesticides on these activities, the degradation of pesticides in the soil, and the movement of pesticides in the soil.

Society of Chemical Industry (1970) Technological Economics of Crop Protection and Pest Control. London: Society of Chemical Industry. 255 pp.

A collection of papers and discussions presented at a conference organized by the British Society of Chemical Industry and the Department of Industrial Science, University of Stirling. Discussion of the economics of pesticide development and use in developed and developing countries.

Study of Critical Environmental Problems (1970) Man's Impact on the Global Environment. Cambridge, Massachusetts MIT Press. 319 pp.

The proceedings and recommendations of an MIT-sponsored summer study. Williamstown, Massachusetts, 1970. Pollution by pesticidal chemicals was a part of the study.

World Health Organization (1970) Control of Pesticides. Geneva: World Health Organization. 150 pp.

A survey of legislation governing the use of pesticides. Covers Czechoslovakia, Denmark, France, Germany (West), Netherlands, New Zealand, Sweden, USSR, United Kingdom, USA, and Venezuela.

Also published in Int. Digest Health Leg. 20:579-726, 1969.

Economic Research Service, U.S. Department of Agriculture (1971) Economic Research on Pesticides for Policy Decision-making. Washington, D.C.: USDA. 172 pp.

Epstein, Samuel S., and Marvin S. Legator (1971) The Mutagenicity of Pesticides, Concepts and Evaluation. Cambridge, Massachusetts: MIT Press. 220 pp.

A slightly expanded version of the section on mutagenicity of pesticides in the Mrak Commission Report.